D0086667

MATRIX STRUCTURAL ANALYSIS

The PWS-KENT Civil Engineering Series List

The Science and Engineering of Materials, Second Edition
Askeland

The Science and Engineering of Materials, Alternate Edition
Askeland

Mechanics of Materials, Second Edition
Bauld

Principles of Foundation Engineering
Das

Principles of Geotechnical Engineering
Das

Mechanics of Materials, Second Edition
Gere/Timoshenko

Matrix Algebra for Engineers, Second Edition
Gere/Weaver

A First Course in the Finite Element Method
Logan

Engineering Mechanics: Statics, Second Edition
McGill/King

Engineering Mechanics: An Introduction to Dynamics, Second Edition
McGill/King

Engineering Mechanics: Statics and an Introduction to Dynamics, Second Edition
McGill/King

Unit Operations and Processes in Environmental Engineering
Reynolds

Matrix Structural Analysis
Sack

Fundamentals of Surveying, Third Edition
Schmidt/Wong

Fundamentals of Structural Steel Design
Segui

Engineering and the Environment
Wanielista/Yousef/Taylor/Cooper

MATRIX STRUCTURAL ANALYSIS

Ronald L. Sack
University of Oklahoma

PWS-KENT Publishing Company
Boston

PWS-KENT
Publishing Company

20 Park Plaza
Boston, Massachusetts 02116

Editor J. Donald Childress, Jr.
Production Editor Carolyn Ingalls
Interior Designer Susan M. C. Caffey
Cover Designer Julia Gecha
Manufacturing Coordinator Marcia A. Locke
Typesetting Santype International Limited
Cover Printing Henry N. Sawyer Co., Inc.
Printing and Binding Book Press, Inc.

PWS-KENT Publishing Company is a division of Wadsworth, Inc.

Printed in the United States of America.

1 2 3 4 5 6 7 8 9—93 92 91 90 89

Library of Congress Cataloging-in-Publication Data

Sack, Ronald L.
Matrix structural analysis.
Bibliography: p.
Includes index.
1. Structural analysis (Engineering)—Matrix methods. I. Title.
TA642.S22 1989 624.1′71 88-32960
ISBN 0-534-91564-7

PREFACE

This book was written for the student and practicing engineer who wish to use matrix methods of structural analysis to predict the static response of structures. The text is introductory, emphasizes the stiffness method, and contains the fundamentals of the flexibility method.

The general theory of the stiffness method is initially derived from the intuitive concepts of the direct solution of the basic equations of equilibrium, compatibility, and material properties. The theory is presented and explained using truss behavior. Thus, at the beginning of the book the reader has the opportunity to observe the method unencumbered by generalized arguments. Subsequently, the principle of virtual work is explained and offered as an alternative theoretical basis for the stiffness method. The flexibility method is similarly derived by direct solution of the basic equations and from the principle of complementary virtual work. Applications of the stiffness method are given for beams, planar frames, space trusses, beam grid works, and space frames. Miscellaneous topics required to complete our coverage of the stiffness method are also described.

The three principal aspects of analyzing structures using matrix methods are: (a) understanding the method (i.e., the theory plus its limitations and applications); (b) developing appropriate computer programs; and (c) solving actual structures on the computer—this involves idealizing the problem, preparing and investigating the data, information processing, and the numerical methods necessary to obtain the solution. Each component is a vital link in implementing and using matrix methods of structural analysis for routine production problems. The material in this text is devoted to an understanding of the method with an appreciation for writing computer programs and using production-level programs to solve actual structures. The stiffness and flexibility methods are cast in a form appropriate for use on the computer, but the details of computer implementation are not discussed. It is occasionally useful for the student to understand the power of the method by investigating structures that require a computer; a sufficient number of exercise problems of this type are provided. We urge the reader to use available software to carry out long tedious computations. For example, programs such as CAL86,[1] used in the mode where only matrices are manipulated, require the user to load the basic matrices and write the appropriate code in the metalanguage of the program to attain the required solution.

The contents of the main body of the book are divided into three categories. The materials in Chapter 1 serve to orient the student to matrix structural analysis and provide a basic introduction and appreciation for the history and scope of the various theorems and methods. The development of the stiffness method in Chapters 2 and 3 using the basic equations and energy meth-

[1]Wilson, E. L. *CAL86 Computer Assisted Learning of Structural Analysis and the CAL/SAP Development System,* Report No. UCB/SESM-86/05, Berkeley, Calif., 1986.

ods, respectively, is strongly tied to practical structures. Chapters 4 and 5 reinforce the theory and give definite applications of the method for various types of structures. The special topics of the stiffness method have been assembled in Chapter 6; thus, the orderly flow of the development is not disrupted, and these important aspects of solving problems are treated together. Chapter 7 on the flexibility method using the solution of the basic equations and the principle of complementary virtual work parallels the stiffness method derivation in Chapters 2 and 3. The theory in this chapter is strongly connected to actual structural problems. The appendix materials complement the main body of the book and give the reader opportunities to study the solution of linear algebraic equations and review elementary matrix operations.

A broad collection of examples demonstrates the principles and assists the reader in developing an active understanding of the concepts. An unlimited number of exercise problems can be obtained through the program shell PANDORAS BOX (available from the author). This program contains a menu of structural groups and subgroups that are illustrated in the text (i.e., appropriate configurations of trusses, beams, and frames will appear at the end of each chapter). The program also contains a random number generator, an executable segment to solve for the structural deflections and forces, plus supporting graphics. An instructor equipped with a disk of PANDORAS BOX can select the category of exercise problem, and the program responds with a graphics display of the structure, the problem data, and the solution. Problem dimensions, loads, etc., are randomly generated; therefore, a different problem can be assigned to each student, and problems need never be repeated from year to year.

The book is designed to be used in aerospace engineering, civil engineering, mechanical engineering, and engineering science curricula. A logical progression of topics with a uniform and continuous flow of information can be obtained by exercising modest discretion in selecting individual chapters or sections. An introductory course in matrix structural analysis with an emphasis on problem-solving skills can be constructed using material from Chapters 1, 2, 4, and 5, with topics such as matrix condensation, release of generalized member nodal forces, and nodal coordinates selected from Chapter 6. By omitting Chapter 5 and including Chapter 3, one obtains a more theoretically oriented presentation. In contrast, an energy orientation toward the subject is obtained by choosing Chapters 1 and 3, plus selections from Chapters 5, 6, and 7. These skeletal outlines can be expanded with additional topics for multicourse sequences, and the book is also arranged so that a cover-to-cover study is possible. It is also envisaged that the book can be used effectively in self-study programs.

Many people have played a role in writing this book. I express my sincere thanks for the valuable suggestions of the reviewers: Dr. Dan Frangopol, University of Colorado; Dr. Daniel L. Garber, University of Maryland; Dr. James K. Nelson, Texas A&M University; and Dr. Jay A. Puckett, University of Wyoming. Finally, I extend special thanks to the students at the University of Idaho and the University of Wyoming for enduring the inconvenience of studying from the manuscript form of this book.

Ronald L. Sack

CONTENTS

SYMBOLS AND NOTATION

Symbols are generally defined where they first appear. Some symbols have been used in different contexts to define several quantities. In general we have used lower-case symbols to indicate quantities associated with element coordinates and capital letters for global quantities. We use **p** (**P**) to indicate nodal applied forces (both loads and reactions) and **u** (**U**) to denote nodal displacements; **k** (**K**) contains the stiffness elements and **f** (**F**) denotes the flexibility matrix. Matrices are shown in bold print, and the elements of a column matrix are written within brackets, { }, to conserve space in the text.

- $\mathbf{a}$ Kinematics matrix (partitioned into $\mathbf{a}_0$, and $\mathbf{a}_1$); matrix of coefficients for polynomial
- A Cross-sectional area of a member
- b Member width
- $\mathbf{b}$ Statics matrix (partitioned into $\mathbf{b}_0$ and $\mathbf{b}_1$)
- $\mathbf{B}$ Matrix relating nodal displacements to element strains
- $\mathbf{d}$ Column matrix of element deformations ($\mathbf{d} = \mathbf{aU}$)
- $\mathbf{d}^o$ Column matrix of initial element deformations
- $\mathbf{e}$ Element force transformation matrix for global displacements
- $\bar{\mathbf{e}}$ Element force transformation matrix for local displacements
- E Modulus of elasticity (i.e., Young's modulus)
- $\mathbf{E}$ Matrix of elastic constants
- $\mathbf{f}$ Element flexibility matrix
- $\mathbf{F}$ Global flexibility matrix
- G Modulus of elasticity in shear
- I Moment of inertia
- $\mathbf{I}$ Identity (unit) matrix
- J St. Venant's torsion constant
- $\mathbf{k}$ Element stiffness matrix with elements k_{ij} expressed in global coordinates
- $\bar{\mathbf{k}}$ Element stiffness matrix with elements $\bar{k}_{ij}$ expressed in local coordinates
- $\mathbf{K}$ Structural stiffness matrix with elements K_{ij} expressed in global coordinates
- L Length
- M Bending moment
- NDOF, NE, NN, NR, NOK, NOS Number of: degrees of freedom; elements; nodes; reactions; kinematic indeterminacies; static indeterminacies
- N Element axial force

$\mathbf{N}$ Column matrix of shape functions

$\mathbf{O}$ Null matrix

$\mathbf{p}$ Column matrix of nodal element forces in global coordinates

$\bar{\mathbf{p}}$ Column matrix of nodal element forces in local coordinates

$\mathbf{p}^o$ Column matrix of initial nodal element forces in global coordinates

$\bar{\mathbf{p}}^{o}$ Column matrix of initial nodal element forces in local coordinates

$\mathbf{P}$ Column matrix of applied nodal forces in global coordinates ($\mathbf{s} = \mathbf{bP} = \mathbf{b}_0\mathbf{P} + \mathbf{b}_1\mathbf{X}$)

$\mathbf{P}_f$ Column matrix of known applied nodal forces in global coordinates

$\mathbf{P}_s$ Column matrix of unknown applied nodal forces in global coordinates

$\mathbf{P}^o$ Column matrix of initial forces in global coordinates

q Distributed load magnitude

$\mathbf{R}$ Column matrix of reaction forces

$\mathbf{s}$ Column matrix of element forces and reactions

T Temperature

$\mathbf{T}$ Transformation matrix

U_i, V_i, W_i Displacements at node i for a structure in the x, y, and z directions, respectively

u, v, w Continuous functions expressing displacements in the x, y, and z directions, respectively

u_i, v_i, w_i Displacements at node i for an element in the x, y, and z directions, respectively

$\bar{u}_i, \bar{v}_i, \bar{w}_i$ Displacements at node i for an element in the $\bar{x}$, $\bar{y}$, and $\bar{z}$ directions, respectively

$\mathbf{u}$ Column matrix of nodal displacements in global coordinates

$\bar{\mathbf{u}}$ Column matrix of nodal displacements in local coordinates

$\mathbf{U}$ Column matrix of nodal displacements for the entire structure in global coordinates

$\mathbf{U}_f$ Column matrix of unknown displacements in global coordinates

$\mathbf{U}_s$ Column matrix of known displacements in global coordinates

V Shear force

W_e, W_e^* Work and complementary work done by external forces

W_i, W_i^* Strain energy and complementary strain energy

$\bar{x}, \bar{y}, \bar{z}$ Orthogonal cartesian global (structural) coordinates

$\bar{x}, \bar{y}, \bar{z}$ Orthogonal cartesian local coordinates

$\mathbf{X}$ Column matrix of redundant forces

SUBSCRIPTS

- i The node (point) associated with the quantity
- f Degrees of freedom with known forces and unknown displacements
- s Degrees of freedom with unknown forces and known displacements
- 0 Quantity associated with the primary structure; used in the flexibility method
- 1 Quantity associated with the structural redundants; used in the flexibility method

SUPERSCRIPTS

- E Force (or moment) that is equivalent in an energy sense to a distributed loading
- F Force (or moment) required to give zero displacement at the point (i.e., a fixed-end force)
- ij The interval (element) associated with the quantity
- o Quantity initially introduced by temperature, fabrication error, precambering, etc.
- T Transpose of a matrix
- -1 Inverse of a matrix

GREEK SYMBOLS

- α (alpha) Coefficient of linear thermal expansion
- α,β,γ Angles measure to a vector from the positive x, y, and z axes, respectively
- γ (gamma) Shear strain
- Γ (gamma) Matrix of direction cosines; an orthogonal transformation
- δ (delta) Deflection; increment of a quantity; first variation of a quantity (a virtual quantity)
- Δ (delta) Deflection; total change in a quantity
- ϵ (epsilon) Translational strain
- θ (theta) Angle; rotation of node with respect to local coordinates
- Θ (theta) Angle; rotation of node with respect to global coordinates
- κ (kappa) Curvature of a beam
- κ_v (kappa) Shear constant
- λ,μ,ν Direction cosines (i.e., cos α, cos β, cos γ, respectively)
- ν (nu) Poisson's ratio
- σ (sigma) Normal stress
- Σ (sigma) Summation of quantities

τ (tau) Shear stress

ϕ (phi) Angle between axis of an element and the global coordinates

GRAPHIC SYMBOLS

Force

or

Moment

Resultant force

Resultant moment

Reactive force

or

Reactive moment

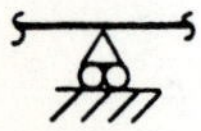

or

Roller support

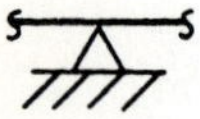

Pinned support

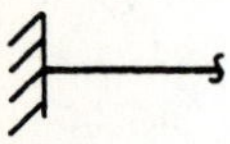

Fixed support

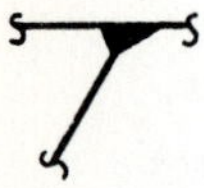

Rigid connection

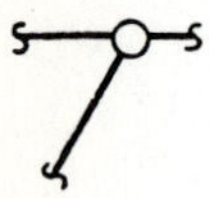

Pinned connection

MATRIX STRUCTURAL ANALYSIS

1 INTRODUCTION

Engineered structures must ensure the safety and welfare of the occupants and general public by performing in a prescribed manner. Strength requirements are accompanied by stiffness constraints to prevent excessive deflections, bouncy floors, outward-tilting walls, uncomfortable structural oscillations, and the like. Thus structural analysis and design are intertwined since behavior is affected by the arrangement of members and distribution of materials. New complex systems require more precise engineering; many major contemporary structures, such as the Boeing 747 aircraft, the Swiss Flesenau Bridge, and the Sears Tower, owe their existence to computer-oriented structural analysis and design. This chapter examines the origins and utility of matrix structural analysis, and its relationship to classical methods.

1.1 HISTORICAL CONTEXT OF MATRIX STRUCTURAL ANALYSIS

The airplane and digital computer are responsible for revolutionizing structural analysis. In the 1940s and 1950s structural engineers were confronted with two highly statically indeterminate systems: the swept-wing and delta-wing aircraft. The governing equations were cast *ab initio* (from the beginning) in matrix format, but this approach required solution of large sets of simultaneous linear algebraic equations. At the time, relaxation methods were used extensively to solve the governing equations of structural behavior; therefore, the requirement to deal with great numbers of algebraic equations was an anathema to the engineer. Fortuitously, the University of Pennsylvania unveiled the 30-ton ENIAC digital computer in 1946. The invention of the transistor in 1947 and the silicon chip in 1959 were pivotal discoveries that accelerated the development of the digital computer and gave impetus to the structural analysis revolution. By embracing this new computing technology, the structural engineers of those two decades completely changed structural analysis. Trusses, beams, and frames were initially investigated, but in the mid-1950s a group at the Boeing Company demonstrated that the procedure could be extended to continua. Common usage now dictates that *matrix structural analysis* designates investigations of structures composed of articulated or discrete components, whereas the *finite element method* denotes analysis of continua.

Structural analysis of the early swept-wing aircraft in 1947 depended upon work from the 1800s by James Clerk Maxwell and Otto Mohr. Their method of *consistent displacements* is an example of a classical *compatibility method* yielding sets of simultaneous linear algebraic equations, with the structural flexibilities as the coefficients and the forces as the unknowns. A new method, called the *flexibility* or *force method*, was formulated and distinguished from traditional compatibility procedures by the fact that all quantities and equations were formulated initially as matrices and manipulated using the associated algebra; therefore, the operations are computer oriented.

In 1953 the delta-wing aircraft was the impetus for a second computer-oriented approach to structural analysis. By broadening the scope of traditional *equilibrium methods* and formulating the equations from inception using matrices, the structural engineers of the 1950s obtained a set of linear algebraic equations with the structural stiffnesses as the coefficients and the displacements as the unknowns. Thus the *stiffness* or *displacement method* was conceived.

Interest in energy methods was also stimulated during this time, but structural mechanics has historically relied upon energy principles. Archimedes (287–212 B.C.), Leonardo da Vinci (1452–1519), and Galileo (1564–1642) each used some form of the work expression to substitute for the equations of equilibrium in lever and pulley systems. Johann Bernoulli (1717) was the first to suggest virtual displacement, and Maupertuis (1740) introduced the concept of measuring equilibrium of rigid bodies by minimizing the total system potential. Leonhard Euler (1744) recognized that energy methods are an alternative approach for solving problems of structural mechanics and used minimization principles to investigate stable equilibrium for deformable bodies; he used expressions for strain energy suggested by Daniel Bernoulli. Lame (1852) derived the principle of conservation of energy and named it for his friend Clapeyron; he used actual forces, stresses, displacements, and strains. James Clerk Maxwell (1864) and Otto Mohr (1874) independently took the results of Lame, and, using a dummy load, investigated statically indeterminate trusses. Thus, the principle of virtual forces is also known as the Maxwell-Mohr method. Castigliano (1873) published the extremum version of Lame's work. Since Lame used actual quantities, Castigliano's theorem, part II, is valid only for linear elastic systems. Crotti (1878) and Engesser (1889) subsequently extended this result, thereby making the minimization principle conform to the principle of complementary virtual work for nonlinear elastic systems.

In 1954 J. Argyris and S. Kelsey formulated matrix structural analysis using energy principles. Matrix structural analysis emanated from physically directed thinking and was derived by satisfying the fundamental equations of structural mechanics; therefore, the application of energy principles was the next logical step in the evolution. Because of the initial popularity of the flexibility method during the 1950s, the corresponding *principle of complementary virtual work* was emphasized. In contrast, the stiffness method arises from the *principle of virtual work*, but early derivations represented this simply as an alternate choice of variables (i.e., displacements as unknowns instead of forces). Subsequent work revealed that the stiffness method, based upon the principle of virtual work, is a numerically efficient procedure for implementing the classical Rayleigh-Ritz method, which was conceived in 1909. The finite element method owes its existence and wide appeal to this fact.

1.2 MATRIX STRUCTURAL ANALYSIS AND CLASSICAL METHODS

Thus matrix structural analysis has come to fruition since the 1940s, but its roots are in classical structural mechanics. Since the computer formulates and solves the equations, large structures can be investigated. We can use either compatibility or equilibrium methods and formulate the method using the fundamental equations of structural mechanics or energy principles. Therefore, it is instructive to recall the basic principles and classical methods of structural analysis and observe their relationship to matrix structural analysis.

Structures must be in equilibrium, with their displacements in a compatible state and material laws satisfied. The structural engineer can investigate these primary behavioral tenets by either: (a) solving the fundamental equations or (b) employing energy principles.

Double integration, the method of elastic weights, and the moment-area method yield structural displacements from the fundamental equations. We use force-displacement relationships to assemble equations of structural response, thereby satisfying equilibrium, compatibility, and material laws. *Compatibility methods* mandate identifying statically indeterminate elements and imposing compatibility requirements, thus producing sets of equations with the structural flexibilities as coefficients and forces as unknowns. The method of consistent displacements and the three-moment equation are examples of the compatibility method.

Alternatively, by invoking equilibrium at points connecting structural elements we can formulate sets of simultaneous linear algebraic equations with the structural stiffnesses as coefficients and displacements as unknowns. This approach begets *equilibrium methods*; the slope-deflection and the moment-distribution methods are two classical procedures in this category.

Energy principles present an alternative approach for investigating structural behavior. The principle of virtual displacements, the unit displacement theorem, and Castigliano's theorem, part I, are examples of energy methods, wherein equilibrium is satisfied implicitly. In contrast, the principle of virtual forces, the unit-load theorem, and Castigliano's theorem, part II, are complementary virtual work theorems that satisfy compatibility implicitly.

Virtual work theorems call for varying the displacement and corresponding strains, whereas *complementary virtual work* theorems require the forces and stresses to undergo variations. The former approach produces equilibrium methods, while the latter gives rise to compatibility procedures. For example, recall the method of least work for linearly elastic systems. By Castigliano's theorem, part II, the partial derivative of the strain energy with respect to a force gives the corresponding displacement. If that displacement is zero (e.g., for a redundant reaction), we obtain what appears to be a minimum principle.

This, of course, is a compatibility method obtained from a complementary virtual work principle.

For systems with many unknowns it is convenient to formulate the equations from the beginning in matrix form; thus, subsequent manipulations are executed using matrix operations, which can be conveniently programmed in a computer language. If we use a compatibility method (either by solving the fundamental equations or invoking a complementary virtual work principle) we obtain sets of simultaneous linear algebraic equations involving structural flexibilities. In contrast, by formulating the solution using either the principle of virtual work or the fundamental equations of structural behavior, we mandate nodal equilibrium and obtain sets of simultaneous linear algebraic equations embodying the structural stiffnesses. The former approach is the *flexibility* or *force method*, whereas the latter is the *stiffness* or *displacement method*.

We can program the force method to automatically identify statically indeterminate components or systems. We formulate the coefficient matrix of system flexibilities using a matrix triple product; one of the basic matrices required for this process expresses system equilibrium, while another simply contains element flexibilities. The computer executes a large number of operations and consumes a great amount of time in formulating the global flexibility matrix. In the formative stages, the stiffness method suffered from engineers obsessed with the duality of the two methods. That is, since the basic equilibrium matrix of the flexibility method can be shown to define the system compatibility equations, we can formulate the global stiffness matrix using a matrix triple product in a fashion resembling that employed in constructing the global flexibility matrix. Equations can elegantly express this duality, but the approach is computationally inefficient. The global stiffness matrix is most efficiently formulated using list processing. By clinging to the duality of equations, the early pioneers nearly rang the death knell for the stiffness method. We now recognize duality for what it is: an interesting fact with few useful computational implications. Today, the stiffness method has almost totally supplanted the flexibility method.

1.3 DISCUSSION

Perhaps in a few years we will not be required to distinguish computer-oriented structural analysis by appending the adjective, matrix. This speculation is strengthened by the capabilities of the available computing hardware and software. In the early days of computers, the user was required to prepare punched cards, learn elaborate access protocol, and struggle with "turn-around time"; the microcomputer has eliminated all of this. In addition, many classical methods can be implemented on a standard spreadsheet, thus qualifying them to be called computer methods. Larger problems simply require more computing power. The supercomputer offers solutions for mammoth systems, whereas intermediate-sized problems can be solved using some component of the array of available equipment between the micro- and supercomputer such as the mini- or mainframe computer.

The spectrum of approaches to structural analysis includes classical, approximate, and computer-oriented methods, and each has its function. By interpreting computer solutions using approximate analysis, the structural engineer can avoid the computer siren that lures acceptance of dubious machine-generated results with the implication that computer output is above question. Since nodes can be misplaced, members inadvertently omitted, and entire systems incorrectly modeled, it is wise to remember the old computer maxim: "garbage in, garbage out."

2 THE STIFFNESS METHOD USING THE BASIC EQUATIONS

Most of today's computer programs for analyzing structural response are based upon the principles of the stiffness method. In this chapter we will formulate this useful method using the basic equations: equilibrium, compatibility, and constitutive (material property relations). These equations must be satisfied within each individual element and throughout the entire structure. The solution of the equations given by the stiffness method yields the displacements of each prescribed point (*node*), and the internal element forces are subsequently obtained by backsubstitution.

The objective of this chapter is to outline the stiffness method and apply it to trusses. We will initially illustrate the method with axial force elements in series and be able to analyze most types of two-dimensional trusses by the end of the chapter. The approach developed in this chapter is extended in Chs. 4 and 5 to the analysis of two- and three-dimensional structures composed of axial force and beam elements. Throughout this book we will study only articulated structures composed of discrete individual elements.

2.1 THE BASIC EQUATIONS

A truss is composed of easily identified individual discrete members, termed *elements*, reaching from joint to joint. The basic equations of structural mechanics must be satisfied within each of the elements. In addition, the truss, composed of the assemblage of all the elements, must also respond in accordance with the basic equations. We can review the basic equations in their most simple form by examining the element in Fig. 2.1a. The element is prismatic with a cross-sectional area A, and it is subjected to forces that are colinear with its centroidal axis. When the forces are applied, the element remains straight, and plane sections normal to the member axis remain plane. We shall assume that the element undergoes sufficiently small displacements so that linear strain-displacement relations are valid, and the equilibrium equations relate to the undeformed state. Also, the element experiences small strains and is made from a material that is homogeneous, isotropic, and linearly elastic; thus, Hooke's law is applicable.

The element is in static equilibrium; hence, the axial forces at each end are equal in magnitude but opposite in direction (we will assume the member

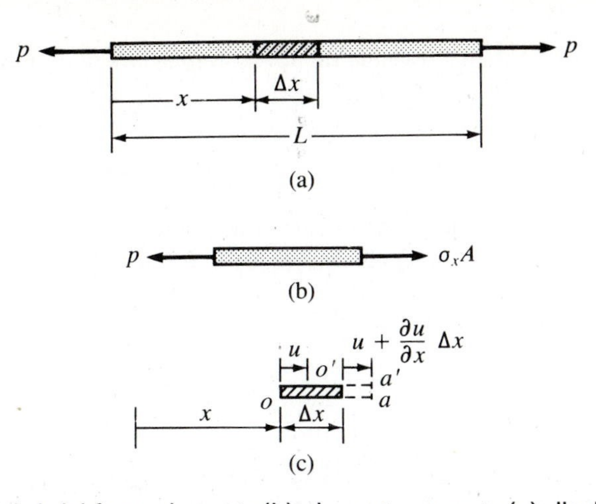

Figure 2.1 (a) Axial force element; (b) element segment; (c) displacement of an element

to be weightless throughout this discussion). The stress, σ_x, is constant over the cross section, and equilibrium must also be satisfied for any portion of the element. We observe from Fig. 2.1b that

$$\sigma_x A = p \tag{2.1}$$

The displacement along the element, u, is a continuous function of x. This ensures that the displacements are geometrically compatible; that is, no rips or tears are introduced within the element by the displacement. The original small length, Δx, moves from points o and a to o' and a', respectively (see Fig. 2.1c). Using the definition for strain gives the following strain-displacement equation:

$$\varepsilon_x = \lim_{\Delta x \to 0} \frac{o'a' - oa}{oa} = \lim_{\Delta x \to 0} \frac{[\Delta x + (\partial u/\partial x)\Delta x] - \Delta x}{\Delta x} = \frac{\partial u}{\partial x} \tag{2.2}$$

The constitutive equation (material behavior relationship) in this case is Hooke's law; that is,

$$\sigma_x = E\varepsilon_x \tag{2.3}$$

where E is the modulus of elasticity.

Combining Eqs. (2.1) through (2.3), and noting that u is a function of x only, yields the following relationship between the force and displacement:

$$\frac{p}{A} = E\,\frac{\partial u}{\partial x} = E\,\frac{du}{dx} \tag{2.4}$$

Rearranging terms gives

$$p\,dx = AE\,du \tag{2.4a}$$

Integrating Eq. (2.4a) yields the force-displacement equation for this axial force element; that is,

$$u = \frac{px}{AE} + C_1 \tag{2.5}$$

where C_1 is a constant of integration. If the member is constrained at the left end, $u(0) = 0$, $C_1 = 0$; furthermore, at $x = L$, $u = \Delta L$, which yields

$$p = \frac{AE}{L} \Delta L \tag{2.6}$$

where ΔL is the total deformation of the element. If the element is incorporated into a truss and both ends are displaced so that $u(0) = u_i$ and $u(L) = u_j$, then $C_1 = u_i$, giving the following force-displacement relationship:

$$p = \frac{AE}{L} (u_j - u_i) \tag{2.6a}$$

Note that $(u_j - u_i)$ is the total deformation of the element, which is equal to ΔL in Eq. (2.6). Thus, by invoking the conditions of equilibrium, compatibility, and stress-strain we have obtained the familiar relationship between force and displacement for an axial force element.

2.2 THE ELEMENT STIFFNESS MATRIX

The axial force element in Fig. 2.2 can deflect in either of the two coordinate directions and rotate about an axis normal to the plane of the paper. Throughout this book a *deflection* indicates the translation of a point on a deformable body, whereas *displacement* is used as a more general term to describe both translations and rotations. Since the element *deformation* (i.e., the relative displacement of points on the element) occurs only in the axial direction, the rotation of the element can be calculated in terms of the deflec-

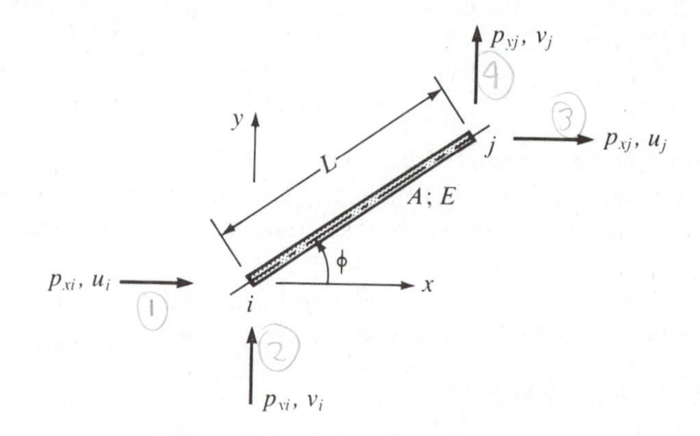

Figure 2.2 The axial force element

tions. If we know the x (u) and y (v) deflections at the ends of the element, the entire response (i.e., deflections, strains, stresses, and forces) can be calculated. Note that the element in Fig. 2.2 has been shown in a general position with respect to the x-y axes. These axes are termed the *global coordinates*. This is in contrast to the system in Fig. 2.1 where the x axis is aligned with the element centroidal axis. The latter is an example of a *local coordinate* system (subsequently denoted with a bar over the symbol). The relationships between quantities in these two systems are discussed in detail in Sec. 2.5.

Each displacement component labeled in Fig. 2.2 is a *degree of freedom*. Generally, the number of degrees of freedom of a system is the number of displacement components or coordinates required to define its position uniquely in space. By this definition, the element in Fig. 2.2 has an infinite number of degrees of freedom since each point can move in two directions. However, from Eq. (2.5) we observe that all deflections along the member can be calculated if the end axial deflections are known. Thus, the axial force element has four independent degrees of freedom: two each in the rectangular cartesian coordinate directions (in all subsequent text in this context, the adjective "independent" will be implied unless it is needed for clarity). Thus, for a truss, the number of degrees of freedom equals twice the number of joints. For the static response of some structures the behavior can be described without including all degreees of freedom. For example, in most rigid frames, the displacements are influenced primarily by flexure, while axial deformations of beams and columns play a minor role. Therefore, it is possible to neglect some nodal degrees of freedom without significant loss of accuracy. A sufficient number of degrees of freedom is required to model a structure so that the actual behavior is closely approximated, but one may not have to include all degrees of freedom in the analysis.

The prismatic axial force element in Fig. 2.2 has a cross-sectional area A, a length L, and it is made from a homogeneous, isotropic, linearly elastic material with a modulus of elasticity E. The deflections at end i in the positive coordinate directions are u_i and v_i, while p_{xi} and p_{yi} are the corresponding forces. The subscript i on the displacements and forces designates the point; thus there are similar quantities at point j. The objective is to relate the displacements, $\mathbf{u} = \{u_i \quad v_i \quad u_j \quad v_j\}$, and the forces, $\mathbf{p} = \{p_{xi} \quad p_{yi} \quad p_{xj} \quad p_{yj}\}$ by systematically deforming the element using four different displacement cases. We will use braces, { }, in the text to designate a column matrix.

First Element Displacement Case:

$$u_i \neq 0; \qquad v_i = u_j = v_j = 0$$

The element in this deformed configuration is shown in Fig. 2.3a. As a result of end i displacing horizontally, the element experiences axial deformation and rotation, as indicated by the dotted lines in Fig. 2.3a. Projecting displaced point i' onto the original element orientation (i.e., point i''), we observe a shortening of the element equal to $u_i \cos \phi$. The implicit assumption is that the displacements are small with respect to the overall element length, and the

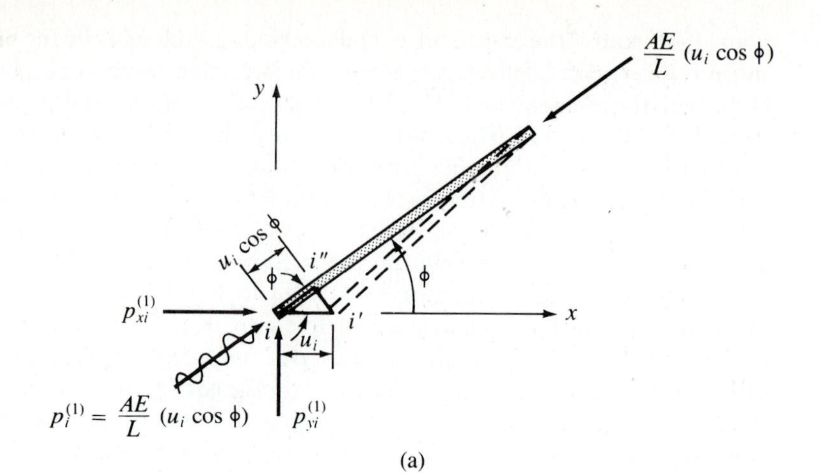

(a)

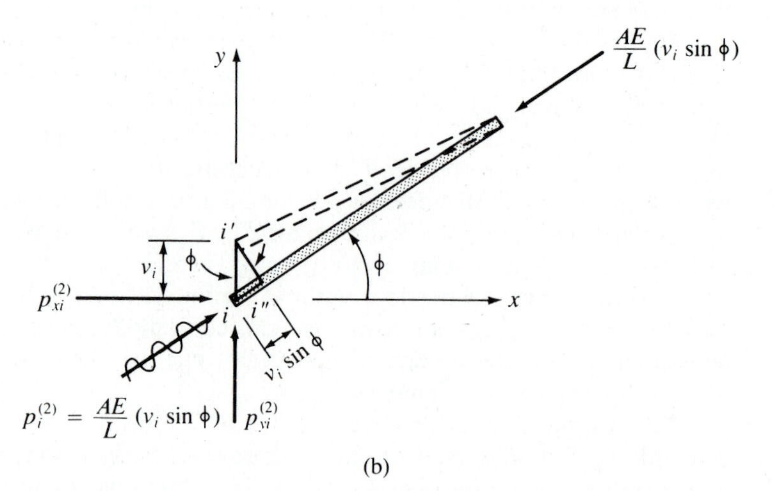

(b)

Figure 2.3 The deformed axial force element: (a) $u_i \neq 0$, $v_i = u_j = v_j = 0$; (b) $v_i \neq 0$, $u_i = u_j = v_j = 0$

deformed geometry approximates the undeformed geometry. Therefore, for the first displacement case the compressive axial force in the member at point i, $p_i^{(\text{case}\,1)}$, is obtained using Eq. (2.6):

$$p_i^{(\text{case}\,1)} = p_i^{(1)} = \frac{AE}{L} u_i \cos \phi \tag{2.7}$$

We will subsequently use only a number in parentheses as a superscript on the force to denote the displacement case number as in Eq. (2.7). The orthogonal cartesian components of the forces at the end of the element are

$$p_{xi}^{(1)} = p_i^{(1)} \cos \phi = \frac{AE}{L} (u_i \cos \phi)(\cos \phi) = \frac{AE}{L} u_i \cos^2 \phi \tag{2.8}$$

and

$$p_{yi}^{(1)} = p_i^{(1)} \sin \phi = \frac{AE}{L} (u_i \cos \phi)(\sin \phi) = \frac{AE}{L} u_i \sin \phi \cos \phi \tag{2.9}$$

For equilibrium of the element, $p_j^{(1)}$ must be equal in magnitude, but opposite in direction, to $p_i^{(1)}$ (see Fig. 2.3a). The two components of force at point j are

$$p_{xj}^{(1)} = -p_{xi}^{(1)} = -\frac{AE}{L} u_i \cos^2 \phi \tag{2.10}$$

and

$$p_{yj}^{(1)} = -p_{yi}^{(1)} = -\frac{AE}{L} u_i \sin \phi \cos \phi \tag{2.11}$$

Second Element Displacement Case:

$$v_i \neq 0; \qquad u_i = u_j = v_j = 0$$

End i of the axial force element is given a displacement in the positive y direction while the other three displacements at the two reference points are not allowed to displace. This configuration is shown in Fig. 2.3b. Since the deformation along the element axis is equal to $v_i \sin \phi$, from Eq. (2.6) the axial compressive force at end i is

$$p_i^{(2)} = \frac{AE}{L} v_i \sin \phi \tag{2.12}$$

The x and y components of the force are

$$p_{xi}^{(2)} = p_i^{(2)} \cos \phi = \frac{AE}{L} v_i \sin \phi \cos \phi \tag{2.13}$$

and

$$p_{yi}^{(2)} = p_i^{(2)} \sin \phi = \frac{AE}{L} v_i \sin^2 \phi \tag{2.14}$$

Equilibrium of the element mandates that

$$p_{xj}^{(2)} = -p_{xi}^{(2)} = -\frac{AE}{L} v_i \sin \phi \cos \phi \tag{2.15}$$

and

$$p_{yj}^{(2)} = -p_{yi}^{(2)} = -\frac{AE}{L} v_i \sin^2 \phi \tag{2.16}$$

Third Element Displacement Case:

$$u_j \neq 0; \qquad u_i = v_i = v_j = 0$$

This deflected configuration of the element will result in an elongation equal to $u_j \cos \phi$. By invoking equilibrium and using Eq. (2.6), similar to what was done in the previous two displacement cases, the forces at the two ends are

$$p_{xj}^{(3)} = -p_{xi}^{(3)} = \frac{AE}{L} u_j \cos^2 \phi \tag{2.17}$$

and

$$p_{yj}^{(3)} = -p_{yi}^{(3)} = \frac{AE}{L} u_j \sin \phi \cos \phi \tag{2.18}$$

Fourth Element Displacement Case:

$$v_j \neq 0; \qquad u_i = v_i = u_j = 0$$

For this last individual case, point j is given a deflection in the y direction while all other deflection components are constrained. Using the equations of equilibrium, along with Eq. (2.6), gives the constraining forces:

$$p_{xj}^{(4)} = -p_{xi}^{(4)} = \frac{AE}{L} v_j \sin \phi \cos \phi \tag{2.19}$$

and

$$p_{yj}^{(4)} = -p_{yi}^{(4)} = \frac{AE}{L} v_j \sin^2 \phi \tag{2.20}$$

General Element Displacement Case:

$$u_i \neq 0; \qquad v_i \neq 0; \qquad u_j \neq 0; \qquad v_j \neq 0$$

We are investigating a linear elastic system undergoing small displacements; therefore, the principle of superposition is valid. Thus if we simultaneously give all four of the displacement components nonzero values, the total constraining forces are the sum of the forces from the four individual displacement cases; that is,

$$\begin{aligned}
p_{xi} &= p_{xi}^{(1)} + p_{xi}^{(2)} + p_{xi}^{(3)} + p_{xi}^{(4)} \\
&= \frac{AE}{L} [(\cos^2 \phi)u_i + (\sin \phi \cos \phi)v_i - (\cos^2 \phi)u_j - (\sin \phi \cos \phi)v_j] \\
p_{yi} &= p_{yi}^{(1)} + p_{yi}^{(2)} + p_{yi}^{(3)} + p_{yi}^{(4)} \\
&= \frac{AE}{L} [(\sin \phi \cos \phi)u_i + (\sin^2 \phi)v_i - (\sin \phi \cos \phi)u_j - (\sin^2 \phi)v_j] \\
p_{xj} &= p_{xj}^{(1)} + p_{xj}^{(2)} + p_{xj}^{(3)} + p_{xj}^{(4)} \\
&= \frac{AE}{L} [-(\cos^2 \phi)u_i - (\sin \phi \cos \phi)v_i + (\cos^2 \phi)u_j + (\sin \phi \cos \phi)v_j] \\
p_{yj} &= p_{yj}^{(1)} + p_{yj}^{(2)} + p_{yj}^{(3)} + p_{yj}^{(4)} \\
&= \frac{AE}{L} [-(\sin \phi \cos \phi)u_i - (\sin^2 \phi)v_i + (\sin \phi \cos \phi)u_j + (\sin^2 \phi)v_j]
\end{aligned} \tag{2.21}$$

These equations can be expressed in matrix form as

$$\underset{\mathbf{p}}{\begin{bmatrix} p_{xi} \\ p_{yi} \\ p_{xj} \\ p_{yj} \end{bmatrix}} = \frac{AE}{L} \underset{\mathbf{k}}{\begin{bmatrix} \cos^2 \phi & \sin \phi \cos \phi & -\cos^2 \phi & -\sin \phi \cos \phi \\ \sin \phi \cos \phi & \sin^2 \phi & -\sin \phi \cos \phi & -\sin^2 \phi \\ -\cos^2 \phi & -\sin \phi \cos \phi & \cos^2 \phi & \sin \phi \cos \phi \\ -\sin \phi \cos \phi & -\sin^2 \phi & \sin \phi \cos \phi & \sin^2 \phi \end{bmatrix}} \underset{\mathbf{u}}{\begin{bmatrix} u_i \\ v_i \\ u_j \\ v_j \end{bmatrix}} \tag{2.22}$$

or

$$\mathbf{p} = \mathbf{ku} \tag{2.22a}$$

Matrix equation (2.22) describes the relationship between the forces and the displacements at the ends of an individual axial force element that is orientated in a general position in two-dimensional space. The forces appear in the same position in the force matrix (**p**) as their associated displacements occur in the displacement matrix (**u**).

The matrix **k** is the element stiffness matrix, and it has a number of unique characteristics that should be noted. A particular column of **k** contains the forces associated with a specific displaced configuration. For example, compare the second column of **k** and the forces computed in the second displacement case described previously. The rules of matrix multiplication mandate this; that is, if we set $v_i \neq 0$, with $u_i = u_j = v_j = 0$, the product **ku** gives the force matrix **p** that contains the forces calculated in the second displacement case. A specific column in **k** represents the set of equilibrating forces necessary to maintain a specific deformed shape; hence the sum of each column is zero. This is a characteristic of the axial force member, but it is not inherent in all element stiffness matrices. Note also that **k** is symmetric (i.e., $k_{ij} = k_{ji}$): a fact that can be demonstrated using the theorem of virtual work (see Sec. 3.2). The reader may wish to defer investigation of this property in detail until reading Ch. 3. Finally, we note that in general **k** has no inverse. That is, there is no unique set of displacements for an arbitrary **p** (unless these forces happen to be in equilibrium). This fact is easily explained by noting the physical problem. No constraints have been specified for the element. Therefore, it is possible for the member to display rigid-body displacements in addition to deformations, and there are no unique displacements for a given set of applied forces. This problem of singularity with respect to solution is alleviated when the element is adequately supported against rigid-body motion (two perpendicular translations and a rotation in two dimensions), or several members are combined into an assemblage that is constrained against rigid-body motion. This point is discussed in Sec. 2.4.

Example 2.1 demonstrates the use of Eq. (2.22) for an individual axial force element that is adequately constrained and loaded at one node.

EXAMPLE 2.1

Calculate the deflections and forces for the single strut in Fig. E2.1 using the stiffness matrix for the axial force member [Eq. (2.22)]; $A = 1600\ \text{mm}^2$ and $E = 200$ GPa. The support symbol at point a can resist horizontal and vertical forces, while that at point b can resist horizontal forces (see "Graphic Symbols" in "Symbols and Notation").

Solution

$$\frac{AE}{L} = \frac{1600 \times 10^{-6}(200 \times 10^{6})}{4} = 80 \times 10^{3}\ \text{kN/m}$$

$$\cos \phi = 0.500$$

$$\sin \phi = 0.866$$

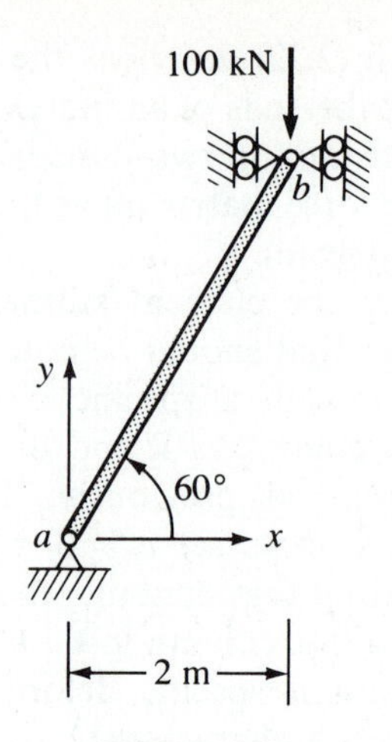

Figure E2.1

From Eq. (2.22),

$$\begin{bmatrix} p_{xa} \\ p_{ya} \\ p_{xb} \\ p_{yb} \end{bmatrix} = 80 \times 10^3 \begin{bmatrix} 0.250 & 0.433 & -0.250 & -0.433 \\ 0.433 & 0.750 & -0.433 & -0.750 \\ -0.250 & -0.433 & 0.250 & 0.433 \\ -0.433 & -0.750 & 0.433 & 0.750 \end{bmatrix} \begin{bmatrix} u_a \\ v_a \\ u_b \\ v_b \end{bmatrix}$$

But $u_a = v_a = u_b = 0$ and $p_{yb} = -100$ kN. From the fourth equation,

$$-100 = 80 \times 10^3(0.750v_b)$$
$$v_b = -1.67 \times 10^{-3} \text{ m}$$

and the first three equations give

$$p_{xa} = 80 \times 10^3(-0.433v_b) = +57.7 \text{ kN}$$
$$p_{ya} = 80 \times 10^3(-0.750v_b) = +100.0 \text{ kN}$$
$$p_{xb} = 80 \times 10^3(0.433v_b) = -57.7 \text{ kN}$$

2.3 NODAL EQUILIBRIUM OF THE STRUCTURE

The basic equations of structural mechanics must be satisfied throughout the structure. In the case of a truss, this is done within each individual element through Eq. (2.6a), but since a truss is an assemblage of several elements, inter-element compatibility and equilibrium must be imposed. We will investigate how this is done by first examining two axial force elements connected in series. The principles involved with this problem will be extended to a two-dimensional truss to study the general form of the procedure.

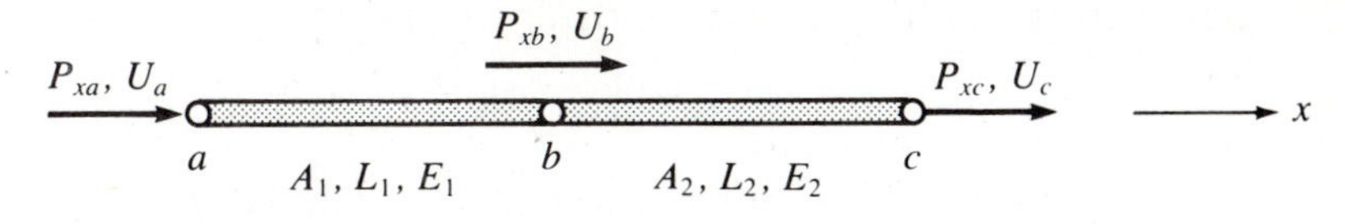

Figure 2.4 The assemblage with two axial force elements

The assemblage in Fig. 2.4 is composed of two axial force elements as described in the previous section [note that $(A_i E_i/L_i) = k_i$ for convenience]. This simple structure is constrained against displacements transverse to the element axes; these restraints are not depicted in Fig. 2.4 for clarity. The relationships between the forces P_{xa}, P_{xb}, and P_{xc} and the corresponding displacements U_{xa}, U_{xb}, and U_{xc} will first be obtained by superposing the forces from three independent displacement cases, wherein each point is selectively given a nonzero displacement while the other points are constrained. This approach is similar to the procedure used for the single element in Sec. 2.2. Note that capital letters are used to designate forces and displacements at the points of the assemblage. This distinguishes these global values from the lower-case quantities for an individual element (i.e., the notation used in Sec. 2.2).

First Assemblage Displacement Case:

$$U_a \neq 0; \qquad U_b = U_c = 0$$

Only element *ab* will be deformed; therefore, using Eq. (2.6a) we calculate the forces on the ends of the element. These are shown in Fig. 2.5a, along with the equilibrating forces exerted on the nodes according to Newton's third law. The superscripts in parentheses on the forces denote the displacement case number.

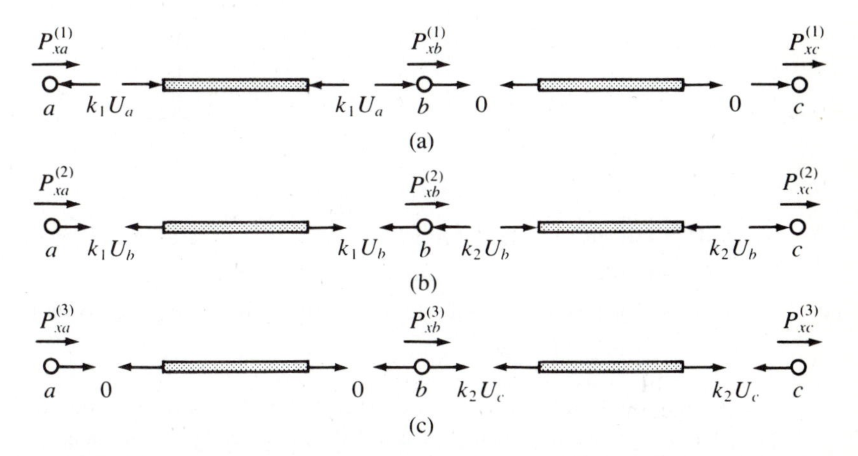

Figure 2.5 Two-element assemblage showing equilibrating forces for nodes and elements for three displacement cases: (a) $U_a \neq 0$; (b) $U_b \neq 0$; (c) $U_c \neq 0$

Second Assemblage Displacement Case:

$$U_b \neq 0; \qquad U_a = U_c = 0$$

Since both elements are attached to node *b*, continuity requires that deformations and hence forces occur in both elements. By invoking Eq. (2.6a) we obtain the forces shown on the elements and nodes in Fig. 2.5b.

Third Assemblage Displacement Case:

$$U_c \neq 0; \qquad U_a = U_b = 0$$

This displaced state is similar to that in the first case in that only one element has been deformed. Applying the appropriate governing equation gives the forces shown in Fig. 2.5c.

General Assemblage Displacement Case:

$$U_a \neq 0; \qquad U_b \neq 0; \qquad U_c \neq 0$$

If nonzero values of U_a, U_b, and U_c are imposed simultaneously, the relationships between these displacements and the corresponding forces are obtained by superposing the results from the three independent displacement cases. We must observe the positive coordinate directions shown in Fig. 2.4, and note that since the elements are connected at node *b*, the total force at that node is obtained by adding the forces from each individual element. This gives

$$\begin{aligned} P_{xa} &= P_{xa}^{(1)} + P_{xa}^{(2)} + P_{xa}^{(3)} = k_1 U_a - k_1 U_b + 0 U_c \\ P_{xb} &= P_{xb}^{(1)} + P_{xb}^{(2)} + P_{xb}^{(3)} = -k_1 U_a + (k_1 + k_2) U_b - k_2 U_c \\ P_{xc} &= P_{xc}^{(1)} + P_{xc}^{(2)} + P_{xc}^{(3)} = 0 U_a - k_1 U_b + k_2 U_c \end{aligned} \tag{2.23}$$

In matrix form these force-displacement equations become

$$\underset{\mathbf{P}}{\begin{bmatrix} P_{xa} \\ P_{xb} \\ P_{xc} \end{bmatrix}} = \underset{\mathbf{K}}{\begin{bmatrix} k_1 & -k_1 & 0 \\ -k_1 & k_1 + k_2 & -k_2 \\ 0 & -k_2 & k_2 \end{bmatrix}} \underset{\mathbf{U}}{\begin{bmatrix} U_a \\ U_b \\ U_c \end{bmatrix}} \tag{2.24}$$

or

$$\mathbf{P} = \mathbf{K}\mathbf{U} \tag{2.25}$$

where $\mathbf{P}$ = column matrix of nodal forces
$\mathbf{U}$ = column matrix of nodal displacements
$\mathbf{K}$ = stiffness matrix of the assemblage (also referred to as the *structural stiffness matrix*)

A perusal of Eq. (2.24) reveals that a given column of the stiffness matrix represents the equilibrating forces for a prescribed displaced shape of the assemblage (i.e., a situation that is analogous to what we observed for the single element in Sec. 2.2). Thus, K_{11} is the force P_{xa} due to a unit displacement of U_a (with $U_b = U_c = 0$), K_{12} is the force P_{xa} due to a unit displacement of U_b (with $U_a = U_c = 0$), etc. Since all forces act in the x direction, a given column must sum to zero for assemblage equilibrium. This suggests that the stiffness matrix can be generated directly from the forces required to maintain equilibrium by imposing selective displacements at each node.

Before proceeding, Eq. (2.22) must be modified for the special case in which the elements are aligned with the x axis. In this case $\phi = 0°$ (see Fig. 2.2). There can be no displacement normal to the element axis because there is no resistance to force (i.e., stiffness) in that direction. Thus, by ignoring the degrees of freedom in the y direction (i.e., p_{yi}, p_{yj}, v_i, and v_j are all zero in this case and all reference to these has been omitted), we have

$$\underset{\mathbf{p}}{\begin{bmatrix} p_{xi} \\ p_{xj} \end{bmatrix}} = \frac{AE}{L} \underset{\mathbf{k}}{\begin{bmatrix} 1 & -1 \\ -1 & 1 \end{bmatrix}} \underset{\mathbf{u}}{\begin{bmatrix} u_i \\ u_j \end{bmatrix}} \tag{2.26}$$

Thus, for the case of elements all aligned with the x axis, $\mathbf{k}$ is the 2 by 2 matrix shown in Eq. (2.26).

Returning to a discussion of the assemblage in Fig. 2.4, we note that Eq. (2.23) shows the equations of nodal equilibrium for the assemblage, and these results indicate a convenient method for assembling the structural stiffness matrix $\mathbf{K}$. Nodal equilibrium is assured by summing the stiffness coefficients from the individual elements that are connected to a given node (this observation will be verified subsequently for more general trusses). This procedure is indicated symbolically in Fig. 2.6, and we note the process can be envisaged as a simple overlay of the individual element stiffness matrices [with the form shown in Eq. (2.26)]. Thus, we obtain $\mathbf{K}$ by starting with a null square matrix, with a size equal to the number of degrees of freedom for the problem, and placing the individual element stiffness matrices, $\mathbf{k}$, in the rows

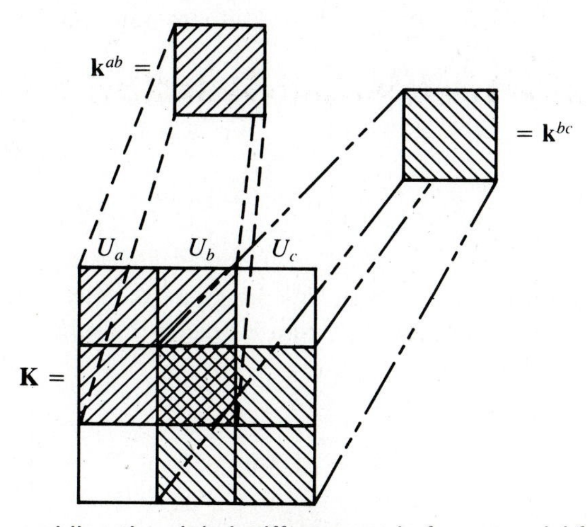

Figure 2.6 Assembling the global stiffness matrix for two axial force elements in series (see Fig. 2.4)

and columns to which they relate. Wherever the **k**'s overlap, addition of the matrix elements is implied, and if no entries from any **k** are overlaid in a particular location of **K**, the entry remains zero. It is important to note that this is not matrix addition in the conventional sense, but it is sometimes referred to as submatrix addition (i.e., **K** is the matrix into which the submatrices, **k**, are added). The combination of the individual element stiffness matrices into the structural stiffness matrix for the entire structure is referred to as the *assembly* or *merge process.* Thus, the assembly process is a convenient method for obtaining **K** which obviates having to calculate numerous equilibrium configurations for the assemblage. Numerical results for a two-element assemblage are presented in Example 2.2.

EXAMPLE 2.2

Obtain the structural stiffness matrix for the two-element assemblage in Fig. E2.2 (displacements are possible only in the x direction); $k_1 = (A_1E_1)/L_1 = 30{,}000$ kN/m and $k_2 = (A_2E_2)/L_2 = 15{,}000$ kN/m.

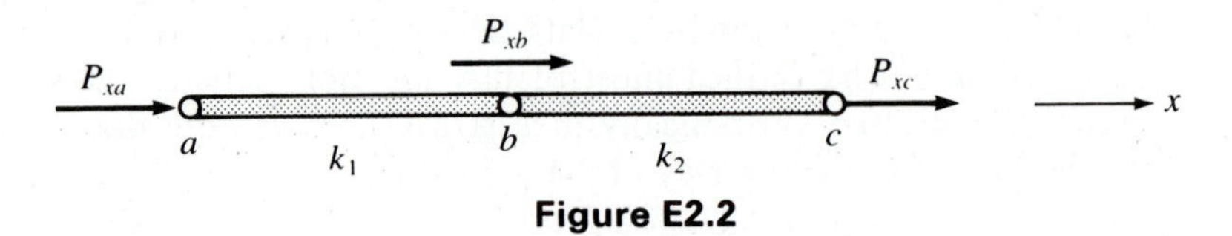

Figure E2.2

Solution

The following individual element stiffness matrices are obtained by substituting into Eq. (2.22):

$$\mathbf{k}^{ab} = 30{,}000\begin{bmatrix} 1 & -1 \\ -1 & 1 \end{bmatrix} = 15{,}000\begin{bmatrix} 2 & -2 \\ -2 & 2 \end{bmatrix}$$

$$\mathbf{k}^{bc} = 15{,}000\begin{bmatrix} 1 & -1 \\ -1 & 1 \end{bmatrix}$$

where the superscripts on the **k**'s designate the element. By starting with a null 3 by 3 matrix and adding $\mathbf{k}^{ab}$ and $\mathbf{k}^{bc}$ using the submatrix procedure implied in Fig. 2.6, we have

$$\mathbf{K} = 15{,}000\begin{bmatrix} 2 & -2 & 0 \\ -2 & 3 & -1 \\ 0 & -1 & 1 \end{bmatrix}$$

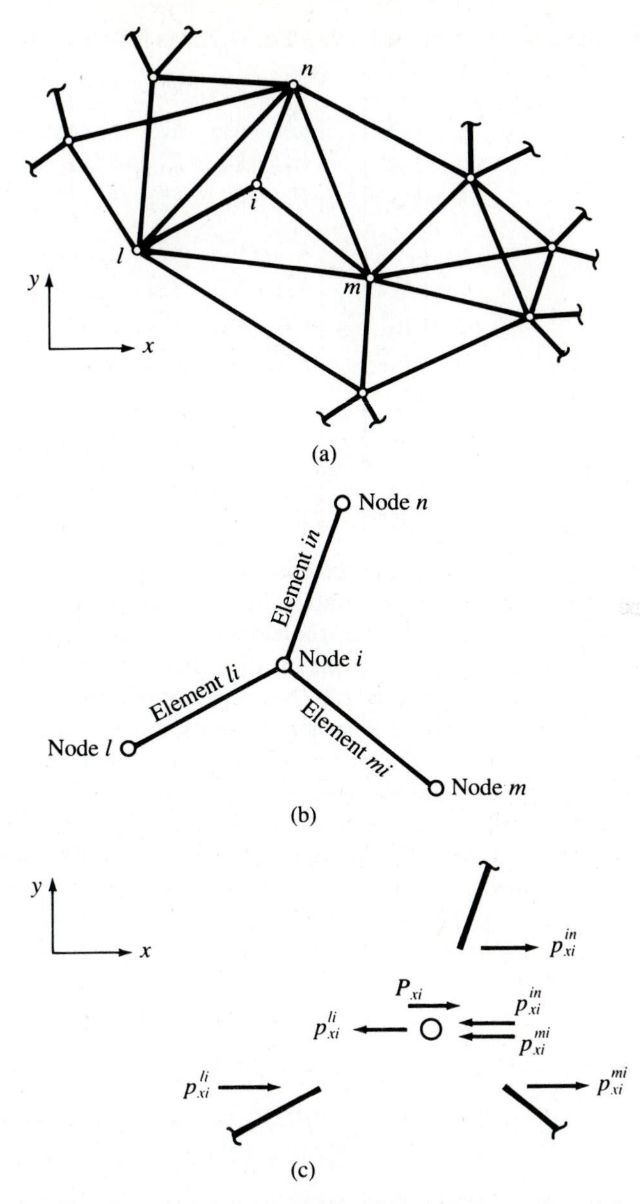

Figure 2.7 (a) Truss showing key nodes; (b) element connectivity at node *i*; (c) free-body diagram of node *i* (for clarity only the *x* force components are shown)

Equilibrium must be enforced for each independent force component that exists at every nodal point of a structure. The truss in Fig. 2.7a has two degrees of freedom per node, and this requires that we use $\mathbf{k}$ as shown in Eq. (2.22). Consider the elements connected to node i in Fig. 2.7b. The force-displacement equations for each of the three axial force elements attached to this node can be expressed using Eq. (2.22) as follows:

$$\begin{bmatrix} p_{xl}^{li} \\ p_{yl}^{li} \\ p_{xi}^{li} \\ p_{yi}^{li} \end{bmatrix} = \begin{bmatrix} k_{11}^{li} & k_{12}^{li} & k_{13}^{li} & k_{14}^{li} \\ k_{21}^{li} & k_{22}^{li} & k_{23}^{li} & k_{24}^{li} \\ k_{31}^{li} & k_{32}^{li} & k_{33}^{li} & k_{34}^{li} \\ k_{41}^{li} & k_{42}^{li} & k_{43}^{li} & k_{44}^{li} \end{bmatrix} \begin{bmatrix} u_l \\ v_l \\ u_i \\ v_i \end{bmatrix} \tag{2.27a}$$

$$\begin{bmatrix} p_{xm}^{mi} \\ p_{ym}^{mi} \\ p_{xi}^{mi} \\ p_{yi}^{mi} \end{bmatrix} = \begin{bmatrix} k_{11}^{mi} & k_{12}^{mi} & k_{13}^{mi} & k_{14}^{mi} \\ k_{21}^{mi} & k_{22}^{mi} & k_{23}^{mi} & k_{24}^{mi} \\ k_{31}^{mi} & k_{32}^{mi} & k_{33}^{mi} & k_{34}^{mi} \\ k_{41}^{mi} & k_{42}^{mi} & k_{43}^{mi} & k_{44}^{mi} \end{bmatrix} \begin{bmatrix} u_m \\ v_m \\ u_i \\ v_i \end{bmatrix} \tag{2.27b}$$

$$\begin{bmatrix} p_{xi}^{in} \\ p_{yi}^{in} \\ p_{xn}^{in} \\ p_{yn}^{in} \end{bmatrix} = \begin{bmatrix} k_{11}^{in} & k_{12}^{in} & k_{13}^{in} & k_{14}^{in} \\ k_{21}^{in} & k_{22}^{in} & k_{23}^{in} & k_{24}^{in} \\ k_{31}^{in} & k_{32}^{in} & k_{33}^{in} & k_{34}^{in} \\ k_{41}^{in} & k_{42}^{in} & k_{43}^{in} & k_{44}^{in} \end{bmatrix} \begin{bmatrix} u_i \\ v_i \\ u_n \\ v_n \end{bmatrix} \tag{2.27c}$$

where the stiffness entries are those from Eq. (2.22); the subscripts denote the position of the entry in the matrix, and the superscripts indicate the end nodes of the element. For example, the element designated by *in* has the local axes directed from node i to node n with the corresponding implied counterclockwise angle ϕ (see Fig. 2.2). No superscripts are required for the displacements since for compatibility all elements must be connected to a common node. For example, $u_i^{li} = u_i^{mi} = u_i^{in} = u_i = U_i$ (since the nodal displacement refers to the global coordinates).

The following describes equilibrium in the x direction only; a similar argument could be made for the y direction. The free-body diagrams for the elements connected to node i are shown in Fig. 2.7c. Only the x forces are shown for clarity. The force applied externally to the node in the x direction is P_{xi}. The equation of equilibrium in the x direction gives

$$P_{xi} = p_{xi}^{li} + p_{xi}^{mi} + p_{xi}^{in}$$

Upon substituting from Eqs. (2.27a) through (2.27c), the equation becomes

$$\begin{aligned} P_{xi} = {} & (k_{33}^{li} + k_{33}^{mi} + k_{11}^{in})U_i + (k_{34}^{li} + k_{34}^{mi} + k_{12}^{in})V_i + k_{31}^{li}U_1 \\ & + k_{32}^{li}V_1 + k_{31}^{mi}U_m + k_{32}^{mi}V_m + k_{13}^{in}U_n + k_{14}^{in}V_n \end{aligned} \tag{2.28}$$

or

$$\begin{aligned} P_{xi} = {} & K_{gg}U_i + K_{gh}V_i + K_{go}U_l + K_{gp}V_l + K_{gq}U_m + K_{gr}V_m \\ & + K_{gs}U_n + K_{gt}V_n \end{aligned} \tag{2.29}$$

where $K_{gg} = k_{33}^{li} + k_{33}^{mi} + k_{11}^{in}$ and $K_{gh} = k_{34}^{li} + k_{34}^{mi} + k_{12}^{in}$

and so forth. That is, the displacement in the x direction at node i (U_i) is considered as the gth degree of freedom for the truss, while the y displacement at node i (V_i) is the hth degree of freedom, etc. The stiffnesses contributed to a given degree of freedom by the various structural elements must be added into the appropriate location in the structural stiffness matrix (**K**). This process is portrayed in Fig. 2.8 and illustrated for a three-member truss in Example 2.3.

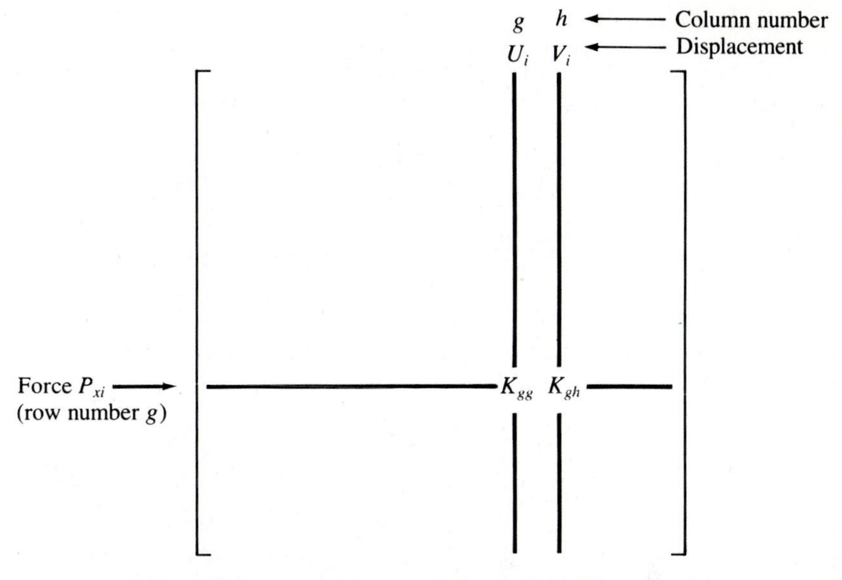

Figure 2.8 Arrangement of the global stiffness matrix

EXAMPLE 2.3

Obtain the structural stiffness matrix for the three-member truss in Fig. E2.3; $A_{ab} = 600\ \text{mm}^2$, $A_{bc} = 1000\ \text{mm}^2$, $A_{ac} = 800\ \text{mm}^2$, and $E = 200$ GPa.

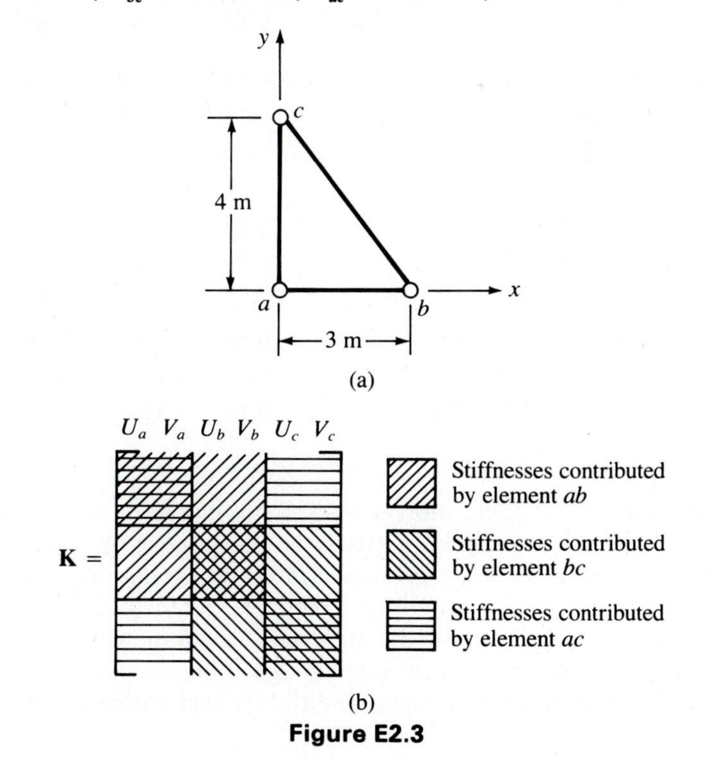

Figure E2.3

Solution

For all elements $AE/L = 40{,}000$ kN/m. For element *ab*, $\phi = 0°$, $\cos\phi = 1.00$, $\sin\phi = 0.00$. Thus

$$k^{ab} = 40{,}000 \begin{bmatrix} \overset{U_a}{1.00} & \overset{V_a}{0.00} & \overset{U_b}{-1.00} & \overset{V_b}{0.00} \\ 0.00 & 0.00 & 0.00 & 0.00 \\ -1.00 & 0.00 & 1.00 & 0.00 \\ 0.00 & 0.00 & 0.00 & 0.00 \end{bmatrix}$$

For element *bc*, $\phi = 126.9°$, $\cos\phi = -0.60$, $\sin\phi = 0.80$. Thus

$$k^{bc} = 40{,}000 \begin{bmatrix} \overset{U_b}{0.36} & \overset{V_b}{-0.48} & \overset{U_c}{-0.36} & \overset{V_c}{0.48} \\ -0.48 & 0.64 & 0.48 & -0.64 \\ -0.36 & 0.48 & 0.36 & -0.48 \\ 0.48 & -0.64 & -0.48 & 0.64 \end{bmatrix}$$

For element *ac*, $\phi = 90°$, $\cos\phi = 0.00$, $\sin\phi = 1.00$, and

$$k^{ac} = 40{,}000 \begin{bmatrix} \overset{U_a}{0.00} & \overset{V_a}{0.00} & \overset{U_c}{0.00} & \overset{V_c}{0.00} \\ 0.00 & 1.00 & 0.00 & -1.00 \\ 0.00 & 0.00 & 0.00 & 0.00 \\ 0.00 & -1.00 & 0.00 & 1.00 \end{bmatrix}$$

Note that the degree of freedom associated with each matrix column has been labeled to help with visualization. Combining the element stiffness matrices as shown in Fig. E2.3b gives the following force-displacement matrix equation for the truss:

$$\begin{bmatrix} P_{xa} \\ P_{ya} \\ P_{xb} \\ P_{yb} \\ P_{xc} \\ P_{yc} \end{bmatrix} = 40{,}000 \begin{bmatrix} 1.00 & 0.00 & -1.00 & 0.00 & 0.00 & 0.00 \\ 0.00 & 1.00 & 0.00 & 0.00 & 0.00 & -1.00 \\ -1.00 & 0.00 & 1.36 & -0.48 & -0.36 & 0.48 \\ 0.00 & 0.00 & -0.48 & 0.64 & 0.48 & -0.64 \\ 0.00 & 0.00 & -0.36 & 0.48 & 0.36 & -0.48 \\ 0.00 & -1.00 & 0.48 & -0.64 & -0.48 & 1.64 \end{bmatrix} \begin{bmatrix} U_a \\ V_a \\ U_b \\ V_b \\ U_c \\ V_c \end{bmatrix}$$

Discussion Each individual stiffness matrix is written using Eq. (2.22), and the structural stiffness matrix is constructed by assembling these three matrices using the logic developed in the foregoing discussion. Note that the assembly of the element stiffness matrices cannot be accomplished by simple matrix addition because the element **k**'s and **K** are not conformable with respect to addition. The graphic assembly representation in Fig. E2.3b assumes that **K** is initially null and the element stiffness matrices are added to the corresponding matrix elements as they are placed in the appropriate locations.

The assembly or merge process lends itself beautifully to computer programming. Typically, a program will number the degrees of freedom of the structure using the nodal designations as a guide. For example, this type of numbering for the truss of Fig. 2.7 could consist of the following (assuming nodes are numbered in ascending alphabetical order):

Displacement	U_i	V_i	U_l	V_l	U_m	V_m	U_n	V_n
Degree of freedom	17	18	23	24	25	26	27	28

Hence Eq. (2.29) is the seventeenth equation of the force-displacement equations, and the structural stiffness coefficients would correspond to the following columns in the seventeenth row of matrix **K** [refer to Eqs. (2.28) and (2.29)]:

Stiffness coefficient	K_{gg}	K_{gh}	K_{go}	K_{gp}	K_{gq}	K_{gr}	K_{gs}	K_{gt}
Location (column)	17	18	23	24	25	26	27	28

Within most computer programs the assembly process is carried out element by element; thus, it is convenient to keep a correspondence table for the degrees of freedom for the structure associated with each individual element. We can envisage this as a rectangular array with the number of rows equal to the total number of degrees of freedom for the structure, and each column represents a particular element. For the truss of Fig. 2.7, the seventeenth through twenty-eighth rows of this array for the illustrated elements appear as follows:

		li	*mi*	*in*	Element number
Global degrees of freedom	17	3	3	1	
	18	4	4	2	
	19	0	0	0	
	20	0	0	0	
	21	0	0	0	
	22	0	0	0	
	23	1	0	0	
	24	2	0	0	
	25	0	1	0	
	26	0	2	0	
	27	0	0	3	
	28	0	0	4	

The numbers in the correspondence table refer to the element degrees of freedom, where u_i, v_i, u_j, v_j are the element degrees of freedom 1, 2, 3, and 4, respectively (Fig. 2.2). This same information can be stored more compactly if we store the degrees of freedom for each member as the rows and enter the global degrees of freedom association with each element degree of freedom. Thus, this improved form of the correspondence table contains only four rows but the same number of columns as before. For the three members *li*, *mi*, and *ni* the table is as follows:

		li	*mi*	*in*	Element number
Element degrees of freedom	u_i	23	25	17	
	v_i	24	26	18	
	u_j	17	17	27	
	v_j	18	18	28	

Since the objective of this book is to teach the theory rather than coding techniques, this argument will not be pursued. We urge the student to use simple matrix computer programs to solve the exercise problems in this book. Therefore, the structural stiffness matrix **K** can be formed by either using submatrix commands or enlarging each of the element **k**'s to be conformable with **K** and using matrix addition. Note that to enlarge **k**, it must be augmented with zeros in the appropriate positions. This latter approach is inefficient for large problems, but it is a tenable method for investigating relatively small structures.

There is another approach for assembling **K** that involves extensive matrix multiplications using the so-called kinematic matrix **a**. This method is extremely inefficient and was abandoned as a viable computational algorithm in the 1960s. It is presented for pedagogical reasons in Sec. 7.6, but we caution the student against adopting this approach for computations.

2.4 NODAL DISPLACEMENTS

The next step in the process requires solution of the equations of nodal equilibrium in the form shown in Eq. (2.25). It appears that if the forces were known and the deflections unknown, the deflections **U** could be obtained by premultiplying the load matrix **P** by the inverse of **K**. In general this is not possible since **P** and **U** both contain a mixture of known and unknown quantities. For example, we note that the two-element assemblage in Fig. 2.4 has not been constrained against motion. Therefore, if an arbitrary set of forces is applied so that equilibrium in the x direction is not attained, i.e.,

$$P_{xa} + P_{xb} + P_{xc} \neq 0$$

the elements will exhibit elastic deformations plus rigid-body translation along the x axis. Since the assemblage does not have at least one node restrained, **K** is singular (i.e., has no inverse); consequently there are no unique displacements for an arbitrary load combination. The reader can verify this fact by attempting to invert **K** as it appears in Eq. (2.24).

If a structure is constrained against rigid-body motion the problem of calculating the deflections is tractable. The displacements are prescribed for the constrained degrees of freedom, and the associated forces (reactions) are unknown. For the other degrees of freedom the forces are specified and the displacements must be calculated. Since both $\mathbf{P}$ and $\mathbf{U}$ contain both known and unknown quantities, we partition the matrices in the force-deflection equations as follows:

$$\begin{bmatrix} \mathbf{P}_f \\ \hline \mathbf{P}_s \end{bmatrix} = \left[\begin{array}{c|c} \mathbf{K}_{ff} & \mathbf{K}_{fs} \\ \hline \mathbf{K}_{sf} & \mathbf{K}_{ss} \end{array}\right] \begin{bmatrix} \mathbf{U}_f \\ \hline \mathbf{U}_s \end{bmatrix} \tag{2.30}$$

where, in general, the subscript f refers to the free or unrestrained nodes and the subscript s designates a supported or constrained node. Hence, $\mathbf{P}_f$ contains applied forces and $\mathbf{U}_f$ is the column matrix of corresponding displacements, while $\mathbf{U}_s$ represents prescribed displacements, and $\mathbf{P}_s$ contains the associated forces. Multiplying the partitioned matrices in Eq. (2.30) gives the two matrix equations

$$\mathbf{P}_f = \mathbf{K}_{ff}\mathbf{U}_f + \mathbf{K}_{fs}\mathbf{U}_s \quad \text{and} \quad \mathbf{P}_s = \mathbf{K}_{sf}\mathbf{U}_f + \mathbf{K}_{ss}\mathbf{U}_s \tag{2.31}$$

When $\mathbf{U}_s$ is null (i.e., rigid supports), then

$$\mathbf{P}_f = \mathbf{K}_{ff}\mathbf{U}_f \tag{2.32}$$

and

$$\mathbf{P}_s = \mathbf{K}_{sf}\mathbf{U}_f \tag{2.33}$$

In this case $\mathbf{P}_s$ contains the reaction forces. Usually it is neither necessary nor useful at this point in the analysis to find the reaction forces; therefore, typically only Eq. (2.32) is used. A comparison of Eqs. (2.32) and (2.30) reveals that three out of four partitions in the matrix $\mathbf{K}$ need not be formed. If the complete $\mathbf{K}$ has been assembled, in view of Eq. (2.32), it is possible to discard all but the $\mathbf{K}_{ff}$ partition in order to calculate the unknown displacements at the constrained nodes (where forces $\mathbf{P}_f$ are specified). The solution process using Eq. (2.32) is

$$\mathbf{U}_f = \mathbf{K}_{ff}^{-1}\mathbf{P}_f \tag{2.34}$$

Constraining node c of the previous two-element assemblage (see Fig. 2.9), we can calculate the displacements using Eq. (2.34). In this case,

$$\begin{aligned}
\mathbf{K}_{ff} &= \begin{bmatrix} k_1 & -k_1 \\ -k_1 & k_1 + k_2 \end{bmatrix} \\
\mathbf{K}_{fs} &= \begin{bmatrix} 0 \\ -k_2 \end{bmatrix} \\
\mathbf{K}_{sf} &= [0 \quad -k_2] \\
\mathbf{K}_{ss} &= [k_2] \\
\mathbf{P}_f &= \begin{bmatrix} P_{xa} \\ P_{xb} \end{bmatrix} \\
\mathbf{P}_s &= [P_{xc}] \\
\mathbf{U}_f &= \begin{bmatrix} U_a \\ U_b \end{bmatrix} \\
\mathbf{U}_s &= [U_c]
\end{aligned} \tag{2.35}$$

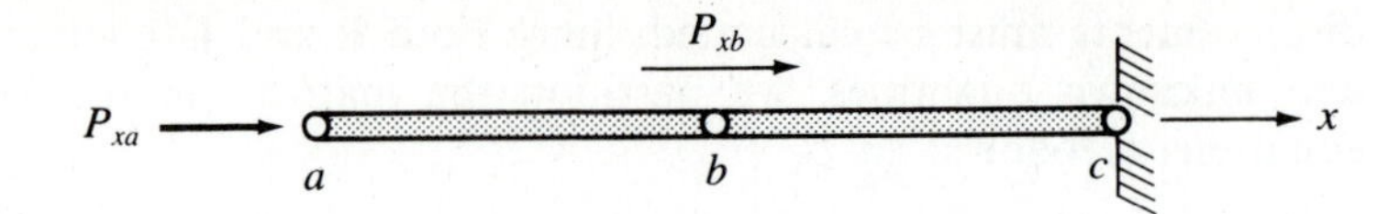

Figure 2.9 Constrained two-element assemblage showing applied nodal forces

Consequently,

$$\mathbf{K}_{ff}^{-1} = \begin{bmatrix} \frac{1}{k_2} + \frac{1}{k_1} & \frac{1}{k_2} \\ \frac{1}{k_2} & \frac{1}{k_2} \end{bmatrix} \tag{2.36}$$

Using Eq. (2.34) we have

$$\begin{bmatrix} U_a \\ U_b \end{bmatrix} = \frac{1}{k_1 k_2} \begin{bmatrix} k_1 + k_2 & k_1 \\ k_1 & k_1 \end{bmatrix} \begin{bmatrix} P_{xa} \\ P_{xb} \end{bmatrix} \tag{2.37}$$

that is,

$$U_a = \left(\frac{1}{k_1} + \frac{1}{k_2}\right) P_{xa} + \frac{1}{k_2} P_{xb} \quad \text{and} \quad U_b = \frac{1}{k_2} P_{xa} + \frac{1}{k_2} P_{xb} \tag{2.38}$$

Observe that the reaction forces $\mathbf{P}_s$ are not explicitly calculated at this point in the analysis. Usually only the partition $\mathbf{K}_{ff}$ is retained and $\mathbf{K}_{fs}$ is discarded. This makes it difficult to return to Eq. (2.33) to obtain these unknown forces. The reaction forces can be obtained after the forces in the elements are known. Calculations for element forces are discussed in Sec. 2.6.

As a matter of interest, it is possible to return to Eq. (2.33) and compute the reaction force P_{xc} for the two-element assemblage in Fig. 2.9. Thus,

$$P_{xc} = [0 \quad -k_2] \begin{bmatrix} \left(\frac{1}{k_2} + \frac{1}{k_1}\right) P_{xa} + \frac{1}{k_2} P_{xb} \\ \frac{1}{k_2} P_{xa} + \frac{1}{k_2} P_{xb} \end{bmatrix} = -P_{xa} - P_{xb}$$

That is, the assemblage is in equilibrium.

The more general partitioning scheme in Eq. (2.31) must be used when both forces and displacements are specified. For example, assume that in Fig. 2.9 P_{xa}, U_b, and U_c are given as nonzero, and we wish to calculate U_a,

P_{xb}, and P_{xc}. In this case, the various partitions are

$$\mathbf{P}_f = P_{xa}$$

$$\mathbf{U}_f = U_a$$

$$\mathbf{P}_s = \begin{bmatrix} P_{xb} \\ P_{xc} \end{bmatrix}$$

$$\mathbf{U}_s = \begin{bmatrix} U_b \\ U_c \end{bmatrix}$$

$$\mathbf{K}_{ff} = k_1$$

$$\mathbf{K}_{fs} = [-k_1 \quad 0]$$

$$\mathbf{K}_{sf} = \begin{bmatrix} -k_1 \\ 0 \end{bmatrix}$$

$$\mathbf{K}_{ss} = \begin{bmatrix} k_1 + k_2 & -k_2 \\ -k_2 & k_2 \end{bmatrix}$$

Equation (2.31) gives

$$P_{xa} = k_1 U_a + [-k_1 \quad 0]\begin{bmatrix} U_b \\ U_c \end{bmatrix} \qquad \text{or} \qquad U_a = \frac{1}{k_1} P_{xa} + U_b$$

and

$$\begin{bmatrix} P_{xb} \\ P_{xc} \end{bmatrix} = \begin{bmatrix} -k_1 \\ 0 \end{bmatrix}\left(\frac{1}{k_1} P_{xa} + U_b\right) + \begin{bmatrix} k_1 + k_2 & -k_2 \\ -k_2 & k_2 \end{bmatrix}\begin{bmatrix} U_b \\ U_c \end{bmatrix}$$
$$= \begin{bmatrix} -P_{xa} + k_2(U_b - U_c) \\ k_2(-U_b + U_c) \end{bmatrix}$$

The system is in equilibrium since $P_{xb} + P_{xc} = -P_{xa}$.

In Examples 2.2 and 2.3 we obtained the structural stiffness matrix **K** for each of the assemblages. In Examples 2.4, 2.5, and 2.6 these same structures are constrained, subjected to forces, and the unknown displacements and forces are calculated using the equations developed in this section. In addition, Example 2.7 illustrates the method applied to a truss with a more complicated arrangement of members (this structure is statically indeterminate).

EXAMPLE 2.4

The assemblage of Example 2.2 is unrestrained at joints a and b and has the following prescribed loads and displacements: $P_{xa} = 120$ kN, $P_{xb} = -75$ kN, and $U_c = 18$ mm. Calculate the unknown nodal displacements and forces.

Solution

Using the partitioning scheme shown in Eq. (2.30) gives

$$\begin{bmatrix} 120 \\ -75 \\ \hline P_{xc} \end{bmatrix} = (15 \times 10^3)\left[\begin{array}{cc|c} 2 & -2 & 0 \\ -2 & 3 & -1 \\ \hline 0 & -1 & 1 \end{array}\right]\begin{bmatrix} U_a \\ U_b \\ \hline 0.018 \end{bmatrix}$$

From Eq. (2.31),

$$\mathbf{U}_f = \mathbf{K}_{ff}^{-1}(\mathbf{P}_f - \mathbf{K}_{fs}\mathbf{U}_s)$$

$$\begin{bmatrix} U_a \\ U_b \end{bmatrix} = \frac{1}{30 \times 10^3}\begin{bmatrix} 3 & 2 \\ 2 & 2 \end{bmatrix}\left[\begin{bmatrix} 120 \\ -75 \end{bmatrix} - (15 \times 10^3)\begin{bmatrix} 0 \\ -1 \end{bmatrix}[0.018]\right]$$

$$= \frac{1}{30 \times 10^3}\begin{bmatrix} 3 & 2 \\ 2 & 2 \end{bmatrix}\left[\begin{bmatrix} 120 \\ -75 \end{bmatrix} - \begin{bmatrix} 0 \\ -270 \end{bmatrix}\right]$$

$$= \frac{1}{30 \times 10^3}\begin{bmatrix} 3 & 2 \\ 2 & 2 \end{bmatrix}\begin{bmatrix} 120 \\ 195 \end{bmatrix} = 10^{-3}\begin{bmatrix} 25 \\ 21 \end{bmatrix} \text{ m}$$

Also from Eq. (2.31),

$$\mathbf{P}_s = \mathbf{K}_{sf}\mathbf{U}_f + \mathbf{K}_{ss}\mathbf{U}_s$$

$$P_{xc} = (15 \times 10^3)[0 \quad -1] \times 10^{-3}\begin{bmatrix} 25 \\ 21 \end{bmatrix} + (15 \times 10^3)[1][0.018]$$

$$= -315 + 270 = -45 \text{ kN}$$

EXAMPLE 2.5

Calculate the displacements and reaction forces for the truss of Example 2.3 if it is supported with $V_b = U_c = V_c = 0$, $P_{xa} = 60$ kN, and $P_{ya} = 90$ kN (Fig. E2.5).

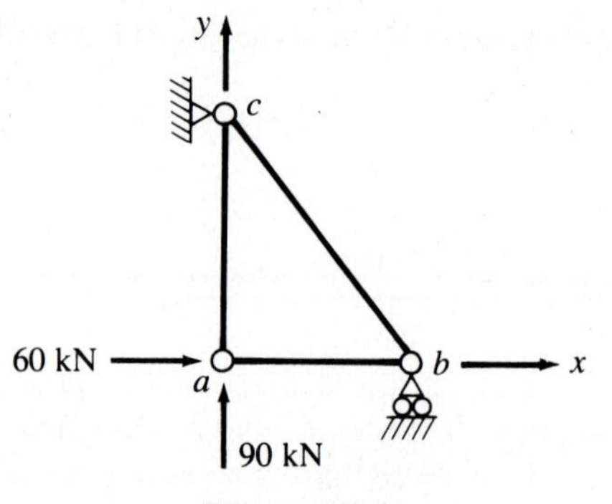

Figure E2.5

Solution

$$\begin{bmatrix} P_{xa}=60 \\ P_{ya}=90 \\ P_{xb}=0 \\ \hdashline P_{yb} \\ P_{xc} \\ P_{yc} \end{bmatrix} = 40{,}000 \left[\begin{array}{ccc:ccc} 1.00 & 0.00 & -1.00 & 0.00 & 0.00 & 0.00 \\ 0.00 & 1.00 & 0.00 & 0.00 & 0.00 & -1.00 \\ -1.00 & 0.00 & 1.36 & -0.48 & -0.36 & 0.48 \\ \hdashline 0.00 & 0.00 & -0.48 & 0.64 & 0.48 & -0.64 \\ 0.00 & 0.00 & -0.36 & 0.48 & 0.36 & -0.48 \\ 0.00 & -1.00 & 0.48 & -0.64 & -0.48 & 1.64 \end{array}\right] \begin{bmatrix} U_a \\ V_a \\ U_b \\ \hdashline V_b=0 \\ U_c=0 \\ V_c=0 \end{bmatrix}$$

From $\mathbf{U}_f = \mathbf{K}_{ff}^{-1}\mathbf{P}_f$ we get

$$\begin{bmatrix} U_a \\ V_a \\ U_b \end{bmatrix} = \frac{1}{14{,}400}\begin{bmatrix} 1.36 & 0.00 & 1.00 \\ 0.00 & 0.36 & 0.00 \\ 1.00 & 0.00 & 1.00 \end{bmatrix}\begin{bmatrix} 60 \\ 90 \\ 0 \end{bmatrix} = 10^{-3}\begin{bmatrix} 5.67 \\ 2.25 \\ 4.17 \end{bmatrix} \text{ m}$$

Using $\mathbf{P}_s = \mathbf{K}_{sf}\mathbf{U}_f$ the reaction forces are

$$\begin{bmatrix} P_{yb} \\ P_{xc} \\ P_{yc} \end{bmatrix} = 40{,}000\begin{bmatrix} 0.00 & 0.00 & -0.48 \\ 0.00 & 0.00 & -0.36 \\ 0.00 & -1.00 & 0.48 \end{bmatrix}(10^{-3})\begin{bmatrix} 5.67 \\ 2.25 \\ 4.17 \end{bmatrix} = \begin{bmatrix} -80 \\ -60 \\ -10 \end{bmatrix} \text{ kN}$$

Discussion Typical input data for loads, geometry, etc., do not warrant carrying out computations to the numerical precision used, but deflections have been shown to three significant digits so the reader can check numerical accuracy. It is not typical to calculate reaction forces in the manner illustrated, but the procedure has been used to demonstrate the use of Eq. (2.31).

EXAMPLE 2.6

Calculate the displacements and reaction forces for the truss of Example 2.3 if it is supported with $U_a = V_a = U_c = 0$, $P_{xb} = 72$ kN, and $P_{yb} = 90$ kN (Fig. E2.6).

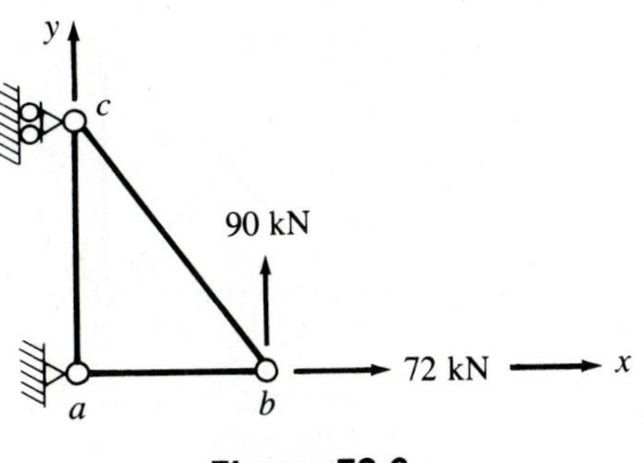

Figure E2.6

Solution

Rearranging the displacements and forces in the matrices of Example 2.5 gives

$$\begin{bmatrix} P_{xb} = 72 \\ P_{yb} = 90 \\ P_{yc} = 0 \\ \hline P_{xa} \\ P_{ya} \\ P_{xc} \end{bmatrix} = 40{,}000 \left[\begin{array}{rrr|rrr} 1.36 & -0.48 & 0.48 & -1.00 & 0.00 & -0.36 \\ -0.48 & 0.64 & -0.64 & 0.00 & 0.00 & 0.48 \\ 0.48 & -0.64 & 1.64 & 0.00 & -1.00 & -0.48 \\ \hline -1.00 & 0.00 & 0.00 & 1.00 & 0.00 & 0.00 \\ 0.00 & 0.00 & -1.00 & 0.00 & 1.00 & 0.00 \\ -0.36 & 0.48 & -0.48 & 0.00 & 0.00 & 0.36 \end{array}\right] \begin{bmatrix} U_b \\ V_b \\ V_c \\ \hline U_a = 0 \\ V_a = 0 \\ U_c = 0 \end{bmatrix}$$

From $\mathbf{U}_f = \mathbf{K}_{ff}^{-1}\mathbf{P}_f$ we get

$$\begin{bmatrix} U_b \\ V_b \\ V_c \end{bmatrix} = \frac{1}{25{,}600} \begin{bmatrix} 0.64 & 0.48 & 0.00 \\ 0.48 & 2.00 & 0.64 \\ 0.00 & 0.64 & 0.64 \end{bmatrix} \begin{bmatrix} 72 \\ 90 \\ 0 \end{bmatrix} = 10^{-3} \begin{bmatrix} 3.49 \\ 8.38 \\ 2.25 \end{bmatrix} \text{m}$$

Using $\mathbf{P}_s = \mathbf{K}_{sf}\,\mathbf{U}_f$ the reaction forces are

$$\begin{bmatrix} P_{xa} \\ P_{ya} \\ P_{xc} \end{bmatrix} = 40{,}000 \begin{bmatrix} -1.00 & 0.00 & 0.00 \\ 0.00 & 0.00 & -1.00 \\ -0.36 & 0.48 & -0.48 \end{bmatrix} 10^{-3} \begin{bmatrix} 3.49 \\ 8.38 \\ 2.25 \end{bmatrix} = \begin{bmatrix} -139.5 \\ -90.0 \\ 67.5 \end{bmatrix} \text{kN}$$

Discussion Partitioning the matrices according to Eq. (2.30) requires a reordering of the equations and displacements as shown. The remainder of the calculations proceed as in Example 2.5.

EXAMPLE 2.7

Obtain the displacements and reaction forces for the truss in Fig. E2.7. For all members $E = 29 \times 10^3$ kips/in.^{2} and $A = 0.993$ in.^{2}.

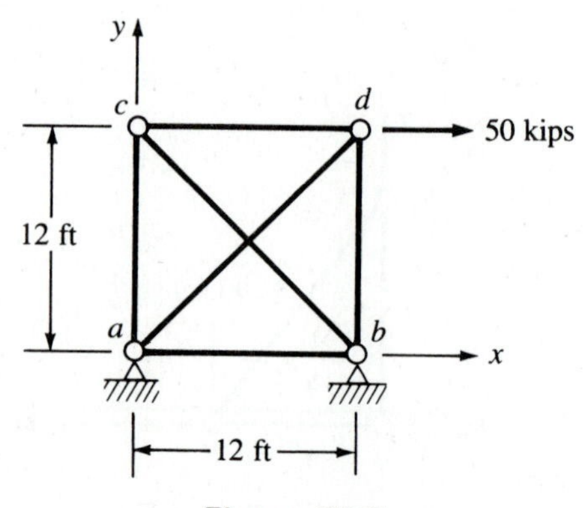

Figure E2.7

Table E2.7

Element	ϕ	cos	sin ϕ
ab, cd	0°	1	0
ac, bd	90°	0	1
ad	45°	$1/\sqrt{2}$	$1/\sqrt{2}$
bc	135°	$-1/\sqrt{2}$	$1/\sqrt{2}$

Solution

Using Eq. (2.22) for the element stiffness matrices: substituting the values from Table E2.7, along with

$$\frac{AE}{L} = \begin{cases} 200.0 \text{ kips/in.} & \text{for elements } ab, cd, ac, bd \\ 141.4 \text{ kips/in.} & \text{for elements } ad, bc \end{cases}$$

we have

$$\mathbf{k}^{ab} = \mathbf{k}^{cd} = 200.0\begin{bmatrix} 1 & 0 & -1 & 0 \\ 0 & 0 & 0 & 0 \\ -1 & 0 & 1 & 0 \\ 0 & 0 & 0 & 0 \end{bmatrix}$$

$$\mathbf{k}^{ac} = \mathbf{k}^{bd} = 200.0\begin{bmatrix} 0 & 0 & 0 & 0 \\ 0 & 1 & 0 & -1 \\ 0 & 0 & 0 & 0 \\ 0 & -1 & 0 & 1 \end{bmatrix}$$

$$\mathbf{k}^{ad} = \frac{141.4}{2}\begin{bmatrix} 1 & 1 & -1 & -1 \\ 1 & 1 & -1 & -1 \\ -1 & -1 & 1 & 1 \\ -1 & -1 & 1 & 1 \end{bmatrix}$$

$$\mathbf{k}^{bc} = \frac{141.4}{2}\begin{bmatrix} 1 & -1 & -1 & 1 \\ -1 & 1 & 1 & -1 \\ -1 & 1 & 1 & -1 \\ 1 & -1 & -1 & 1 \end{bmatrix}$$

$$\mathbf{K} = 200.0 \begin{array}{c} \begin{matrix} U_a & V_a & U_b & V_b & U_c & V_c & U_d & V_d \end{matrix} \\ \left[\begin{array}{cccc|cccc} 1.354 & 0.354 & -1.000 & 0.000 & 0.000 & 0.000 & -0.354 & -0.354 \\ 0.354 & 1.354 & 0.000 & 0.000 & 0.000 & -1.000 & -0.354 & -0.354 \\ -1.000 & 0.000 & 1.354 & -0.354 & -0.354 & 0.354 & 0.000 & 0.000 \\ 0.000 & 0.000 & -0.354 & 1.354 & 0.354 & -0.354 & 0.000 & -1.000 \\ \hline 0.000 & 0.000 & -0.354 & 0.354 & 1.354 & -0.354 & -1.000 & 0.000 \\ 0.000 & -1.000 & 0.354 & -0.354 & -0.354 & 1.354 & 0.000 & 0.000 \\ -0.354 & -0.354 & 0.000 & 0.000 & -1.000 & 0.000 & 1.354 & 0.354 \\ -0.354 & -0.354 & 0.000 & -1.000 & 0.000 & 0.000 & 0.354 & 1.354 \end{array}\right] \end{array}$$

Noting that $\mathbf{U}_s$ is null and using Eq. (2.31) gives

$$\begin{bmatrix} U_c \\ V_c \\ U_d \\ V_d \end{bmatrix} = \frac{1}{200.0}\begin{bmatrix} 1.354 & -0.354 & -1.000 & 0.000 \\ -0.354 & 1.354 & 0.000 & 0.000 \\ -1.000 & 0.000 & 1.354 & 0.354 \\ 0.000 & 0.000 & 0.354 & 1.354 \end{bmatrix}^{-1} \begin{bmatrix} 0 \\ 0 \\ 50 \\ 0 \end{bmatrix}$$

$$\begin{bmatrix} U_c \\ V_c \\ U_d \\ V_d \end{bmatrix} = \frac{1}{200.0}\begin{bmatrix} 2.135 & 0.558 & 1.693 & -0.442 \\ 0.558 & 0.884 & 0.442 & -0.116 \\ 1.693 & 0.442 & 2.135 & -0.558 \\ -0.442 & -0.116 & -0.558 & 0.884 \end{bmatrix} \begin{bmatrix} 0 \\ 0 \\ 50 \\ 0 \end{bmatrix} = \begin{bmatrix} 0.423 \\ 0.110 \\ 0.534 \\ -0.140 \end{bmatrix} \text{in.}$$

and from $\mathbf{P}_s = \mathbf{K}_{sf}\,\mathbf{U}_f$,

$$\begin{bmatrix} P_{xa} \\ P_{ya} \\ P_{xb} \\ P_{yb} \end{bmatrix} = 200.0\begin{bmatrix} 0.000 & 0.000 & -0.354 & -0.354 \\ 0.000 & -1.000 & -0.354 & -0.354 \\ -0.354 & 0.354 & 0.000 & 0.000 \\ 0.354 & -0.354 & 0.000 & -1.000 \end{bmatrix} \begin{bmatrix} 0.423 \\ 0.110 \\ 0.534 \\ -0.140 \end{bmatrix} = \begin{bmatrix} -27.9 \\ -50.0 \\ -22.1 \\ 50.0 \end{bmatrix} \text{kips}$$

Discussion Although this truss is statically indeterminate to the second degree, the analysis proceeds in exactly the same way as for the statically determinate trusses of the previous two examples; in the stiffness method the statical indeterminacy of the structure is not a consideration. A partial explanation can be obtained by recalling that statical indeterminacy is significant when one of the so-called compatibility methods is used but not when equilibrium methods are used. Since the stiffness method belongs to this second group, this consideration is suppressed during the investigation. It is sufficient for the reader to note at this point that all the pertinent equations (equilibrium, compatibility, and constitutive) are satisfied; thus the solution is valid. Static and kinematic indeterminacy are discussed in Sec. 2.8 and further investigated in Sec. 7.6.

After merging the individual element stiffness matrices, the structural stiffness matrix is obtained as shown; therefore, it must be reordered to attain the form suggested by Eq. (2.30). For this truss, it is a simple matter to extract the required partitions $\mathbf{K}_{ff}$ and $\mathbf{K}_{sf}$ without going through the actual reordering.

There are other methods for enforcing the boundary (support) conditions and obtaining displacements that are computationally more efficient than the approach implied by Eq. (2.31). It is possible to program the assembly process for $\mathbf{K}$ so that only the partition $\mathbf{K}_{ff}$ is produced. At the end of Sec. 2.3 we discussed a tabular array for denoting how members are connected together to form the structure. This array can be used in the programming logic for creating the global stiffness matrix. By noting in this array that a degree of freedom does not exist in $\mathbf{K}_{ff}$, the corresponding terms from the element stiffness matrices will not be included in the final assembled global stiffness

Table 2.1 Member connectivity array for Example E2.6

		ab	*bc*	*ac*	*Element*
	u_i	0	1	0	
Element degrees	v_i	0	2	0	
of freedom	u_j	1	0	0	
	v_j	2	3	3	
Global degrees of freedom: $U_b = 1$, $V_b = 2$, and $V_c = 3$					

matrix. For example, such an array for the truss of Example E2.6 is shown in Table 2.1. Thus the first, second, and third columns (and rows) of $\mathbf{K}_{ff}$ refer to U_b, V_b, and V_c, respectively. The element degrees of freedom correspond to those in Eq. (2.22). During the assembly when processing element *ab*, the lower right 2 by 2 matrix of Eq. (2.22) [i.e., entries (3,3), (3,4), (4,3), and (4,4)] will be placed into the upper left corner [i.e., positions (1,1), (1,2), (2,1), and (2,2)] of $\mathbf{K}_{ff}$. The remainder of this element stiffness matrix will not be used. For element *bc*, the third rows and columns of $\mathbf{k}$ will not be included in $\mathbf{K}_{ff}$, and for element *ac*, only the entry (4,4) will be added into $\mathbf{K}_{ff}$.

The boundary conditions can also be imposed upon the equilibrium equations without partitioning each of the matrices. This is done by using a computational approach that can best be explained by returning to Example 2.6. We will use $\mathbf{K}$ from Example 2.5 in which the rows and columns are ordered with the degrees of freedom in sequence. The first, second, and fifth degrees of freedom are constrained; therefore, we will replace these rows and columns by zeros, with ones for the first, second, and fifth diagonal elements. In addition, the corresponding forces will be set to zero in $\mathbf{P}$. This gives the following:

$$\begin{bmatrix} \underline{0} \\ \underline{0} \\ 72 \\ 90 \\ \underline{0} \\ 0 \end{bmatrix} = 40{,}000 \begin{bmatrix} \underline{1} & \underline{0} & \underline{0} & \underline{0} & \underline{0} & \underline{0} \\ \underline{0} & \underline{1} & \underline{0} & \underline{0} & \underline{0} & \underline{0} \\ \underline{0} & \underline{0} & 1.36 & -0.48 & \underline{0} & 0.48 \\ \underline{0} & \underline{0} & -0.48 & 0.64 & \underline{0} & -0.64 \\ \underline{0} & \underline{0} & \underline{0} & \underline{0} & \underline{1} & \underline{0} \\ \underline{0} & \underline{0} & 0.48 & -0.64 & \underline{0} & 1.64 \end{bmatrix} \begin{bmatrix} U_a \\ V_a \\ U_b \\ V_b \\ U_c \\ V_c \end{bmatrix}$$

The quantities that have replaced the original values are underlined. Note that the first, second, and fifth equations now reflect the support conditions: $U_a = 0$; $V_a = 0$; and $U_c = 0$. These constrained degrees of freedom are multiplied by zero in the third, fourth, and sixth equations. The unknown displacements are obtained by solution using all six equations. This gives the same values calculated in Example 2.6. It is important to note that the reaction forces (P_{xa}, P_{ya}, and P_{xc}) are not zero as stated in the matrix equations; we have simply set these quantities to zero for computational expediency.

A third method for enforcing support conditions also involves modifying the equations associated with the constrained degrees of freedom. Again using the structure from Example 2.6 and the equilibrium equations in the order shown in Example 2.5, we change the first, second, and fifth equations as follows:

$$P_{xa} = 40{,}000[10^8 U_a + 0.00V_a - 1.00U_b + 0.00V_b + 0.00U_c + 0.00V_c]$$
$$P_{ya} = 40{,}000[0.00U_a + 10^8 V_a + 0.00U_b + 0.00V_b + 0.00U_c - 1.00V_c]$$
$$P_{xc} = 40{,}000[0.00U_a + 0.00V_a - 0.36U_b + 0.48V_b + 10^8 U_c - 0.48V_c]$$

Note that the large number 10^8 has replaced the original stiffness coefficients located on the diagonal in **K**. The choice of the arbitrary magnitude of this value will depend upon the computer being used. If we divide the first equation by the coefficient of U_a, all the resulting coefficients in that equation, plus the left side, approximate zero. This results in a value for U_a that is approximately equal to zero (within the precision of the computer being used). By manipulating the other two equations similarly, we obtain zero displacements for the remaining constrained degrees of freedom. Rather than isolating the equations for the support forces, all the equilibrium equations are solved as a system (i.e., $\mathbf{U} = \mathbf{K}^{-1}\mathbf{P}$). In this case, large values are placed on the diagonal of **K** for the equations that are associated with the constraint forces.

While both of the above numerical procedures appear to be rational approaches, they can lead to incorrect solutions. The problem is associated with the numbers that are used to replace the original stiffness coefficients on the diagonal. If these values are too dissimilar in magnitude (either larger or smaller) from the other stiffness coefficients, the system of equations can be ill conditioned with respect to solution. The various methods commonly used for solving the equilibrium equations are described in App. A; solution errors are discussed in Sec. A.2.

2.5 COORDINATE TRANSFORMATIONS

During the analysis of a structure it is necessary to specify information in various coordinate systems. For example, the horizontal and vertical displacements of the nodes of a truss are typically calculated; but, in addition, we need the axial force in each element to carry out the subsequent design computations. For the truss in Fig. 2.10a, the overall *global coordinate system* is used to describe the nodal displacements, and an individual *local coordinate system* is needed to define the axial force in each element. Additional coordinate systems may also be necessary for some situations. For example, *nodal coordinates* are required to analyze structures with supports that are not aligned with the global coordinates; these coordinates are described in Sec. 6.2. The end forces and displacements are shown in Fig. 2.10b on element *ij* for both coordinate systems. Quantities in local coordinates are designated with an overbar, while the global coordinate values have no overbar. We will continue to use lower-case letters for quantities associated with the elements and capital letters for displacements, forces, etc., referenced to the assembled structure.

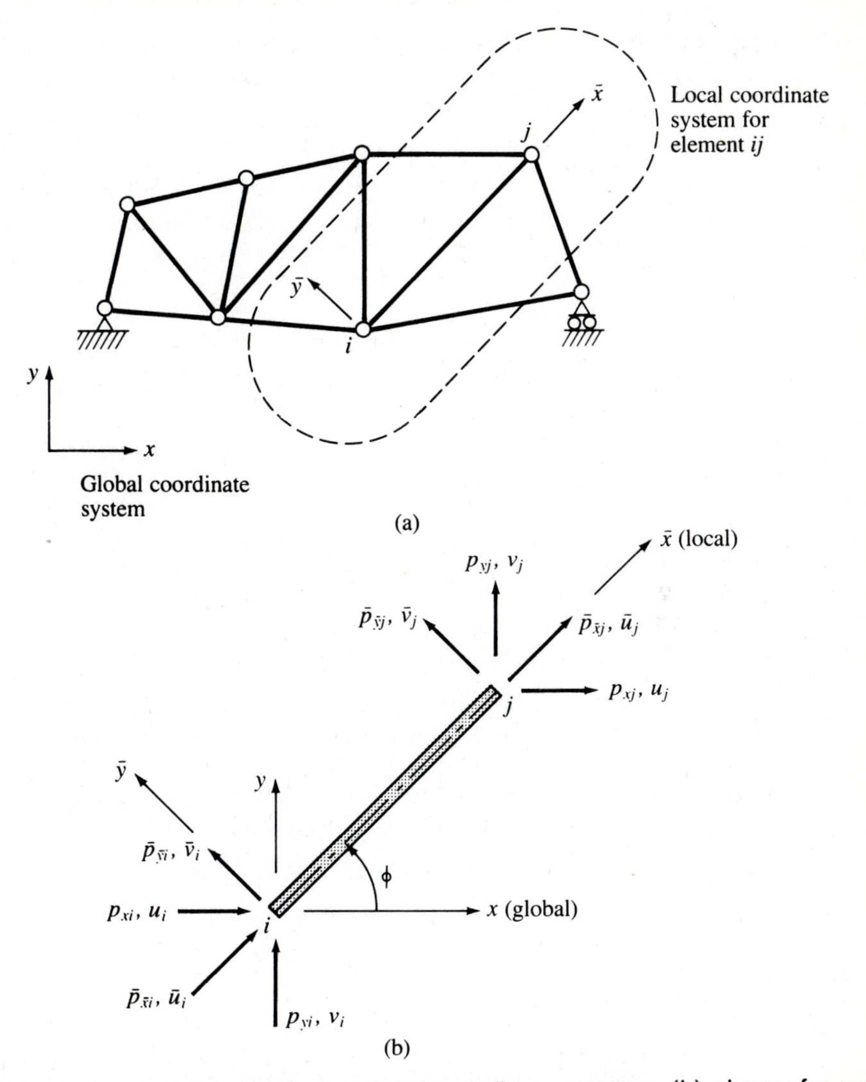

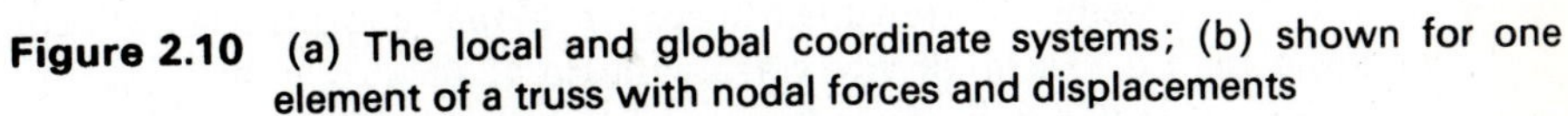

Figure 2.10 (a) The local and global coordinate systems; (b) shown for one element of a truss with nodal forces and displacements

Transforming quantities from one coordinate system to another is done using the appropriate coordinate transformations. For example, the local and global forces at node i in Fig. 2.10b are related as follows:

$$\bar{p}_{\bar{x}i} = p_{xi} \cos \phi + p_{yi} \sin \phi \tag{2.39}$$

and

$$\bar{p}_{\bar{y}i} = -p_{xi} \sin \phi + p_{yi} \cos \phi \tag{2.40}$$

Using similar trigonometric relations at node j gives

$$\bar{p}_{\bar{x}j} = p_{xj} \cos \phi + p_{yj} \sin \phi \tag{2.41}$$

and

$$\bar{p}_{\bar{y}j} = -p_{xj} \sin \phi + p_{yj} \cos \phi \tag{2.42}$$

Combining Eqs. (2.39) through (2.42) yields the following force transformation:

$$\underset{\bar{\mathbf{p}}}{\begin{bmatrix} \bar{p}_{\bar{x}i} \\ \bar{p}_{\bar{y}i} \\ \bar{p}_{\bar{x}j} \\ \bar{p}_{\bar{y}j} \end{bmatrix}} = \underset{\mathbf{T}}{\begin{bmatrix} \cos\phi & \sin\phi & 0 & 0 \\ -\sin\phi & \cos\phi & 0 & 0 \\ 0 & 0 & \cos\phi & \sin\phi \\ 0 & 0 & -\sin\phi & \cos\phi \end{bmatrix}} \underset{\mathbf{p}}{\begin{bmatrix} p_{xi} \\ p_{yi} \\ p_{xj} \\ p_{yj} \end{bmatrix}} \tag{2.43}$$

or

$$\bar{\mathbf{p}} = \mathbf{T}\mathbf{p} \tag{2.43a}$$

where $\bar{\mathbf{p}}$ = column matrix containing nodal force components in the local coordinate system

$\mathbf{p}$ = column matrix containing nodal force components in the global coordinate system

$\mathbf{T}$ = coordinate transformation matrix

The matrix **T** is an orthogonal matrix: a square matrix whose inverse equals its transpose; that is,

$$\mathbf{T}^T = \mathbf{T}^{-1} \tag{2.44}$$

We leave it as an exercise for the reader to demonstrate that $\mathbf{T}^T\mathbf{T}$ (or $\mathbf{T}\mathbf{T}^T$) equals the identity matrix **I**. This consequence of orthogonality will be useful when the transformation of the stiffness matrix is described. All coordinate transformation matrices need be neither square nor orthogonal; the more general transformation laws are investigated in Sec. 3.3.

Premultiplying both sides of Eq. (2.43a) by $\mathbf{T}^{-1}$ gives

$$\mathbf{T}^{-1}\bar{\mathbf{p}} = \mathbf{T}^{-1}\mathbf{T}\mathbf{p} = \mathbf{I}\mathbf{p} = \mathbf{p}$$

Using Eq. (2.44) yields

$$\mathbf{p} = \mathbf{T}^T\bar{\mathbf{p}} \tag{2.45}$$

Thus, it is possible to transform forces in global coordinates to local coordinates using Eq. (2.43a) or vice versa with Eq. (2.45).

Equation (2.43) is a transformation law for two vectors (column matrices). The relationship between nodal displacements in local and global coordinates can also be expressed using this same transformation since **u** and $\bar{\mathbf{u}}$ are two vectors related in the same fashion as **p** and $\bar{\mathbf{p}}$. Hence

$$\underset{\bar{\mathbf{u}}}{\begin{bmatrix} \bar{u}_i \\ \bar{v}_i \\ \bar{u}_j \\ \bar{v}_j \end{bmatrix}} = \underset{\mathbf{T}}{\begin{bmatrix} \cos\phi & \sin\phi & 0 & 0 \\ -\sin\phi & \cos\phi & 0 & 0 \\ 0 & 0 & \cos\phi & \sin\phi \\ 0 & 0 & -\sin\phi & \cos\phi \end{bmatrix}} \underset{\mathbf{u}}{\begin{bmatrix} u_i \\ v_i \\ u_j \\ v_j \end{bmatrix}} \tag{2.46}$$

or

$$\bar{\mathbf{u}} = \mathbf{T}\mathbf{u} \tag{2.46a}$$

where $\mathbf{u}$ and $\bar{\mathbf{u}}$ are the column matrices containing the displacements expressed in the global and local coordinates, respectively. The inverse transformation is obtained by premultiplying both sides of Eq. (2.46a) by $\mathbf{T}^{-1}$ and using Eq. (2.44), giving

$$\mathbf{u} = \mathbf{T}^T\bar{\mathbf{u}} \tag{2.47}$$

Equation (2.22a) expressed the force-displacement relationship for an element in global coordinates as follows:

$$\mathbf{p} = \mathbf{k}\mathbf{u}$$

Substituting Eqs. (2.45) and (2.47) into the above equation gives

$$\mathbf{T}^T\bar{\mathbf{p}} = \mathbf{k}\mathbf{T}^T\bar{\mathbf{u}} \tag{2.48}$$

Premultiplying both sides by $\mathbf{T}$ and making use of the orthogonality property of $\mathbf{T}$ yields

$$\bar{\mathbf{p}} = \mathbf{T}\mathbf{k}\mathbf{T}^T\bar{\mathbf{u}} \tag{2.49}$$

or

$$\bar{\mathbf{p}} = \bar{\mathbf{k}}\bar{\mathbf{u}} \tag{2.50}$$

That is, since the forces and displacements in local coordinates are related as shown in Eq. (2.49), we can conclude that

$$\bar{\mathbf{k}} = \mathbf{T}\mathbf{k}\mathbf{T}^T \tag{2.51}$$

Executing the product in Eq. (2.51) gives the element stiffness matrix in local coordinates as follows:

$$\bar{k} = \frac{AE}{L}\begin{bmatrix} 1 & 0 & -1 & 0 \\ 0 & 0 & 0 & 0 \\ -1 & 0 & 1 & 0 \\ 0 & 0 & 0 & 0 \end{bmatrix} \tag{2.52}$$

Hence, the complete force-displacement equation for the element is

$$\begin{bmatrix} \bar{p}_{\bar{x}i} \\ \bar{p}_{\bar{y}i} \\ \bar{p}_{\bar{x}j} \\ \bar{p}_{\bar{y}j} \end{bmatrix} = \frac{AE}{L}\begin{bmatrix} 1 & 0 & -1 & 0 \\ 0 & 0 & 0 & 0 \\ -1 & 0 & 1 & 0 \\ 0 & 0 & 0 & 0 \end{bmatrix}\begin{bmatrix} \bar{u}_i \\ \bar{v}_i \\ \bar{u}_j \\ \bar{v}_j \end{bmatrix} \tag{2.53}$$

The information contained in Eq. (2.53) is identical to that in Eq. (2.26); it describes the basic force-deflection relationships for an axial force element where equilibrium, compatibility, and the material relationships have been satisfied [see Eq. (2.6a)].

The stiffness matrix can also be transformed from local to global coordinates. Starting with Eq. (2.50) and using the force and displacement transformations [Eqs. (2.43a) and (2.46a)], along with the orthogonality condition for $\mathbf{T}$, gives

$$\mathbf{p} = \mathbf{T}^T\bar{\mathbf{k}}\mathbf{T}\mathbf{u} \tag{2.54}$$

Therefore,

$$\mathbf{k} = \mathbf{T}^T\bar{\mathbf{k}}\mathbf{T} \tag{2.55}$$

The stiffness transformation in Eq. (2.51) was derived initially since the matrix $\mathbf{k}$ had already been obtained, but Eq. (2.55) is probably the more frequently used of the two transformations. In using a matrix interpretative program, the stiffness matrix for each individual element [in local coordinates, as shown in Eq. (2.52)] can be conveniently programmed and stored, along with the appropriate transformation matrix $\mathbf{T}$. Subsequently, the transformation from local to global coordinates can be executed using the matrix triple product in Eq. (2.55). Alternatively, for a production-level program, it is probably more efficient to generate $\mathbf{k}$ explicitly in the form of Eq. (2.22) using a subprogram.

2.6 ELEMENT FORCES

After calculating the nodal displacements, the forces in each of the individual elements can be computed; various options exist for doing this. The computation must proceed element by element. Initially, the displacement matrix, $\mathbf{u}$, for each element must be constructed by selecting the appropriate entries from the matrix of global displacements, $\mathbf{U}$ (or $\mathbf{U}_f$). If this is being done within a production-level program, the search of $\mathbf{U}$ is performed using information indicating how each member is connected to the various nodes (e.g., the member connectivity array in Table 2.1).

We can obtain forces at the ends of the element in global coordinates using Eq. (2.22). That is, $\mathbf{u}$ for each element is premultiplied by the respective element stiffness matrix. Unfortunately, generally none of these four force components are aligned with the element axis; the element axial force is of primary interest to the structural engineer.

The element forces in local coordinates, $\bar{\mathbf{p}}$, can be calculated using Eq. (2.50). This computation requires the matrix of element displacements in local coordinates, $\bar{\mathbf{u}}$, which must be obtained using Eq. (2.46a); that is,

$$\bar{\mathbf{p}} = \bar{\mathbf{k}}\bar{\mathbf{u}} = \bar{\mathbf{k}}\mathbf{T}\mathbf{u} \tag{2.56}$$

Since $\bar{\mathbf{p}} = \{\bar{p}_{\bar{x}i} \quad \bar{p}_{\bar{y}i} \quad \bar{p}_{\bar{x}j} \quad \bar{p}_{\bar{y}j}\}$, most of the information obtained by the computation of Eq. (2.56) is either redundant or of little value. Note that $\bar{p}_{\bar{y}i} = \bar{p}_{\bar{y}j} = 0$, since the element can sustain no forces in the $\bar{y}$ direction. Also, the element must be in equilibrium; therefore, $\bar{p}_{\bar{x}i} = -\bar{p}_{\bar{x}j}$. There is only one independent force component for the axial force element, and it can be obtained as follows:

$$s^{ij} = \underbrace{\frac{AE}{L}[-1 \quad 0 \quad 1 \quad 0]}_{\bar{\mathbf{e}}} \underbrace{\begin{bmatrix} \bar{u}_i \\ \bar{v}_i \\ \bar{u}_j \\ \bar{v}_j \end{bmatrix}}_{\bar{\mathbf{u}}^{ij}} \tag{2.57}$$

or
$$s^{ij} = \bar{\mathbf{e}}\bar{\mathbf{u}}^{ij} \tag{2.57a}$$

where s^{ij} is the force in element ij and the row matrix $\bar{\mathbf{e}}$ is the local element force-transformation matrix. Note that $\bar{\mathbf{e}}$ is the third row of $\bar{\mathbf{k}}$; hence s^{ij} is the element force $\bar{p}_{\bar{x}j}$. A positive value of s^{ij} indicates that the element is in tension, while compression is denoted by a negative value.

Alternatively, the element force can be calculated as follows:

$$s^{ij} = \bar{\mathbf{e}}\bar{\mathbf{u}}^{ij} = \bar{\mathbf{e}}\mathbf{T}\mathbf{u}^{ij} = \mathbf{e}\mathbf{u}^{ij} \tag{2.58}$$

where

$$\mathbf{e} = \bar{\mathbf{e}}\mathbf{T}$$

$$= \frac{AE}{L}[-1 \quad 0 \quad 1 \quad 0]\begin{bmatrix} \cos\phi & \sin\phi & 0 & 0 \\ -\sin\phi & \cos\phi & 0 & 0 \\ 0 & 0 & \cos\phi & \sin\phi \\ 0 & 0 & -\sin\phi & \cos\phi \end{bmatrix}$$

$$= \frac{AE}{L}[-\cos\phi \quad -\sin\phi \quad \cos\phi \quad \sin\phi] \tag{2.59}$$

where $\mathbf{e}$ is the global element force-transformation matrix. The approach suggested by Eq. (2.58) is more efficient than that of Eq. (2.57a) because it involves fewer computations. Examples 2.8 and 2.9 illustrate the element force calculations for Examples 2.6 and 2.7, respectively. Equation (2.57a) has been used in these examples so that the reader can clearly visualize the transformation. The reader may wish to calculate the element forces using Eq. (2.58) to compare the two approaches.

EXAMPLE 2.8

Calculate the element forces for the truss of Example 2.6.

Solution

For element ab (all displacements in m),

$$\mathbf{T} = \mathbf{I}\bar{\mathbf{u}}^{ab} = \mathbf{u}^{ab} = \{U_a \quad V_a \quad U_b \quad V_b\}$$
$$= 10^{-3}\{0.000 \quad 0.000 \quad 3.488 \quad 8.381\}$$

$$s^{ab} = \bar{\mathbf{e}}\mathbf{u}^{ab} = 40{,}000[-1 \quad 0 \quad 1 \quad 0](10^{-3})\begin{bmatrix} 0.000 \\ 0.000 \\ 3.488 \\ 8.381 \end{bmatrix} = 139.5 \text{ kN (t)}$$

For element bc,

$$\mathbf{u}^{bc} = \{U_b \quad V_b \quad U_c \quad V_c\} = 10^{-3}\{3.488 \quad 8.381 \quad 0.000 \quad 2.250\}$$

$$\bar{\mathbf{u}}^{bc} = \mathbf{T}\mathbf{u}^{bc} = \begin{bmatrix} -0.6 & 0.8 & 0.0 & 0.0 \\ -0.8 & -0.6 & 0.0 & 0.0 \\ 0.0 & 0.0 & -0.6 & 0.8 \\ 0.0 & 0.0 & -0.8 & -0.6 \end{bmatrix} (10^{-3}) \begin{bmatrix} 3.488 \\ 8.381 \\ 0.000 \\ 2.250 \end{bmatrix} = 10^{-3} \begin{bmatrix} 4.612 \\ -7.819 \\ 1.800 \\ -1.350 \end{bmatrix}$$

$$s^{bc} = \bar{\mathbf{e}}\bar{\mathbf{u}}^{bc} = 40{,}000[-1 \quad 0 \quad 1 \quad 0](10^{-3}) \begin{bmatrix} 4.612 \\ -7.819 \\ 1.800 \\ -1.350 \end{bmatrix} = -112.5 \text{ kN (c)}$$

For element ac,

$$\mathbf{u}^{ac} = \{U_a \quad V_a \quad U_c \quad V_c\} = 10^{-3}\{0.000 \quad 0.000 \quad 0.000 \quad 2.250\}$$

$$\bar{u}^{ac} = \mathbf{T}\mathbf{u}^{ac} = \begin{bmatrix} 0.0 & 1.0 & 0.0 & 0.0 \\ -1.0 & 0.0 & 0.0 & 0.0 \\ 0.0 & 0.0 & 0.0 & 1.0 \\ 0.0 & 0.0 & -1.0 & 0.0 \end{bmatrix} (10^{-3}) \begin{bmatrix} 0.000 \\ 0.000 \\ 0.000 \\ 2.250 \end{bmatrix} = 10^{-3} \begin{bmatrix} 0.000 \\ 0.000 \\ 2.250 \\ 0.000 \end{bmatrix}$$

$$s^{ac} = \bar{\mathbf{e}}\bar{\mathbf{u}}^{ac} = 40{,}000[-1 \quad 0 \quad 1 \quad 0](10^{-3}) \begin{bmatrix} 0.000 \\ 0.000 \\ 2.250 \\ 0.000 \end{bmatrix} = 90.0 \text{ kN (t)}$$

Discussion The symbols (t) and (c) following the element forces designate tension and compression, respectively. Note that the nodal displacements are shown to four significant digits as contrasted with three digits in Example 2.6. In that previous example, the displacements were the final results, whereas in the computations of this example they are intermediate results. Generally, throughout this book final results are shown to three significant digits, and four significant digits are used for intermediate computations. The applied forces are given to two significant digits; thus the member forces should also be given to this numerical precision. The additional digit has been included to allow the reader to check the results with hand calculations using static equilibrium of the truss.

EXAMPLE 2.9

Calculate the element forces for the truss of Example 2.7.

Solution

For elements *ab* and *cd* (all displacements in in.),

$$\mathbf{T} = \mathbf{I} \qquad \text{and} \qquad \bar{\mathbf{u}} = \mathbf{u}$$

$$s^{ab} = 200.0[-1 \quad 0 \quad 1 \quad 0]\begin{bmatrix} 0.000 \\ 0.000 \\ 0.000 \\ 0.000 \end{bmatrix} = 0.0 \text{ kip}$$

$$s^{cd} = 200.0[-1 \quad 0 \quad 1 \quad 0]\begin{bmatrix} 0.423 \\ 0.110 \\ 0.534 \\ -0.140 \end{bmatrix} = 22.0 \text{ kips (t)}$$

For elements *ac* and *bd*,

$$\mathbf{T} = \begin{bmatrix} 0.0 & 1.0 & 0.0 & 0.0 \\ -1.0 & 0.0 & 0.0 & 0.0 \\ 0.0 & 0.0 & 0.0 & 1.0 \\ 0.0 & 0.0 & -1.0 & 0.0 \end{bmatrix}$$

$$s^{ac} = 200.0[-1 \quad 0 \quad 1 \quad 0]\begin{bmatrix} 0.000 \\ 0.000 \\ 0.110 \\ -0.423 \end{bmatrix} = 22.0 \text{ kips (t)}$$

$$s^{bd} = 200.0[-1 \quad 0 \quad 1 \quad 0]\begin{bmatrix} 0.000 \\ 0.000 \\ -0.140 \\ -0.534 \end{bmatrix} = -28.0 \text{ kips (c)}$$

For element *ad*,

$$\mathbf{u}^{ad} = \{0.000 \quad 0.000 \quad 0.534 \quad -0.140\}$$

$$\bar{\mathbf{u}}^{ad} = \mathbf{T}\mathbf{u}^{ad} = \frac{1}{\sqrt{2}}\begin{bmatrix} 1 & 1 & 0 & 0 \\ -1 & 1 & 0 & 0 \\ 0 & 0 & 1 & 1 \\ 0 & 0 & -1 & 1 \end{bmatrix}\begin{bmatrix} 0.000 \\ 0.000 \\ 0.534 \\ -0.140 \end{bmatrix} = \begin{bmatrix} 0.000 \\ 0.000 \\ 0.279 \\ -0.479 \end{bmatrix}$$

$$s^{ad} = 141.4[-1 \quad 0 \quad 1 \quad 0]\begin{bmatrix} 0.000 \\ 0.000 \\ 0.279 \\ -0.477 \end{bmatrix} = 39.4 \text{ kips (t)}$$

For element bc,

$$\mathbf{u}^{bc} = \{0.000 \quad 0.000 \quad 0.423 \quad 0.110\}$$

$$\bar{\mathbf{u}}^{bc} = \mathbf{T}\mathbf{u}^{bc} = \frac{1}{\sqrt{2}}\begin{bmatrix} -1 & 1 & 0 & 0 \\ -1 & -1 & 0 & 0 \\ 0 & 0 & -1 & 1 \\ 0 & 0 & -1 & -1 \end{bmatrix}\begin{bmatrix} 0.000 \\ 0.000 \\ 0.423 \\ 0.110 \end{bmatrix} = \begin{bmatrix} 0.000 \\ 0.000 \\ -0.221 \\ -0.377 \end{bmatrix}$$

$$s^{bc} = 141.4[-1 \quad 0 \quad 1 \quad 0]\begin{bmatrix} 0.000 \\ 0.000 \\ -0.221 \\ -0.377 \end{bmatrix} = -31.2 \text{ kips (c)}$$

2.7 SETTLEMENT, INITIAL, AND THERMAL STRAINS

A truss with settlement of the supports can be analyzed using the nodal equilibrium equations in Eq. (2.31). In this case $\mathbf{U}_s$ is not null, and the nodal displacements with prescribed forces are obtained by solving the first equation to yield

$$\mathbf{U}_f = \mathbf{K}_{ff}^{-1}(\mathbf{P}_f - \mathbf{K}_{fs}\mathbf{U}_s) \tag{2.60}$$

The second part of Eq. (2.31) yields the forces associated with the prescribed displacements, that is,

$$\mathbf{P}_s = \mathbf{K}_{sf}\mathbf{U}_f + \mathbf{K}_{ss}\mathbf{U}_s \tag{2.61}$$

The element forces are calculated using one of the approaches described in Sec. 2.6; that is, by using either Eq. (2.56), (2.57a), or (2.58). Examples 2.10 and 2.11 demonstrate this analysis procedure for a statically determinate and also for a statically indeterminate truss, respectively.

EXAMPLE 2.10

The forces at node b are removed from the truss in Example 2.6 and the structure is subjected to the following support displacements: $U_a = 15$ mm; $V_a = 20$ mm; and $U_c = 27$ mm. Calculate: (a) the unknown nodal displacements $\mathbf{U}_f$; (b) the forces, $\mathbf{P}_s$, associated with the prescribed displacements; and (c) the forces in the three elements.

Solution

(a) Using the first of the nodal equilibrium equations [Eq. (2.31)] and the matrix partitions in Example 2.6,

$$\begin{bmatrix} 0 \\ 0 \\ 0 \end{bmatrix} = \mathbf{K}_{ff}\mathbf{U}_f + 40{,}000\begin{bmatrix} -1.00 & 0.00 & -0.36 \\ 0.00 & 0.00 & 0.48 \\ 0.00 & -1.00 & -0.48 \end{bmatrix}\begin{bmatrix} 0.015 \\ 0.020 \\ 0.027 \end{bmatrix}$$

$\mathbf{K}_{ff}$ is the same as shown in Example 2.6; therefore, solving for $\mathbf{U}_f$ using the previously obtained matrix $\mathbf{K}_{ff}^{-1}$ we have

$$\begin{bmatrix} U_b \\ V_b \\ V_c \end{bmatrix} = \frac{1}{25{,}600}\begin{bmatrix} 0.64 & 0.48 & 0.00 \\ 0.48 & 2.00 & 0.64 \\ 0.00 & 0.64 & 0.64 \end{bmatrix}\begin{bmatrix} 988.8 \\ -518.4 \\ 1318.4 \end{bmatrix} = \begin{bmatrix} 0.015 \\ 0.011 \\ 0.020 \end{bmatrix} \text{ m}$$

(b) The forces, $\mathbf{P}_s$, associated with the known displacements are obtained using Eq. (2.61) and the appropriate matrix partitions from Example 2.6:

$$\mathbf{P}_s = \begin{bmatrix} P_{xa} \\ P_{ya} \\ P_{xc} \end{bmatrix}$$

$$= 40{,}000\begin{bmatrix} -1.00 & 0.00 & 0.00 \\ 0.00 & 0.00 & -1.00 \\ -0.36 & 0.48 & -0.48 \end{bmatrix}\begin{bmatrix} 0.015 \\ 0.011 \\ 0.020 \end{bmatrix}$$

$$+ 40{,}000\begin{bmatrix} 1.00 & 0.00 & 0.00 \\ 0.00 & 1.00 & 0.00 \\ 0.00 & 0.00 & 0.36 \end{bmatrix}\begin{bmatrix} 0.015 \\ 0.020 \\ 0.027 \end{bmatrix}$$

$$= \begin{bmatrix} -600 \\ -800 \\ -389 \end{bmatrix} + \begin{bmatrix} 600 \\ 800 \\ 389 \end{bmatrix} = \begin{bmatrix} 0 \\ 0 \\ 0 \end{bmatrix} \text{ kN}$$

(c) See the basic matrices in Example 2.6. For element *ab*,

$$\bar{\mathbf{u}}^{ab} = \mathbf{u}^{ab}$$

$$s^{ab} = 40{,}000[-1 \quad 0 \quad 1 \quad 0]\begin{bmatrix} 0.015 \\ 0.020 \\ 0.015 \\ 0.011 \end{bmatrix} = 0 \text{ kN}$$

For element *bc*, transforming the displacements to local coordinates and substituting into Eq. (2.58) gives

$$s^{bc} = 40{,}000[-1 \quad 0 \quad 1 \quad 0]\begin{bmatrix} -0.0002 \\ -0.0186 \\ -0.0002 \\ -0.0336 \end{bmatrix} = 0 \text{ kN}$$

For element *ac*, transforming the displacements to local coordinates and substituting into Eq. (2.58) gives

$$s^{ac} = 40{,}000[-1 \quad 0 \quad 1 \quad 0]\begin{bmatrix} 0.020 \\ -0.015 \\ 0.020 \\ -0.027 \end{bmatrix} = 0 \text{ kN}$$

....

Discussion In general, a simple statically determinate truss such as this does not warrant an analysis using the stiffness method, since we could anticipate that neither reaction forces nor element forces will be induced by the support movements. The various nodal displacements can be checked using simple kinematics; that is, the truss experiences rigid-body translations and rotation. The point of the example is to demonstrate the use of the stiffness method when the support movements are prescribed. Note that the effects of support movement can be superposed with the response caused by applied forces. Thus, if the forces of Example 2.6 were applied simultaneously with the support movements, displacements and element forces would be additive.

EXAMPLE 2.11

The force at node d is removed from the truss in Example 2.7 and the structure is subjected to the following support displacements: $U_a = -0.25$ in.; $U_b = -0.50$ in.; and $V_b = 0.25$ in. Calculate: (a) the displacements $\mathbf{U}_f$; (b) the forces, $\mathbf{P}_s$, associated with the prescribed displacements; and (c) the forces in all elements.

Solution

(a) Using the first of the nodal equilibrium equations [Eq. (2.31)] and the matrix partitions in Example 2.7 gives

$$\begin{bmatrix} 0 \\ 0 \\ 0 \\ 0 \end{bmatrix} = \mathbf{K}_{ff}\mathbf{U}_f + 200.0 \begin{bmatrix} 0.000 & 0.000 & -0.354 & 0.354 \\ 0.000 & -1.000 & 0.354 & -0.354 \\ -0.354 & -0.354 & 0.000 & 0.000 \\ -0.354 & -0.354 & 0.000 & -1.000 \end{bmatrix} \begin{bmatrix} -0.25 \\ 0.00 \\ -0.50 \\ 0.25 \end{bmatrix}$$

$\mathbf{K}_{ff}$ is the same as shown in Example 2.7; therefore, solving for $\mathbf{U}_f$ using the previously obtained matrix $\mathbf{K}_{ff}^{-1}$ we have

$$\begin{bmatrix} U_c \\ V_c \\ U_d \\ V_d \end{bmatrix} = \frac{1}{200.0} \begin{bmatrix} 2.135 & 0.558 & 1.693 & -0.422 \\ 0.558 & 0.884 & 0.442 & -0.116 \\ 1.693 & 0.442 & 2.135 & -0.558 \\ -0.442 & -0.116 & -0.558 & 0.884 \end{bmatrix} \begin{bmatrix} -53.10 \\ 53.10 \\ -17.10 \\ 32.30 \end{bmatrix}$$

$$= \begin{bmatrix} -0.6399 \\ 0.0287 \\ -0.6112 \\ 0.2787 \end{bmatrix} \text{in.}$$

(b) The forces, $\mathbf{P}_s$, associated with the known displacements are obtained using Eq. (2.61) and the appropriate matrix partitions from Example 2.7:

$$\begin{bmatrix} P_{xa} \\ P_{ya} \\ P_{xb} \\ P_{yb} \end{bmatrix} = 200.0 \begin{bmatrix} 0.000 & 0.000 & -0.354 & -0.354 \\ 0.000 & -1.000 & -0.354 & -0.354 \\ -0.354 & 0.354 & 0.000 & 0.000 \\ 0.354 & -0.354 & 0.000 & -1.000 \end{bmatrix} \begin{bmatrix} -0.6399 \\ 0.0287 \\ -0.6112 \\ 0.2787 \end{bmatrix}$$

$$+200.0 \begin{bmatrix} 1.354 & 0.354 & -1.000 & 0.000 \\ 0.354 & 1.354 & 0.000 & 0.000 \\ -1.000 & 0.000 & 1.354 & -0.354 \\ 0.000 & 0.000 & -0.354 & 1.354 \end{bmatrix} \begin{bmatrix} -0.25 \\ 0.00 \\ -0.50 \\ 0.25 \end{bmatrix}$$

$$= \begin{bmatrix} 55.74 \\ 0.00 \\ -55.74 \\ 0.00 \end{bmatrix} \text{kips}$$

(c) See the basic matrices and transformations in Example 2.7:

$$\begin{aligned}
s^{ab} &= \bar{\mathbf{e}}\{-0.2500 \quad 0.0000 \quad -0.5000 \quad 0.2500\} = -50 \text{ kips (c)} \\
s^{cd} &= \bar{\mathbf{e}}\{-0.6399 \quad 0.0287 \quad -0.6112 \quad 0.2787\} = 5.74 \text{ kips (t)} \\
s^{ac} &= \bar{\mathbf{e}}\{\ 0.0000 \quad 0.2500 \quad 0.0287 \quad 0.6399\} = 5.74 \text{ kips (t)} \\
s^{bd} &= \bar{\mathbf{e}}\{\ 0.2500 \quad 0.5000 \quad 0.2787 \quad 0.6112\} = 5.74 \text{ kips (t)} \\
s^{ad} &= \bar{\mathbf{e}}\{-0.1768 \quad 0.1768 \quad -0.2351 \quad 0.6293\} = -8.12 \text{ kips (c)} \\
s^{bc} &= \bar{\mathbf{e}}\{\ 0.5303 \quad 0.1768 \quad 0.4728 \quad 0.4322\} = -8.12 \text{ kips (c)}
\end{aligned}$$

where

$$\bar{\mathbf{e}} = \frac{AE}{L}[-1 \quad 0 \quad 1 \quad 0]$$

Discussion This example illustrates the use of the stiffness method for support movements of a statically indeterminate truss. In contrast to the response of a statically determinate truss, the element forces and reaction forces for a statically indeterminate structure are generally nonzero. The solution procedure is identical to that used for the statically determinate truss of Example 2.10, but here the numerous computations probably warrant use of the stiffness method. Nodal equilibrium is implicitly satisfied in the stiffness method, but the reader may wish to check independently that all the nodes are in equilibrium. This procedure can be used to review the numerical accuracy of the calculations (see Sec. A.2 for a complete discussion of this point).

The effects of support settlement can also be analyzed using an approach similar to that employed for temperature changes, fabrication errors, and precambering. This procedure will be explained in the subsequent paragraphs. Adapting this method for support movement is briefly described at the end of this section. A comparison of matrix partitioning and the alternate method is contained in Example 2.15 at the end of this chapter.

When a truss is subjected to temperature changes, fabrication errors, or other initial strains, the principle of superposition can be used to obtain the solution. For example, assume that member *de* of the truss in Fig. 2.11a experiences a temperature increase above the ambient conditions; the element will increase in length as shown. This in turn will introduce nodal deflections. The structure in Fig. 2.11 is statically indeterminate so we anticipate that forces will be introduced into the elements and the reaction forces will be nonzero. The solution procedure can be envisaged as a three-step process shown in Fig. 2.11 and described as follows:

1. Apply initial fixed-end forces to the element (or elements in the more general case) subjected to thermal strains; this will result in zero nodal displacements (see Fig. 2.11b).
2. Remove the forces described in step 1 and introduce forces at all nodes that are equal in magnitude but opposite in direction to the initial fixed-

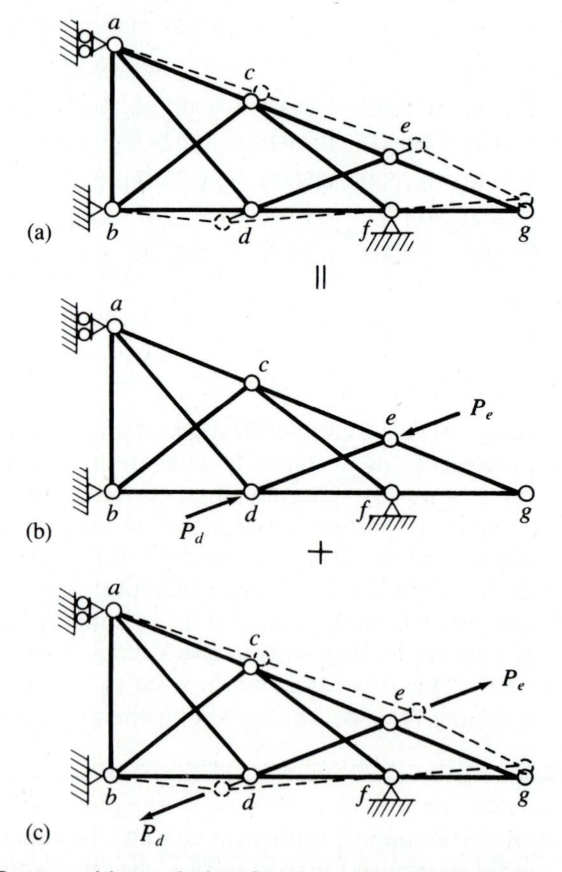

Figure 2.11 Superposition solution for thermal and initial effects: (a) displacements with elongation of member *de*; (b) all nodes fixed (step 1); (c) application of equivalent forces (step 2)

end forces of step 1 (see Fig. 2.11c). These are the so-called *equivalent forces*. Calculate the nodal displacements.

3. Superpose the results from steps 1 and 2 to obtain the actual displacements and element forces experienced by the truss (see Fig. 2.11a).

The force-displacement relation for an individual axial force element subjected to initial forces is

$$\bar{\mathbf{p}} = \bar{\mathbf{k}}\bar{\mathbf{u}} + \bar{\mathbf{p}}^o \tag{2.62}$$

where $\bar{\mathbf{p}}^o$ is the column matrix of forces (in local coordinates) that are necessary to constrain the member so that nodal displacements are zero when it experiences thermal changes, fabrication errors, etc. For example, if the axial force element of Fig. 2.10 is subjected to a temperature increase equal to ΔT, the strain is $\alpha\Delta T$, the stress is $E\alpha\Delta T$, and

$$\bar{\mathbf{p}}^o = AE\alpha\Delta T \begin{bmatrix} 1 \\ 0 \\ -1 \\ 0 \end{bmatrix} \tag{2.63}$$

where A = element cross-sectional area
E = modulus of elasticity
α = coefficient of linear thermal expansion

With an increase in temperature, the element will expand longitudinally. By noting directions of the applied forces (in local coordinates) as indicated by Eq. (2.63), we can observe that the element will be in compression.

If the element is fabricated ΔL too long, the strain is $\Delta L/L$, the stress is $E(\Delta L/L)$, and

$$\bar{\mathbf{p}}^o = \frac{AE\Delta L}{L} \begin{bmatrix} 1 \\ 0 \\ -1 \\ 0 \end{bmatrix} \tag{2.64}$$

As in the case of a temperature change, the axial forces are those necessary to constrain the element with zero nodal displacements; therefore, the element will be in compression with $\bar{p}_{xi}$ directed in the positive local coordinate direction and $\bar{p}_{xj}$ in the negative direction.

Note that the forces in Eq. (2.64) typically occur when the structure is erected, while those in Eq. (2.63) could be encountered anytime during the lifetime of the truss. In spite of this, for convenience we will designate both of these as *initial force matrices*. To analyze the response of the total truss for the effects of temperature, fabrication errors, etc., the initial force matrix for each element must be transformed into global coordinates using the transformation of Eq. (2.45). These forces are then combined at the nodal points when equilibrium of the assemblage is enforced, and this gives the following matrix equation relating forces and displacements in global coordinates for the total structure

$$\mathbf{P} = \mathbf{K}\mathbf{U} + \mathbf{P}^o \tag{2.65}$$

where $\mathbf{P}^o$ is the column matrix of the initial forces obtained by adding the nodal forces contributed by each of the elements. These matrix equations can be partitioned with the subscripting designations used in Eq. (2.30) to yield

$$\begin{bmatrix} \mathbf{P}_f \\ \hline \mathbf{P}_s \end{bmatrix} = \left[\begin{array}{c|c} \mathbf{K}_{ff} & \mathbf{K}_{fs} \\ \hline \mathbf{K}_{sf} & \mathbf{K}_{ss} \end{array}\right] \begin{bmatrix} \mathbf{U}_f \\ \hline \mathbf{U}_s \end{bmatrix} + \begin{bmatrix} \mathbf{P}_f^o \\ \hline \mathbf{P}_s^o \end{bmatrix} \tag{2.66}$$

where $\mathbf{P}_f^o$ and $\mathbf{P}_s^o$ are the column matrices of the initial forces at the nodes corresponding to $\mathbf{P}_f$ and $\mathbf{P}_s$, respectively. Multiplication gives

$$\begin{aligned} \mathbf{P}_f &= \mathbf{K}_{ff}\mathbf{U}_f + \mathbf{K}_{fs}\mathbf{U}_s + \mathbf{P}_f^o \\ \text{and} \quad \mathbf{P}_s &= \mathbf{K}_{sf}\mathbf{U}_f + \mathbf{K}_{ss}\mathbf{U}_s + \mathbf{P}_s^o \end{aligned} \tag{2.67}$$

If the supports are rigid ($\mathbf{U}_s = \mathbf{0}$), the unknown displacements are obtained by solving the first of the matrix equations in Eq. (2.67) as follows:

$$\mathbf{P}_f = \mathbf{K}_{ff}\mathbf{U}_f + \mathbf{P}_f^o \tag{2.68}$$

or

$$\mathbf{P}_f - \mathbf{P}_f^o = \mathbf{P}_f + \mathbf{P}_f^E = \mathbf{K}_{ff}\mathbf{U}_f \tag{2.68a}$$

Thus

$$\mathbf{U}_f = \mathbf{K}_{ff}^{-1}(\mathbf{P}_f - \mathbf{P}_f^o) \tag{2.68b}$$

The physical interpretation of subtracting $\mathbf{P}_f^o$ from both sides of the equation is that we are applying the initial fixed-end forces to the truss in the direction opposite to the one in which they were originally directed; this is step 2 of the process described above. Note that $-\mathbf{P}_f^o = \mathbf{P}_f^E$ where the latter set of forces are the *equivalent forces* suggested in step 2.

The solution of Eqs. (2.67) [or Eq. (2.68) if the supports are rigid] produces the nodal displacements induced by the initial forces. Now the element forces can be calculated; this is suggested in step 3 of the procedure described above. It is imperative to use all the terms of Eq. (2.62) for this computation. That is, some of the elements are initially subjected to end forces, $\bar{\mathbf{p}}^o$, and these must be included in the calculations. The first term of Eq. (2.62) ($\bar{\mathbf{k}}\bar{\mathbf{u}}$) produces the forces caused by nodal displacements. Just as in the usual computation of element forces, various options exist for performing the calculations with initial forces. The matrix $\bar{\mathbf{p}}$ in Eq. (2.62) contains redundant information: the forces normal to the element axis are zero and the end longitudinal forces are equal in magnitude and opposite in direction. By discarding all but the third equation of this set, we have

$$s^{ij} = \bar{\mathbf{e}}\bar{\mathbf{u}} + \bar{s}^o \tag{2.69}$$

where $\bar{s}^o$ is the initial element force (in local coordinates) and $\bar{\mathbf{e}}$ is given in Eq. (2.57). Recall that the force-transformation matrix $\bar{\mathbf{e}}$ was formed using the third row from the matrix $\bar{\mathbf{k}}$ (the equation that yields the local force $\bar{\mathbf{p}}_{xj}$);

therefore, $\bar{s}^o$ is the (3,1) entry of the matrix $\bar{\mathbf{p}}^o$. Hence, for a temperature change ΔT,

$$\bar{s}^o = -AE\alpha\Delta T \tag{2.69a}$$

and for an initial strain $\Delta L/L$,

$$\bar{s}^o = -\frac{AE\Delta L}{L} \tag{2.69b}$$

Alternatively, the element forces can be obtained as follows:

$$s^{ij} + \bar{\mathbf{e}}\bar{\mathbf{u}} + \bar{s}^o = \mathbf{eu} + \bar{s}^o \tag{2.70}$$

where $\mathbf{e} = \bar{\mathbf{e}}\mathbf{T}$. Equation (2.70) is computationally more efficient than Eq. (2.69). Examples 2.12 and 2.13 illustrate the calculations involving initial forces for a statically determinate and a statically indeterminate truss, respectively. Equation (2.69) has been used in these examples so that the reader can visualize the transformation; the reader may wish to carry out the element force calculations using Eq. (2.70).

EXAMPLE 2.12

The forces at node b are removed from the truss in Example 2.6; element bc is fabricated 15 mm too long. Calculate: (a) the nodal displacements; (b) the reaction forces; and (c) the element forces.

Solution

(a) The matrix of initial forces for element bc in local coordinates is obtained using Eq. (2.64) with $E = 200$ GPa and $A_{bc} = 1000\ \text{mm}^2$ (from Example 2.3). Thus we have

$$\bar{\mathbf{p}}^o = \begin{bmatrix} \bar{p}^o_{\bar{x}b} \\ \bar{p}^o_{\bar{y}b} \\ \bar{p}^o_{\bar{x}c} \\ \bar{p}^o_{\bar{y}c} \end{bmatrix} = \begin{bmatrix} 600 \\ 0 \\ -600 \\ 0 \end{bmatrix} \text{kN}$$

These forces are transformed into global coordinates as follows:

$$\mathbf{p}^o = \mathbf{T}^T\bar{\mathbf{p}}^o = \begin{bmatrix} p^o_{xb} \\ p^o_{yb} \\ p^o_{xc} \\ p^o_{yc} \end{bmatrix} = \begin{bmatrix} -0.6 & -0.8 & 0.0 & 0.0 \\ 0.8 & -0.6 & 0.0 & 0.0 \\ 0.0 & 0.0 & -0.6 & -0.8 \\ 0.0 & 0.0 & 0.8 & -0.6 \end{bmatrix} \begin{bmatrix} 600 \\ 0 \\ -600 \\ 0 \end{bmatrix} = \begin{bmatrix} -360 \\ 480 \\ 360 \\ -480 \end{bmatrix} \text{kN}$$

Using the first of the matrix equations in Eq. (2.67) and noting that $\mathbf{P}_f$ and $\mathbf{U}_s$ are both null column matrices, we have

$$\mathbf{U}_f = \mathbf{K}_{ff}^{-1}(-\mathbf{P}_f^o)$$

where $\mathbf{P}_f^o$ is obtained in this case only from $\mathbf{p}^o$ of element bc since only one element has initial forces; therefore,

$$\begin{bmatrix} U_b \\ V_b \\ V_c \end{bmatrix} = \frac{1}{25{,}600}\begin{bmatrix} 0.64 & 0.48 & 0.00 \\ 0.48 & 2.00 & 0.64 \\ 0.00 & 0.64 & 0.64 \end{bmatrix}\begin{bmatrix} 360 \\ -480 \\ 480 \end{bmatrix} = 10^{-3}\begin{bmatrix} 0.00 \\ -18.75 \\ 0.00 \end{bmatrix} \text{m}$$

(b) The force matrix $\mathbf{P}_s$ is obtained using the second matrix equation in Eq. (2.67) and the appropriate matrix partitions from Example 2.6. Note that $\mathbf{P}_s^o \neq 0$:

$$\mathbf{P}_s = 40{,}000\begin{bmatrix} -1.00 & 0.00 & 0.00 \\ 0.00 & 0.00 & -1.00 \\ -0.36 & 0.48 & -0.48 \end{bmatrix}[10^{-3}]\begin{bmatrix} 0.00 \\ -18.75 \\ 0.00 \end{bmatrix} + \begin{bmatrix} 0 \\ 0 \\ 360 \end{bmatrix} = \begin{bmatrix} 0 \\ 0 \\ 0 \end{bmatrix}$$

(c) See the basic matrices and transformations in Example 2.6:

$$\begin{aligned} s^{ab} &= \bar{\mathbf{e}}[10^{-3}]\{0.00 \quad 0.00 \quad 0.00 \quad -18.75\} = 0 \text{ kN} \\ s^{bc} &= \bar{\mathbf{e}}[10^{-3}]\{-15.00 \quad 11.25 \quad 0.00 \quad 0.00\} + [-600] = 0 \text{ kN} \\ s^{ac} &= \bar{\mathbf{e}}[10^{-3}]\{0.00 \quad 0.00 \quad 0.00 \quad 0.00\} = 0 \text{ kN} \end{aligned}$$

Discussion Again, as in Example 2.10, this simple statically determinate truss would not warrant an analysis using the stiffness method, but the point of this example is to demonstrate the application of the stiffness method when initial forces are present. Note that the fabrication error results in neither support reactions nor forces in the element since the truss is statically determinate. After using Eq. (2.67) to calculate $\mathbf{U}_f$ and $\mathbf{P}_s$, it is important to include any initial forces in the element force computations [i.e., Eq. (2.69) must be used]. For element bc, $s^o = -600$ kN, but this quantity is zero for the other two elements.

EXAMPLE 2.13

The force at node d is removed from the truss in Example 2.7; element cd is heated 100°F above the temperature of the other elements. Calculate: (a) the nodal displacements; (b) the reaction forces; and (c) the element forces.

Solution

(a) The matrix of initial forces in local coordinates is obtained using Eq. (2.63). From Example 2.7, $E = 29 \times 10^6$ lb/in.^{2} and $A_{cd} = 0.993$ in.^{2}; furthermore, let

$\alpha = 6.5 \times 10^{-6}$ in./(in. · °F). For element cd,

$$\bar{\mathbf{p}}^o = \mathbf{p}^o = \begin{bmatrix} p^o_{xc} \\ p^o_{yc} \\ p^o_{xd} \\ p^o_{yd} \end{bmatrix} = \begin{bmatrix} 18.72 \\ 0.00 \\ -18.72 \\ 0.00 \end{bmatrix} \text{kips}$$

Using the first of the matrix equations in Eq. (2.67) and noting that $\mathbf{P}_f$ and $\mathbf{U}_s$ are both null, we get

$$\mathbf{U}_f = \mathbf{K}_{ff}^{-1}(-\mathbf{P}^o_f)$$

where $\mathbf{P}^o_f$ is in general obtained by combining the initial force matrices contributed by each of the elements. Since in this case only one element is heated, $\mathbf{P}^o_f = \mathbf{p}^{o(cd)}$ because the displacements in $\mathbf{U}_f$ are also the displacements of element cd. Thus,

$$\begin{bmatrix} U_c \\ V_c \\ U_d \\ V_d \end{bmatrix} = \frac{1}{200.0} \begin{bmatrix} 2.135 & 0.558 & 1.693 & -0.442 \\ 0.558 & 0.884 & 0.442 & -0.116 \\ 1.693 & 0.442 & 2.135 & -0.558 \\ -0.442 & 0.116 & -0.558 & 0.884 \end{bmatrix} \begin{bmatrix} -18.72 \\ 0.00 \\ 18.72 \\ 0.00 \end{bmatrix}$$

$$= \begin{bmatrix} -0.0414 \\ -0.0108 \\ 0.0414 \\ -0.0108 \end{bmatrix} \text{in.}$$

(b) The force matrix $\mathbf{P}_s$ is obtained using the second matrix equation in Eq. (2.67) and the appropriate matrix partitions from Example 2.7. Note that $\mathbf{P}^o_s = \mathbf{0}$, which yields

$$\mathbf{P}_s = \begin{bmatrix} P_{xa} \\ P_{ya} \\ P_{xb} \\ P_{yb} \end{bmatrix} = 200.0 \begin{bmatrix} 0.000 & 0.000 & -0.354 & -0.354 \\ 0.000 & -1.000 & -0.354 & -0.354 \\ -0.354 & 0.354 & 0.000 & 0.000 \\ 0.354 & -0.354 & 0.000 & -1.000 \end{bmatrix} \begin{bmatrix} -0.0414 \\ -0.0108 \\ 0.0414 \\ -0.0108 \end{bmatrix}$$

$$= \begin{bmatrix} -2.16 \\ 0.00 \\ 2.16 \\ 0.00 \end{bmatrix} \text{kips}$$

(c) See the basic matrices and transformations in Example 2.7:

$$s^{ab} = 0 \text{ kips}$$

$$s^{cd} = \bar{\mathbf{e}}\{-0.0414 \quad -0.0108 \quad 0.0414 \quad -0.0108\} + [-18.72]$$
$$= 16.56 \quad -18.72 = -2.16 \text{ kips (c)}$$

$$s^{bd} = s^{ac} = \bar{\mathbf{e}}\{0.0000 \quad 0.0000 \quad -0.0108 \quad 0.0414\} = -2.16 \text{ kips (c)}$$

$$s^{ad} = s^{bc} = \bar{\mathbf{e}}\{0.0000 \quad 0.0000 \quad 0.0216 \quad -0.0369\} = 3.05 \text{ kips (t)}$$

....

Discussion The modest temperature increase in one element of this statically indeterminate truss results in nonzero forces in the elements and the two horizontal reactions. The solution differs from that of Example 2.12 in that $\mathbf{P}_s^o$ is null and hence does not enter into the calculations for $\mathbf{P}_s$. Since only element *cd* has initial forces s^o, Eq. (2.69) reduces to Eq. (2.57a) for all other elements. Because the truss and the thermal expansion are both geometrically symmetric, it is to be expected that the deflections and element forces will also be geometrically symmetric.

As an alternate to the matrix partitioning of Eqs. (2.60) and (2.61), the three-step approach described for temperature effects, fabrication errors, and precambering can be used for support movements. The initial element force matrices, $\mathbf{p}^o$, contain the forces required to constrain all nodes (except those with prescribed displacements) from moving. In Example 2.10, all elements will experience initial forces, and these are computed as follows:

$$\begin{aligned}
\mathbf{p}^{o(ab)} &= \mathbf{k}^{ab}\{0.015 \quad 0.020 \quad 0.000 \quad 0.000\} \\
&= \{600.0 \quad 0.0 \quad -600.0 \quad 0.0\}\ \text{kN} \\
\mathbf{p}^{o(ac)} &= \mathbf{k}^{ac}\{0.015 \quad 0.020 \quad 0.000 \quad 0.000\} \\
&= \{0.0 \quad 800.0 \quad 0.0 \quad -800.0\}\ \text{kN} \\
\mathbf{p}^{o(bc)} &= \mathbf{k}^{bc}\{0.000 \quad 0.000 \quad 0.027 \quad 0.000\} \\
&= \{-388.8 \quad 518.4 \quad 388.8 \quad -518.4\}\ \text{kN}
\end{aligned}$$

where

$$\mathbf{p}^{o(ij)} = \{p_{xi}^o \quad p_{yi}^o \quad p_{xj}^o \quad p_{yj}^o\}$$

In using Eq. (2.68), $\mathbf{P}_f$ is null and $\mathbf{P}_f^o = \{P_{xb}^o \quad P_{yb}^o \quad P_{yc}^o\}$ is obtained by combining forces obtained from $\mathbf{p}^{o(ab)}$, $\mathbf{p}^{o(ac)}$, and $\mathbf{p}^{o(bc)}$, which gives

$$\mathbf{P}^o = \{-988.8 \quad 518.4 \quad -1318.4\}\ \text{kN}$$

Note that $\mathbf{P}^o$ is the same column matrix obtained from the product $\mathbf{K}_{fs}\mathbf{U}_s$ in Example 2.10; that is, these forces will yield the identical displacements calculated previously. Recall that for calculating element forces in Example 2.10 we used Eq. (2.61), but now we must employ Eq. (2.70) where $\bar{s}^{o(ab)} = -600$ kN (c), $\bar{s}^{o(ac)} = -800$ kN (c), and $\bar{s}^{o(bc)} = -648$ kN (c). This procedure will show that all element forces are zero, which is the same conclusion reached in Example 2.10. The settlement problem of Example 2.11 could also be solved using this alternative approach; Example 2.15 is another illustration of the use of initial forces for investigating support movements.

2.8 STATIC AND KINEMATIC INDETERMINACY

The indeterminacy of a structure is a property that influences structural response. For discrete systems the *static indeterminacy* equals the number of unknown forces in excess of the available equations of static equilibrium, whereas the *kinematic indeterminacy* is the number of degrees of freedom required to completely describe the movement of the structure. For compatibility methods of analysis, static indeterminacy is the parameter of interest, while kinematic indeterminacy is the corresponding measure for equilibrium methods. Note that the truss in Example 2.6 is statically determinate because the number of unknown forces equals the number of equations of static equilibrium. That is, there are three unknown reaction forces and three additional member forces to be calculated, but at each of the three nodes there are two equilibrium equations available. In contrast, the statically indeterminate truss of Example 2.7 has ten unknown forces (six element forces and four reactions) to be calculated, but there are only eight equations of static equilibrium. Both structures were analyzed using the stiffness method with no immediate need to know the static indeterminacy of the structure. Note that Example 2.6 required three displacements to describe the response and Example 2.7 needed four. These are the number of global force-displacement equations required in each problem, and they are also the kinematic indeterminacy of the two trusses.

Typically, the first step in the static analysis of a truss is to establish whether it can be solved using only the equations of static equilibrium. If NE is the number of elements (truss members) and NR is the number of independent reaction forces, the number of unknown forces is NE + NR. Writing the two nontrivial equations of equilibrium at each node gives 2NN equations, where NN is the number of nodes. Thus, the criteria for classification can be summarized as follows: if

$$\text{NE} + \text{NR} - 2\text{NN}\begin{cases} = 0, & \text{the truss is statically determinate.} \\ > 0, & \text{the truss is statically indeterminate.} \\ < 0, & \text{the truss is a mechanism.} \end{cases} \tag{2.71}$$

These criteria include the degree of static indeterminacy, or redundancy, of both the internal (i.e., elements) and the external (i.e., reactions) forces, but they do not ensure that the structure will be stable. That is, the first two criteria are *necessary conditions* for static indeterminacy, but they are *not sufficient* to state that the structure is stable. The degree of static indeterminacy equals the number of additional equations of displacement that must be used to augment the equations of static equilibrium when using one of the compatibility methods. For example, the truss in Fig. 2.11a is statically indeterminate to the second degree. If this structure were being investigated using the method of consistent displacements, we would first reduce the structure to a statically determinate one by removing two support forces (e.g., those at node f) and obtain the member forces plus the horizontal and vertical displacements at node f for this primary structure. Subsequently, we would calculate the forces necessary to ensure compatibility (zero displacements at node f in this case). This compatibility method could also be used if the truss had more than the minimum number of elements necessary to maintain internal equilibrium. The three-moment equation is an example of another compatibility method that is used to solve statically indeterminate beams. The flexibility method (also called the force method) is based upon compatibility arguments; it is described in detail in Ch. 7.

In contrast, the strategies used in equilibrium methods of analysis do not depend upon a conscious knowledge of static indeterminacy. The stiffness method is an equilibrium method; this designation stems from the fact that equilibrium is invoked at the nodal points in establishing the force-displacement equations. The reader may recall from previous experience that the slope-deflection and moment-distribution methods are both equilibrium methods for analyzing beams and frames. Kinematic and static indeterminacy are dual measures for the equilibrium and compatibility methods, respectively. Establishing kinematic indeterminacy is not a crucial aspect of the stiffness method, but it is a measure of the complexity of the problem. Kinematic determinacy is the condition where all nodal displacements are constrained to zero. Thus, in the stiffness method, in effect we initially reduce the structure to a kinematically determinate state by envisaging that all unconstrained nodes are restrained to zero. The kinematic indeterminacy (redundancy) is the number of degrees of freedom that must be imaginatively restrained to accomplish this. Following this, overall structural equilibrium is restored at each node; this is described in Sec. 2.3 [e.g., see Eqs. (2.27) through (2.29)]. The kinematic redundancy is the number of global force-displacement equations that must be solved and equals the size of the partition $\mathbf{K}_{ff}$. The trusses of Examples 2.6 and 2.7 have three and four kinematic redundancies, respectively.

The static and kinematic indeterminacy of various trusses are shown in Fig. 2.12. In general, for trusses, the kinematic indeterminacy is greater than the static indeterminacy; this is not the case for all types of structures. We note that in general

$$\text{NOS} = \text{NE} + \text{NR} - 2\text{NN} \tag{2.72a}$$

and

$$\text{NOK} = 2\text{NN} - \text{NR} \tag{2.72b}$$

where NOS is the number of static redundancies and NOK is the number of kinematic redundancies. Adding the above two equations gives

$$\text{NOS} + \text{NOK} = \text{NE}$$

In Fig. 2.12a and b we have the same truss with different supports. In this case NE = 15, and we note that increasing the static indeterminacy of the truss decreases the kinematic indeterminacy. Similarly, for the truss of Fig. 2.12c and d, additional supports in the latter structure imply a smaller set of force-displacement equations to be solved. Computational efficiency for the stiffness and flexibility methods cannot simply be measured by comparing the kinematic and static indeterminacies of a given structure. The number of simultaneous linear algebraic equations to be solved is one measure of solution effort, but the method and ease of assembling the equations must also be considered.

The topic of indeterminacy is discussed for each different type of structure encountered in the subsequent chapters. A more thorough investigation of the

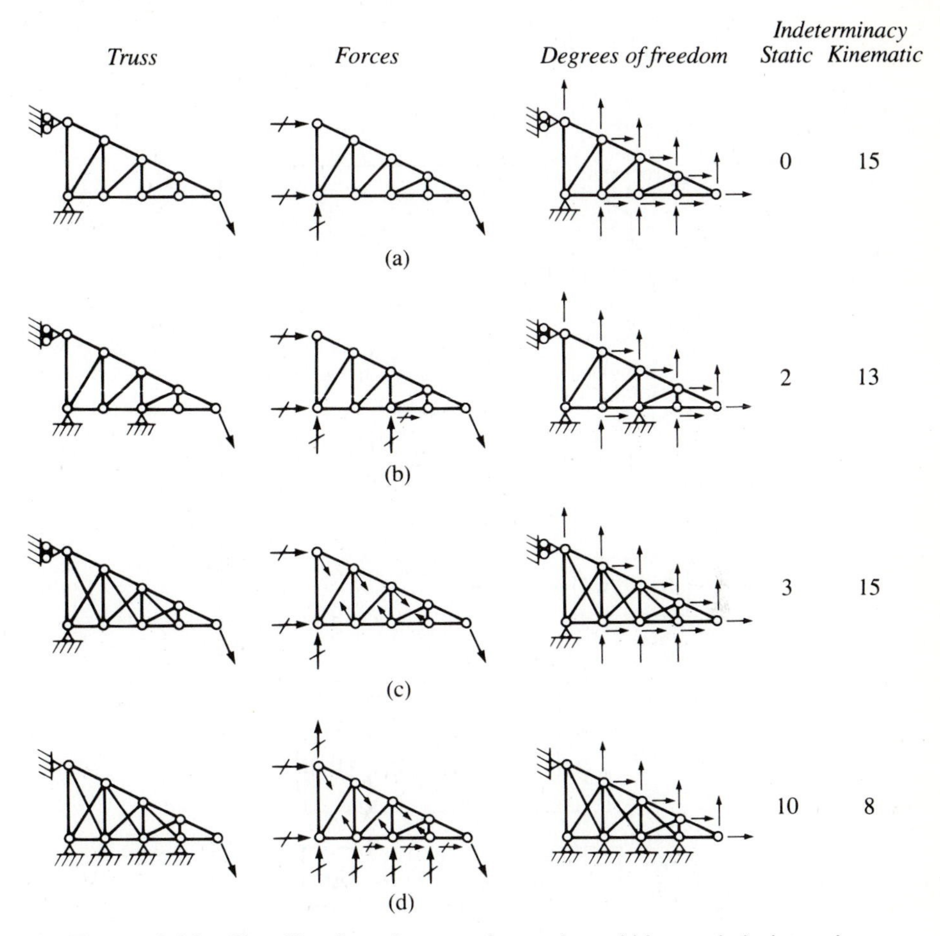

Figure 2.12 Classification of trusses for static and kinematic indeterminacy

basic concepts is contained in Ch. 7 where the duality of the flexibility and stiffness methods is examined.

2.9 SUMMARY

The basic tenets of the matrix stiffness method have been described and illustrated for planar trusses. The fundamental processes of the method are summarized in Fig. 2.13, and pertinent equation numbers are indicated for convenience. Before executing the analysis, the structure must be defined, and this requires three broad categories of input data: (a) nodal; (b) element; and (c) load. The location of each node in global coordinates must be specified. In addition, we must prescribe whether or not the associated degrees of freedom are constrained; this information is subsequently used in partitioning the global stiffness matrix. Each element is designated by the nodes at its ends, its

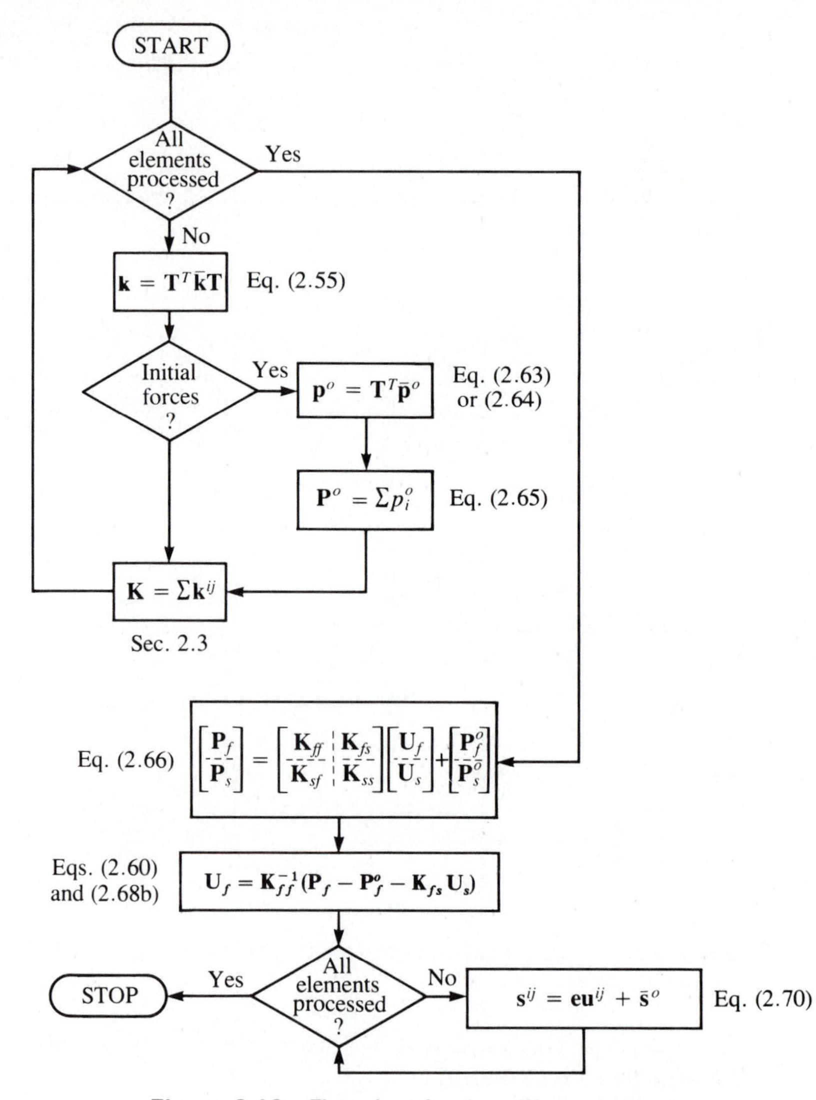

Figure 2.13 Flowchart for the stiffness method

cross-sectional area (geometric property), and the material properties. Also, initial forces (i.e., forces induced by thermal changes, fabrication errors, and/or precambering) are typically indicated when describing the elements. The applied nodal forces must also be prescribed as part of the input data.

A major task in the stiffness method is the formation of the global stiffness matrix ($\mathbf{K}$). Note in Fig. 2.13 that assembling $\mathbf{K}$ is done by creating the stiffness matrix for each element in global coordinates ($\mathbf{k}$) and adding these to the global stiffness matrix. This is the meaning of the symbol $\sum \mathbf{k}^{ij}$, where $\mathbf{k}^{ij}$ is the stiffness matrix of the element connected to nodes i and j. $\mathbf{K}$ is not formed by simple matrix addition since $\mathbf{K}$ and $\mathbf{k}^{ij}$, in general, are not conformable for addition; hence the summation sign is symbolic of the assembly process. The nodal forces associated with initial effects ($\mathbf{p}^o$) are calculated while looping through the elements. The $\mathbf{p}^o$ contributed by each element are added to form the column matrix of initial forces ($\mathbf{P}^o$). Again the summation sign is used to symbolize the process.

The nodal input data are then used to partition the global force-displacement equations according to Eq. (2.66). Most programs do not form all four partitions of $\mathbf{K}$ and then select the pertinent parts as indicated in Fig. 2.13. Rather, if the structure is constrained by rigid supports so that $\mathbf{U}_s$ is null, only $\mathbf{K}_{ff}$ is assembled and retained. The solution for the global displacements is the most time-consuming computer operation of the stiffness method. This is typically done by one of the elimination methods, which are fully described in Sec. A.1.

The final step of the stiffness method consists of calculating the forces in each of the elements. Again, as in assembling $\mathbf{K}$, the computations are performed element by element. In executing Eq. (2.70), $\mathbf{u}$ for each element must be formed by searching and selecting the correct entries from $\mathbf{U}_f$ (or $\mathbf{U}$ if the supports are not rigid). If initial forces are involved, it is imperative that $\bar{\mathbf{s}}^o$ be added in as prescribed by Eq. (2.70) since this type of analysis is basically a three-step process (see Sec. 2.7).

Examples 2.14 and 2.15 apply the stiffness method in its entirety for two trusses. Both structures are subjected to applied forces; in addition, the former has a fabrication error, while the latter is subjected to support settlement. It may be instructive to consult the flowchart of Fig. 2.13 while reviewing the computations of these examples.

EXAMPLE 2.14

Calculate the nodal displacements and element forces for the truss in Fig. E2.14: (a) for the illustrated loading and (b) if the loads shown are

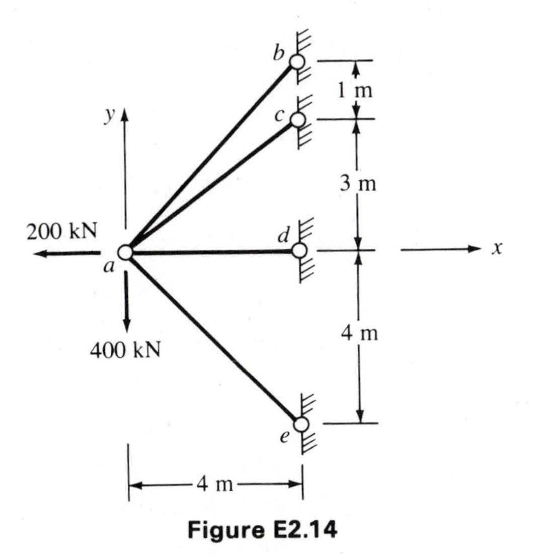

Figure E2.14

removed and element ac is fabricated 5 mm too short. $E = 200$ GPa, $A_{ab} = A_{ae} = 1131\ \text{mm}^2$, $A_{ac} = 500\ \text{mm}^2$, and $A_{ad} = 400\ \text{mm}^2$.

Table E2.14

Element	ϕ	$\cos\phi$	$\sin\phi$
ab	45.00°	0.707	0.707
ac	36.87°	0.800	0.600
ad	0.00°	1.000	0.000
ae	−45.00°	0.707	−0.707

Solution

(a) Using Eq. (2.22) for the element stiffness matrices and substituting the values from Table E2.14, along with

$$\frac{AE}{L} = \begin{cases} 2 \times 10^4\ \text{kN/m}^2 & \text{for elements } ac \text{ and } ad \\ 4 \times 10^4\ \text{kN/m}^2 & \text{for elements } ab \text{ and } ae \end{cases}$$

we have

$$\mathbf{k}^{ab} = (4 \times 10^4)\begin{bmatrix} 0.50 & 0.50 \\ 0.50 & 0.50 \end{bmatrix}$$

$$\mathbf{k}^{ae} = (4 \times 10^4)\begin{bmatrix} 0.50 & -0.50 \\ -0.50 & 0.50 \end{bmatrix}$$

$$\mathbf{k}^{ac} = (2 \times 10^4)\begin{bmatrix} 0.64 & 0.48 \\ 0.48 & 0.36 \end{bmatrix}$$

$$\mathbf{k}^{ad} = (2 \times 10^4)\begin{bmatrix} 1.00 & 0.00 \\ 0.00 & 0.00 \end{bmatrix}$$

Note that since only node a is unconstrained for all elements, only the 2 by 2 partitions of the element stiffness matrices that will be merged into $\mathbf{K}_{ff}$ are shown. Therefore,

$$\mathbf{K}_{ff} = (2 \times 10^4)\begin{bmatrix} 3.64 & 0.48 \\ 0.48 & 2.36 \end{bmatrix} \quad \text{and} \quad \mathbf{K}_{ff}^{-1} = (10^{-4})\begin{bmatrix} 0.1412 & -0.0287 \\ -0.0287 & 0.2177 \end{bmatrix}$$

Since $\mathbf{P}_f = \{-200 \quad -400\}$ kN, Eq. (2.34) gives

$$\mathbf{U}_f = 10^{-3}\{-1.675 \quad -8.134\}\ \text{m}$$

The element forces are calculated using Eqs. (2.58) and (2.59), and these give

$$s^{ab} = 277.4\ \text{kN (t)}$$
$$s^{ac} = 124.4\ \text{kN (t)}$$
$$s^{ad} = 33.5\ \text{kN (t)}$$

and

$$s^{ae} = -182.7\ \text{kN (c)}$$

(b) If element *ac* is fabricated 5 mm too short,

$$AE\,\frac{\Delta L}{L} = (500 \times 10^{-6})(200 \times 10^{6})\,\frac{5 \times 10^{3}}{5} = 100 \text{ kN}$$

Thus $\bar{\mathbf{p}}^o = \{-100 \quad 0 \quad 100 \quad 0\}$ kN. Using Eqs. (2.45) and (2.46),

$$\mathbf{p}^o = \mathbf{T}^T\bar{\mathbf{p}}^o = \{-80 \quad -60 \quad 80 \quad 60\} \text{ kN}$$

$\mathbf{K}_{ff}$ and $\mathbf{K}_{ff}^{-1}$ are the same as in part (a). Using Eq. (2.67) with $\mathbf{P}_f = \mathbf{0}$, $\mathbf{U}_s = \mathbf{0}$, and $\mathbf{P}_f^o = \{-80 \quad -60\}$, gives

$$\mathbf{U}_f = 10^{-3}\{0.957 \quad 1.077\} \text{ m}$$

The element forces are calculated using Eq. (2.58) for all elements with the exception of element *ac*, where Eq. (2.70) must be applied since an initial force is involved. This gives

$$s^{ab} = -57.5 \text{ kN (c)}$$

$$s^{ac} = -28.2 + 100.0 = 71.8 \text{ kN (t)}$$

$$s^{ad} = -19.1 \text{ kN (c)}$$

and

$$s^{ae} = 3.4 \text{ kN (t)}$$

Discussion The **k**'s for each element as given by Eq. (2.22) are 4 by 4, but we note that only those 2 by 2 partitions associated with node *a* will occur in the partition $\mathbf{K}_{ff}$. To conserve space only the retained partitions have been shown. The element force-transformation matrix **e** [Eq. (2.59)] has been used. Since the nonzero displacements in global coordinates associated with each element are contained in $\mathbf{U}_f$, we need only use the first two entries of **e**. For the case with a fabrication error the general analysis is similar to that of part (a); the forces necessary to constrain the nodal points ($\mathbf{p}^o$) yield structural displacements. The reader should note that if an element is initially loaded with $\bar{\mathbf{p}}^o$, then it is mandatory that these forces be included in computing the final element forces s^{ij}. For this example only element *ac* required this consideration. If the forces of part (a) were applied simultaneously with the fabrication error of part (b), the total displacements and element forces would be obtained by superposing the two individual solutions.

EXAMPLE 2.15

Calculate the nodal displacements and element forces for the symmetrical truss in Fig. E2.15: (a) for the applied forces and (b) if the illustrated forces are removed and support *b* settles 0.50 in. $E = 29 \times 10^3$ kips/in.^{2}, $A_{ac} = A_{bd} = 3$ in.^{2}, $A_{bc} = A_{ad} = 5$ in.^{2}, and $A_{cd} = 2$ in.^{2}.

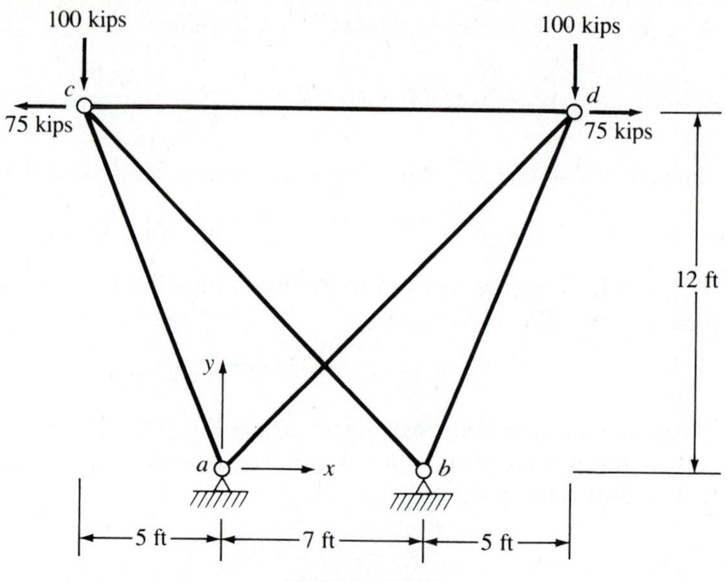

Figure E2.15

Table E2.15

Element	ϕ	$\cos\phi$	$\sin\phi$
ac	112.62°	−0.3846	0.9231
bd	67.38°	0.3846	0.9231
bc	135.00°	−0.7071	0.7071
ad	45.00°	0.7071	0.7071
cd	0.00°	1.0000	0.0000

Solution

(a) Using Eq. (2.22) and Table E2.15, the element stiffness matrices are as follows:

$$\mathbf{k}^{ac} = 557.7\begin{bmatrix} 0.1479 & -0.3550 & -0.1479 & 0.3550 \\ -0.3550 & 0.8521 & 0.3550 & -0.8521 \\ -0.1479 & 0.3550 & 0.1479 & -0.3550 \\ 0.3550 & -0.8521 & -0.3550 & 0.8521 \end{bmatrix}$$

$$\mathbf{k}^{bd} = 557.7\begin{bmatrix} 0.1479 & 0.3550 & -0.1479 & -0.3550 \\ 0.3550 & 0.8521 & -0.3550 & -0.8521 \\ -0.1479 & -0.3550 & 0.1479 & 0.3550 \\ -0.3550 & -0.8521 & 0.3550 & 0.8521 \end{bmatrix}$$

$$\mathbf{k}^{bc} = 712.1\begin{bmatrix} 0.5000 & -0.5000 & -0.5000 & 0.5000 \\ -0.5000 & 0.5000 & 0.5000 & -0.5000 \\ -0.5000 & 0.5000 & 0.5000 & -0.5000 \\ 0.5000 & -0.5000 & -0.5000 & 0.5000 \end{bmatrix}$$

$$\mathbf{k}^{ad} = 712.1\begin{bmatrix} 0.5000 & 0.5000 & -0.5000 & -0.5000 \\ 0.5000 & 0.5000 & -0.5000 & -0.5000 \\ -0.5000 & -0.5000 & 0.5000 & 0.5000 \\ -0.5000 & -0.5000 & 0.5000 & 0.5000 \end{bmatrix}$$

$$\mathbf{k}^{cd} = 284.3\begin{bmatrix} 1.0000 & 0.0000 & -1.0000 & 0.0000 \\ 0.0000 & 0.0000 & 0.0000 & 0.0000 \\ -1.0000 & 0.0000 & 1.0000 & 0.0000 \\ 0.0000 & 0.0000 & 0.0000 & 0.0000 \end{bmatrix}$$

These give

$$\mathbf{K}_{ff} = \begin{bmatrix} 722.83 & -554.03 & -284.30 & 000.00 \\ -554.03 & 831.27 & 000.00 & 000.00 \\ -284.30 & 000.00 & 722.83 & 554.03 \\ 000.00 & 000.00 & 554.03 & 831.27 \end{bmatrix}$$

From Eq. (2.34),

$$\begin{bmatrix} U_c \\ V_c \\ U_d \\ V_d \end{bmatrix} = (10^{-2})\begin{bmatrix} 0.80016 & 0.53330 & 0.64339 & -0.42881 \\ 0.53330 & 0.47574 & 0.42881 & -0.28580 \\ 0.64339 & 0.42881 & 0.80016 & -0.53330 \\ -0.42881 & -0.28580 & -0.53330 & 0.47574 \end{bmatrix}\begin{bmatrix} -75 \\ -100 \\ 75 \\ -100 \end{bmatrix}$$

$$= \{-0.2221 \quad -0.2683 \quad 0.2221 \quad -0.2683\} \text{ in.}$$

Using Eq. (2.58) we have the element forces

$$s^{ac} = s^{bd} = -90.5 \text{ kips (c)}$$
$$s^{ad} = s^{cd} = -23.3 \text{ kips (c)}$$
$$s^{cd} = 126.3 \text{ kips (t)}$$

(b) In the case of settlement we must use Eq. (2.31), with

$$\mathbf{U}_s = \{U_a \quad V_a \quad U_b \quad V_b\} = \{0.00 \quad 0.00 \quad 0.00 \quad -0.50\} \text{ in.}$$

and $$\mathbf{K}_{fs} = \begin{bmatrix} -82.50 & 198.00 & -356.01 & 356.01 \\ 198.00 & -475.19 & 356.01 & -356.01 \\ -356.01 & -356.01 & -82.50 & -198.00 \\ -356.01 & -356.01 & -198.00 & -475.19 \end{bmatrix}$$

$\mathbf{P}_f$ is null and $\mathbf{K}_{fs}\mathbf{U}_s = \{-178 \quad 178 \quad 100 \quad 238\}$ kips.

Using $\mathbf{K}_{ff}$ from part (a) and solving Eq. (2.31) yields

$$\{U_c \quad V_c \quad U_d \quad V_d\} = \{0.8571 \quad 0.3571 \quad 0.8571 \quad -0.8571\} \text{ in.}$$

Equation (2.59) reveals that all element forces are zero.

(b) *Alternate solution using initial end forces.* This can be envisaged as a three-step process similar to that used for thermal effects, fabrication errors, and precambering. Since node b moves down 0.50 in., it will be necessary to initially constrain nodes c and d with the appropriate nodal forces, which are

$$\begin{aligned}\mathbf{p}^{o(bc)} &= \mathbf{k}^{bc}\{0.00 \quad -0.50 \quad 0.00 \quad 0.00\}\\ &= \{178 \quad -178 \quad -178 \quad 178\} \text{ kips}\\ \mathbf{p}^{o(bd)} &= \mathbf{k}^{bd}\{0.00 \quad -0.50 \quad 0.00 \quad 0.00\}\\ &= \{-100 \quad -238 \quad 100 \quad 238\} \text{ kips}\end{aligned}$$

where $$\mathbf{p}^{o(ij)} = \{p_{xi}^{o} \quad p_{yi}^{o} \quad p_{xj}^{o} \quad p_{yj}^{o}\}$$

Combining these element initial forces gives

$$\mathbf{P}_f^o = \{P_{xc}^o\, P_{yc}^o\, P_{xd}^o\, P_{yd}^o\} = \{-178 \quad 178 \quad 100 \quad 238\}$$

Note that these forces are identical to the product $\mathbf{K}_{fs}\mathbf{U}_s$ obtained in the first solution of the settlement problem. Therefore the displacements and element forces would also be the same as those obtained previously.

Discussion Since the truss and its loading in part (a) are symmetrical, we could anticipate symmetrical results, which in fact is the situation. This is a statically indeterminate truss with one redundancy; therefore, the fact that no element forces are induced by differential support settlement in part (b) requires some reflection. The static redundancy is introduced by the supports, and the vertical movement of node b can be envisaged as though the structure were constructed with a slide mechanism at this point that does not resist vertical motion. With this portrayal it is evident that the movement will not cause element forces. Whether the settlement problem is treated by matrix partitioning of $\mathbf{K}$ or by initial forces usually depends upon the preferences of the structural engineer and the capabilities of the programs being used. One needs to be careful in computing the element forces with these two approaches. Using matrix partitioning, we treat the settlement as a known displacement, and this must be acknowledged in computing element forces; that is, for elements bc and bd the associated nodal displacement matrices contain $V_b = -0.50$. In contrast, when initial forces are used, we visualize that all nodes are fixed in position until $\mathbf{P}^o$ is applied. Therefore forces for elements bc and bd must be computed using Eq. (2.70) with values of $\bar{s}^o$ equal to 252 and 257 kips, respectively.

The stiffness method can be extended to other types of structures, and this is done in subsequent chapters. Beams and frames in two dimensions, space trusses, grid works, and space frames are all amenable to analysis by the stiffness method. Based upon the experience gained in this chapter, one can envisage that the overall approach can be adapted to structures composed of discrete elements simply by furnishing the appropriate basic matrices for element stiffness, initial forces, and element force transformation. A number of special topics associated with the stiffness method have not been addressed in this chapter, and we leave the discussion of these to Ch. 6.

The stiffness method has been described in this chapter by using the basic equilibrium, compatibility, and material constitutive equations. The method can be described for simple structures by satisfying these fundamental relationships. Another approach that must be used in some cases consists of examining the energy of the system. The principle of virtual work is applied to formulate the equations of the stiffness method in the following chapter. The reader is urged to study Ch. 3; a mastery of energy methods helps one to understand the bases for the stiffness method procedure.

2.10 Problems

2.1 Describe four properties of the element stiffness matrix for an axial force element and explain the basis for each property.

2.2 The strut in Fig. P2.2 has $A = 1.125 \times 10^{-3}$ m^2, $E = 200$ GPa and is supported with $U_a = V_a = U_b = 0$.

(a) Calculate V_b if it is loaded with $P_{yb} = -108$ kN.

(b) Calculate the force P_{yb} if $V_b = 20$ mm.

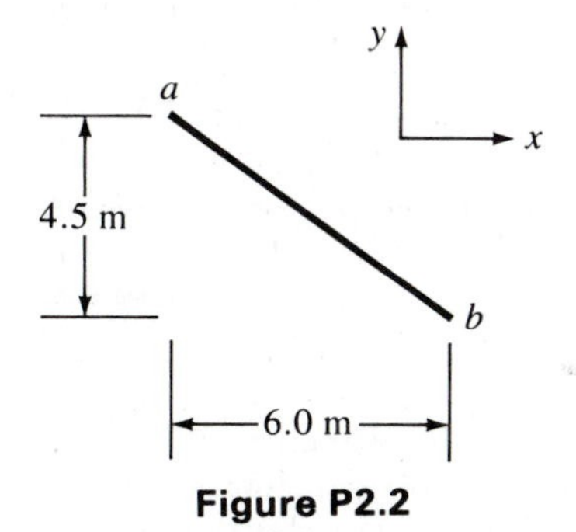

Figure P2.2

2.3 The strut in Fig. P2.3 has $A = 3.51$ in.^{2}, $E = 29 \times 10^3$ ksi and is supported with $U_a = U_b = V_b = 0$.

(a) Calculate V_a if it is loaded with $P_{ya} = 120$ kips.

(b) Calculate the force P_{ya} if $V_a = 0.55$ in.

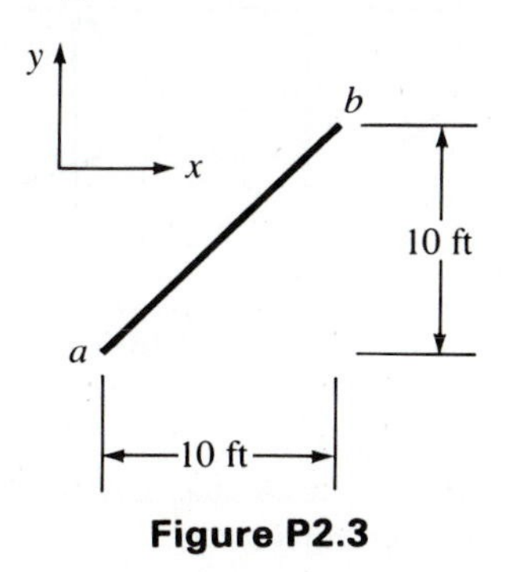

Figure P2.3

2.4 For the three-element assemblage in Fig. P2.4 $(AE/L)_{ab} = 48 \times 10^3$ kN/m; $(AE/L)_{bc} = 24 \times 10^3$ kN/m; and $(AE/L)_{cd} = 63 \times 10^3$ kN/m. Nodes can displace only in the x direction.

(a) Obtain the stiffness matrix for the assemblage by imposing the four possible distinct displacement cases. Note in the calculations where equilibrium, continuity, and the element force-displacement relations are imposed.

(b) Form the stiffness matrix for the assemblage by superimposing the stiffness matrices for the individual elements.

(c) The assemblage has node a fixed in position, and it is subjected to the applied forces: $P_{xb} = P_{xc} = P_{xd} = 192$ kN. Using the stiffness method, calculate the unknown displacements at nodes b, c, and d, and the reaction force at a.

(d) The assemblage has nodes a and d fixed in position, and it is subjected to the forces: $P_{xb} = P_{xc} = 192$ kN. Using the stiffness method, calculate the displacements at nodes b and c and the reactions at nodes a and d.

(e) The assemblage has nodes a and d fixed in position. If node b is given a displacement of 24 mm (in the positive x direction), and $P_{xc} = 192$ kN, calculate the displacement at c, the force at b, and the unknown reactions using the stiffness method.

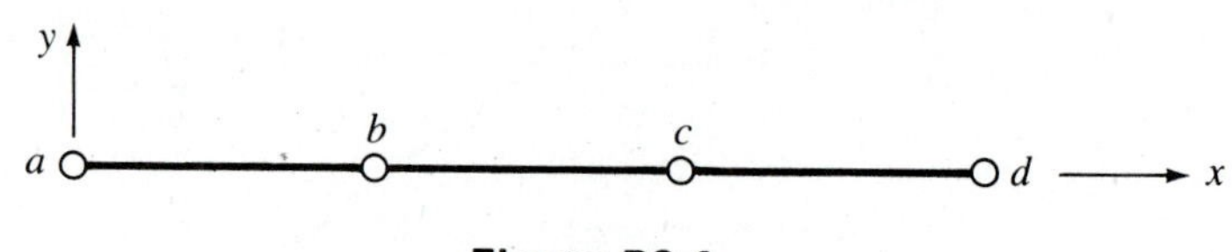

Figure P2.4

2.5 Sketch an array with an appropriate number of blank entries to accommodate the unconstrained structural stiffness matrix (**K**) for the truss in Fig. P2.5. Place x's in the locations of the matrix that have nonzero entries. Assume that all elements have a 4 by 4 stiffness matrix with nonzero entries as shown in Eq. (2.22).

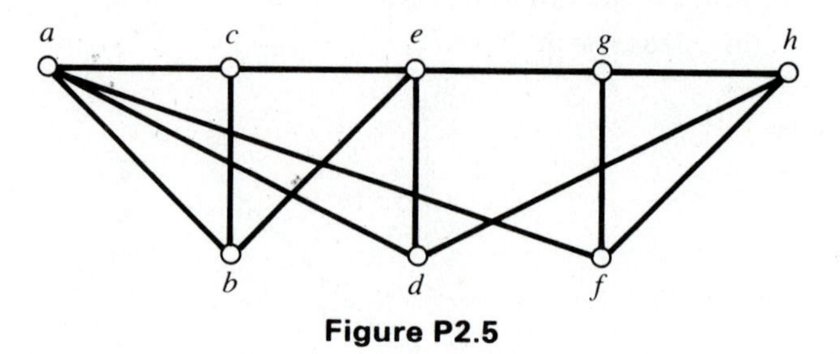

Figure P2.5

2.6 Sketch the approximate truss that has the following structural stiffness matrix. The x's indicate nonzero quantities.

$$
\mathbf{K} = \begin{bmatrix}
x & x & x & x & 0 & 0 & 0 & 0 & x & x \\
x & x & x & x & 0 & 0 & 0 & 0 & x & x \\
x & x & x & x & x & x & 0 & 0 & x & x \\
x & x & x & x & x & x & 0 & 0 & x & x \\
0 & 0 & x & x & x & x & x & x & x & x \\
0 & 0 & x & x & x & x & x & x & x & x \\
0 & 0 & 0 & 0 & x & x & x & x & x & x \\
0 & 0 & 0 & 0 & x & x & x & x & x & x \\
x & x & x & x & x & x & x & x & x & x \\
x & x & x & x & x & x & x & x & x & x
\end{bmatrix} \qquad
\mathbf{U} = \begin{bmatrix} 0 \\ 0 \\ x \\ x \\ x \\ x \\ 0 \\ 0 \\ x \\ x \end{bmatrix}
\begin{matrix} 1 \\ 2 \\ 3 \\ 4 \\ 5 \\ 6 \\ 7 \\ 8 \\ 9 \\ 10 \end{matrix}
$$

(columns of $\mathbf{K}$ numbered 1 2 3 4 5 6 7 8 9 10)

2.7 Calculate **K** for the truss shown in Fig. P2.7. $AE/L = 2 \times 10^4$ kN/m for all elements.

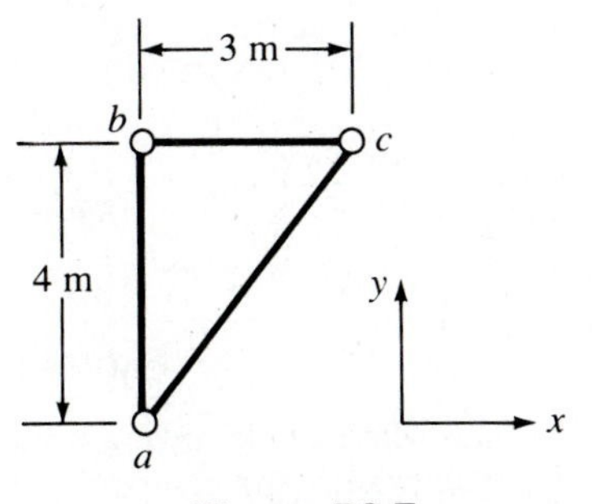

Figure P2.7

2.8 The structural stiffness matrix (**K**) for the truss shown in Fig. P2.8, with AE = constant for all elements, is

$$
\mathbf{K} = \frac{AE}{L}\begin{bmatrix}
1.00 & 0.00 & -1.00 & 0.00 & 0.00 & 0.00 \\
0.00 & 1.00 & 0.00 & 0.00 & 0.00 & -1.00 \\
-1.00 & 0.00 & 1.35 & -0.35 & -0.35 & 0.35 \\
0.00 & 0.00 & -0.35 & 0.35 & 0.35 & -0.35 \\
0.00 & 0.00 & -0.35 & 0.35 & 0.35 & -0.35 \\
0.00 & -1.00 & 0.35 & -0.35 & -0.35 & 1.35
\end{bmatrix}
$$

(a) Why do the columns of **K** add to zero?

(b) What do the numbers in column 3 represent?

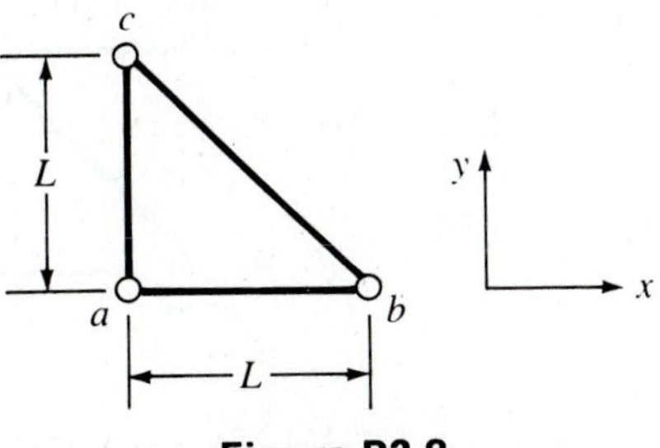

Figure P2.8

(c) Why are all the terms on the main diagonal positive?

(d) Why doesn't $\mathbf{K}^{-1}$ exist?

2.9 The truss in Fig. P2.9 has $\mathbf{K}_{ff}$ as shown. What forces are required at node a to deform the truss so that $V_a = 1$, while $U_a = 0$? What are the forces in each element if AE/L = constant?

$$\mathbf{K}_{ff} = \frac{AE}{2L}\begin{bmatrix} 4 & \sqrt{3} \\ \sqrt{3} & 4 \end{bmatrix}$$

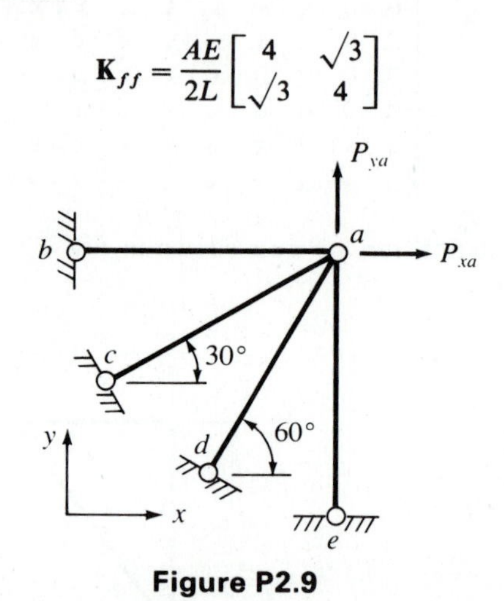

Figure P2.9

2.10 Describe how and where the following concepts are incorporated into the stiffness method: **(a)** equilibrium; **(b)** compatibility; **(c)** stress-strain relations; and **(d)** rigid-body translations and rotations.

2.11 Illustrate the physical significance of the stiffness coefficients corresponding to the forces shown in Fig. P2.11.

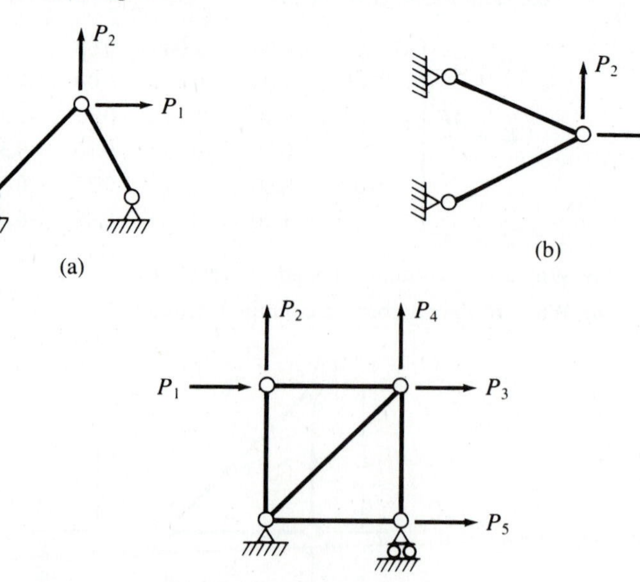

Figure P2.11

2.12 Calculate $\mathbf{K}_{ff}$ for the truss in Fig. P2.12. $A = 0.96$ in.^{2} and $E = 30 \times 10^3$ ksi for all elements.

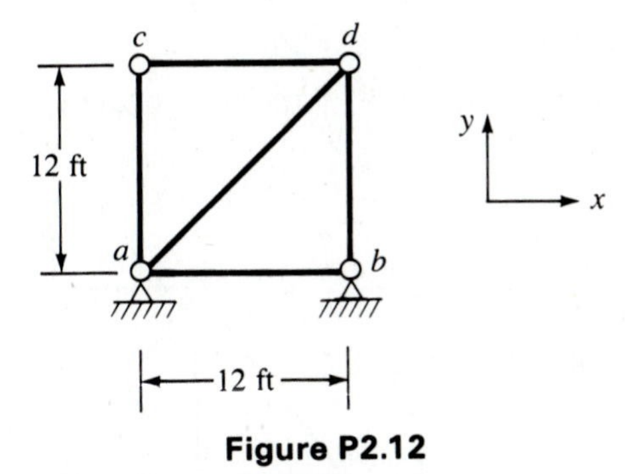

Figure P2.12

2.13 (a) Calculate $\mathbf{K}_{ff}$ for the truss in Fig. P2.13. Cross-sectional areas of elements: diagonals 500 mm^2; horizontals 600 mm^2. $E = 200$ GPa.

(b) The truss of part (a) has an element added between nodes a and d with an area of 984.9 mm^2 and $E = 200$ GPa. Calculate $\mathbf{K}_{ff}$.

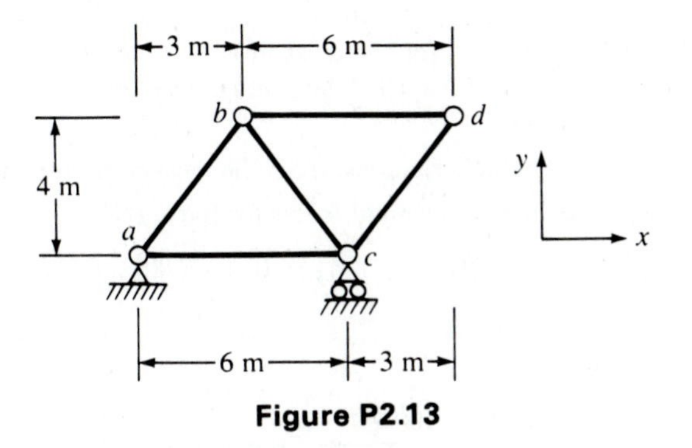

Figure P2.13

2.14 Calculate $\mathbf{K}_{ff}$ and $\mathbf{U}_f$ for the truss in Fig. P2.14. Diagonal elements have cross-sectional areas of $A\sqrt{2}$; all others have an area of A. E is constant.

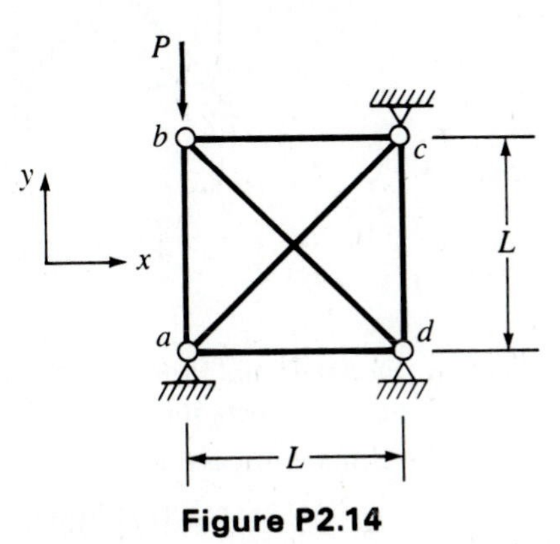

Figure P2.14

2.15 Cross-sectional areas of elements: diagonals 5 in.^{2}, verticals 4 in.^{2}; horizontals 3 in.^{2}. $E = 29 \times 10^3$ ksi. Calculate forces in elements of the truss in Fig. P2.15 if we know that

$$\{U_b \quad V_b \quad U_c \quad V_c \quad U_d \quad V_d\}$$
$$= \{0.1294 \quad -0.0970 \quad -0.0466 \quad -0.2213 \quad 0.2226 \quad -0.6172\} \text{ in.}$$

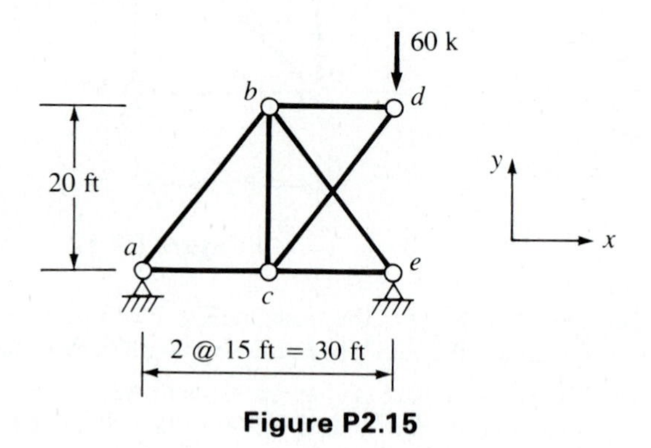

Figure P2.15

2.16 (a) Calculate $\mathbf{K}_{ff}$ for the truss in Fig. P2.16. All horizontal and vertical elements have $A = 12 \times 10^{-4}$ m^2; diagonals have $A = 16.97 \times 10^{-4}$ m^2. $E = 200$ GPa.

(b) Element *ad* is removed from the truss of part (a). Calculate $\mathbf{K}_{ff}$.

(c) Calculate the element forces for part (a) if

$$\{U_b \quad V_b \quad U_c \quad V_c\} = 10^{-3}\{5.786 \quad 1.929 \quad 5.464 \quad -4.821\} \text{ m}$$

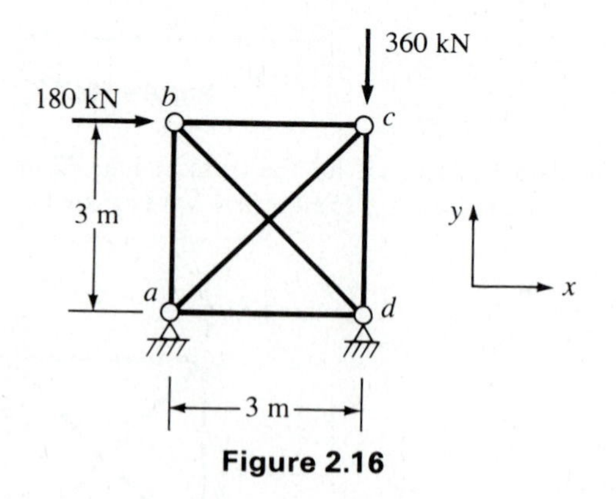

Figure 2.16

2.17 (a) Calculate $\mathbf{K}_{ff}$ for the truss in Fig. P2.17. Elements intersect at right angles at node *c*, and AE/L = constant for all elements.

(b) Calculate the element forces for part (a) if

$$\{U_c \quad V_c \quad V_d \quad U_e\} = (L/AE)\{47.5 \quad 22.5 \quad -8.33 \quad 61.6\}$$

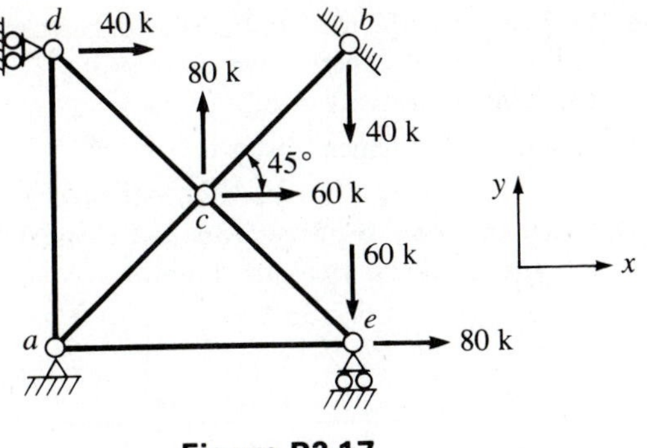

Figure P2.17

2.18 (a) Calculate $\mathbf{K}_{ff}$ for the truss in Fig. P2.18. Elements intersect at right angles at node a, and for all elements $E = 1.5 \times 10^3$ kips/in.^{2} and $L = 5$ ft. $A = 0.60$ in.^{2} (ab and ac) and $A = 0.40$ in.^{2} (ad and ae).

(b) Calculate the element forces for part (a) if $\{U_a \quad V_a\} = \{1.131 \quad 0.566\}$ in.

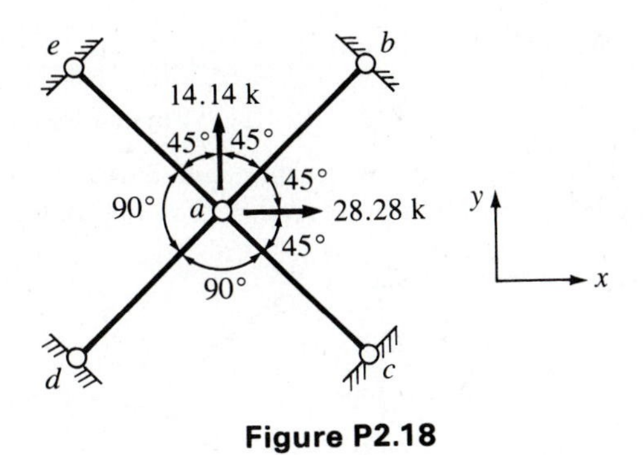

Figure P2.18

2.19 Calculate the nodal displacements and element forces for the truss shown in Fig. P2.19 if all elements have $AE/L = 2 \times 10^4$ kN/m.

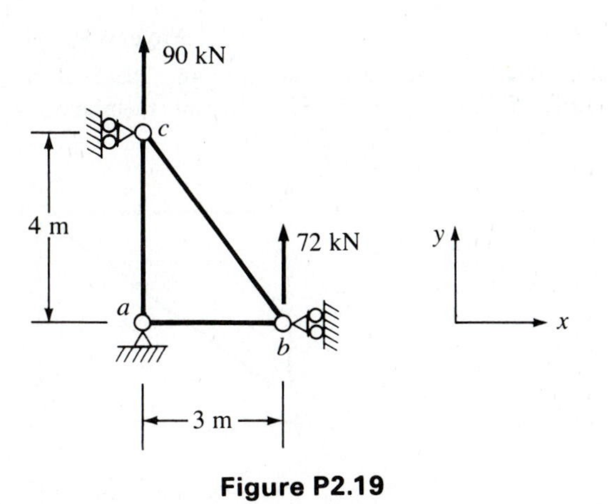

Figure P2.19

2.20 **(a)** Form $\mathbf{K}_{ff}$ for the truss in Fig. P2.20. Elements *ac* and *bd* have cross-sectional areas $= A$; all others have areas $= A\sqrt{2}$. Material for all elements has a modulus of elasticity $= E$.

(b) Calculate the element forces if

$$\{U_b \quad V_b \quad U_d \quad V_d\} = (PL/AE)\{1 \quad -9/4 \quad 0 \quad -3/4\}$$

(c) Calculate nodal displacements and element forces for the truss of part (a) if all cross-sectional areas are doubled.

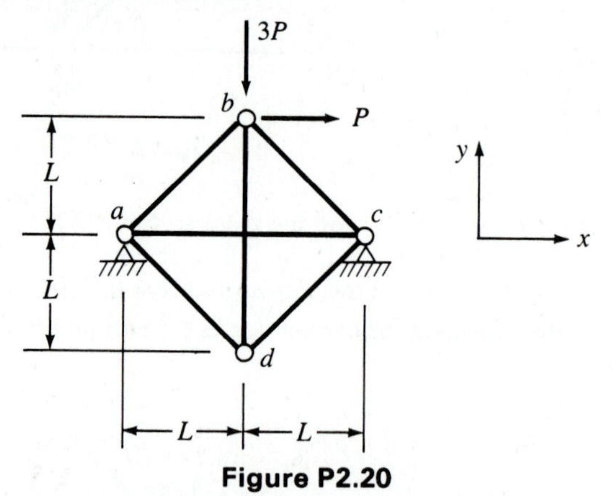

Figure P2.20

2.21 Calculate nodal displacements and element forces for the truss shown in Fig. P2.21. $A = 0.50$ in.^{2} and $E = 29 \times 10^3$ ksi for all members.

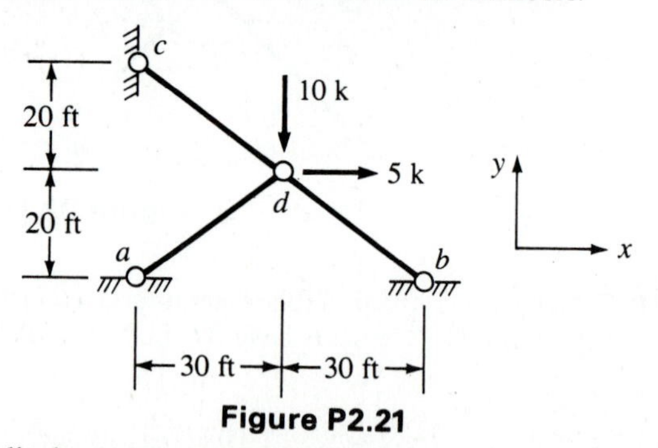

Figure P2.21

2.22 Calculate nodal displacements and element forces for the truss shown in Fig. P2.22. $E = 200$ GPa for all members; $A_{ab} = A_{cd} = 3000$ mm^2, $A_{bc} =$ 4000 mm^2, and $A_{ac} = 5000$ mm^2.

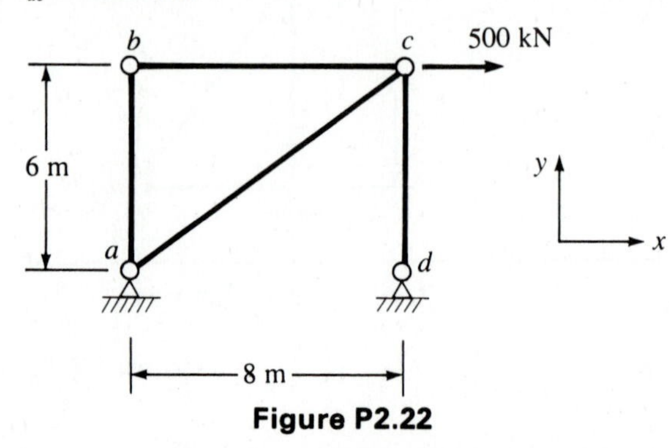

Figure P2.22

2.23 Calculate nodal displacements and element forces for the truss in Fig. P2.23; AE/L is constant for all members.

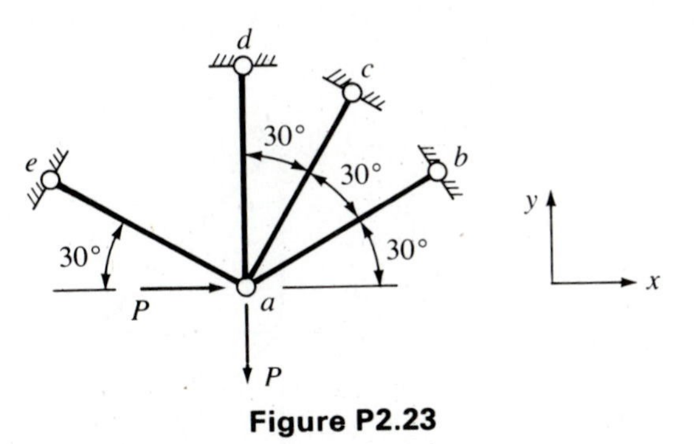

Figure P2.23

2.24 Calculate nodal displacements and element forces for the truss in Fig. P2.24; $E = 29 \times 10^3$ ksi and $A = 1.5$ in.^{2} for all members. Nodal coordinates in feet.

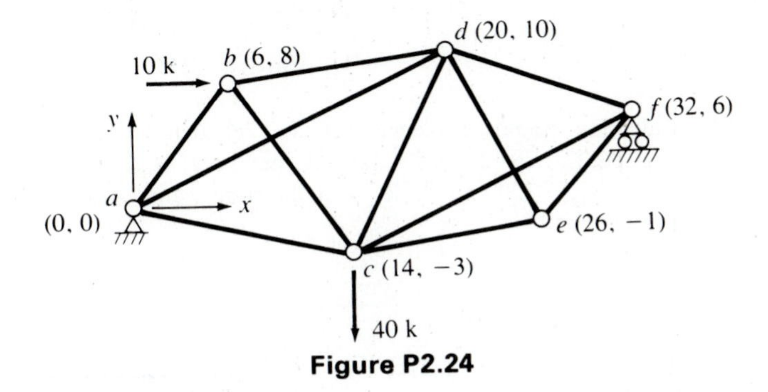

Figure P2.24

2.25 Calculate nodal displacements and element forces for the truss in Fig. P2.25; $E = 29 \times 10^3$ ksi and member areas are: 0.83 in.^{2} for the lower chord; 1.03 in.^{2} for the upper chord; and 1.18 in.^{2} for all web members. Applied forces (1 kip each) have the following orientation: $P_1(260°)$; $P_2(270°)$; and $P_3(80°)$, where angles are referenced to the coordinate axes.

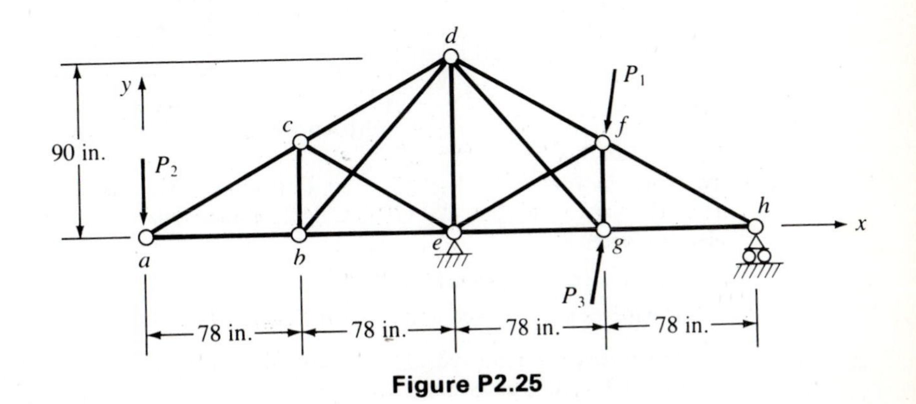

Figure P2.25

2.26 Calculate nodal displacements and element forces for the truss in Fig. P2.26; $E = 29 \times 10^3$ ksi and member areas are: 0.76 in.^{2} for verticals; 0.43 in.^{2} for horizontals; and 1.01 in.^{2} for diagonals. Applied forces (1 kip each) have the following orientation: $P_1(290°)$; $P_2(120°)$; and $P_3(350°)$, where angles are referenced to the coordinate axes.

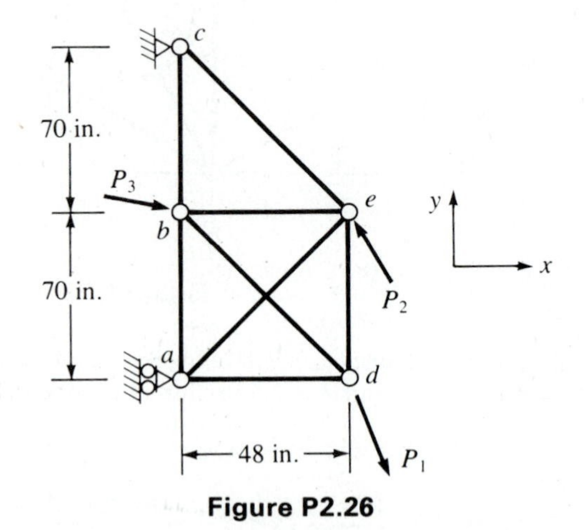

Figure P2.26

2.27 Calculate nodal displacements and element forces for the truss shown in Fig. P2.27. $E = 200$ GPa for all members; $A_{ab} = 1600$ mm^2, $A_{bc} = 1300$ mm^2, $A_{ce} = A_{ad} = 3200$ mm^2, $A_{ae} = 3900$ mm^2, and $A_{bd} = 2600$ mm^2 (the truss is symmetrical).

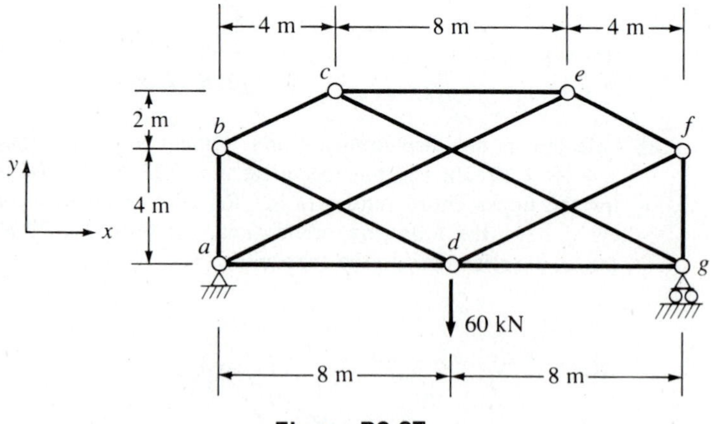

Figure P2.27

2.28 The truss in Fig. P2.28 has $A_{ab} = 30 \times 10^{-4}$ m^2; $A_{ac} = 20 \times 10^{-4}$ m^2; $A_{ad} = 15 \times 10^{-4}$ m^2; and $E = 200$ GPa. Calculate the nodal displacements and the element forces if:

(a) The illustrated forces are imposed.

(b) The forces are removed and support *b* moves 7.07 mm downward.

(c) The forces are removed and element *ab* increases in temperature by 45°C [$\alpha = 1.2 \times 10^{-5}$ mm/(mm · °C)].

(d) The forces are removed and elements *ab* and *ad* have been fabricated with length errors of +5 mm and −5 mm, respectively.

(e) All element areas are doubled in part (a).

(f) The area of only element *ab* in part (a) is doubled.

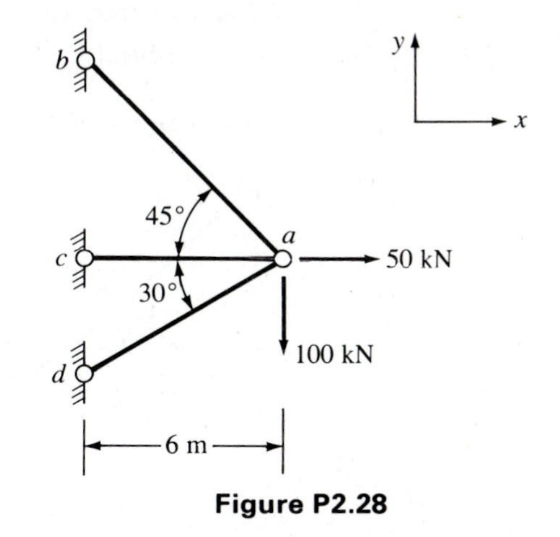

Figure P2.28

2.29 The support at *e* for the truss of Prob. 2.15 is replaced by a roller so that $U_e \neq 0$. Calculate the element forces and the rotation of element *ac* due to:

(a) The illustrated force.

(b) A temperature decrease for elements *ac* and *ce* of 30°F and 100°F for element *ab* [$\alpha = 6.5 \times 10^{-6}$ in./(in. · °F)]—applied force removed.

(c) A fabrication error in elements *ab* and *be* of +0.50 in.—applied force removed.

2.30 The truss shown in Fig. P2.30 has $A_{ac} = 4$ in.^{2}, $A_{ad} = A_{ab} = 2\sqrt{2}(A_{ac})$, and $E = 29 \times 10^3$ ksi. Calculate the nodal displacements and element forces if:

(a) $P_{xa} = 100$ kips.

(b) Element *ac* is fabricated 0.06 in. too long.

(c) Support *a* settles 0.05 in.

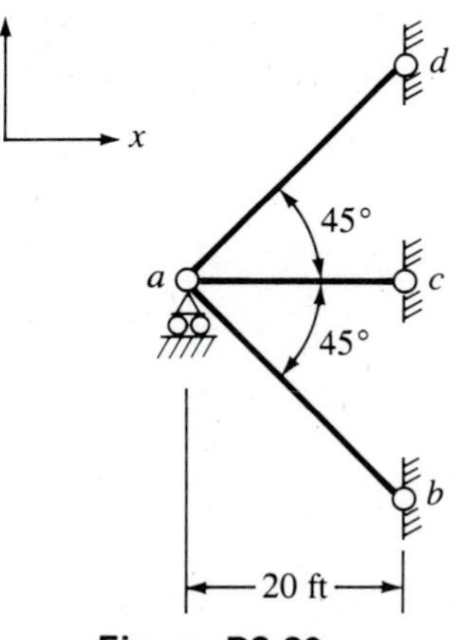

Figure P2.30

2.31 The truss shown in Fig. P2.31 has $E = 29 \times 10^3$ ksi and cross-sectional areas of: 6.21 in.^{2} (*ae* and *ed*) and 1.76 in.^{2} (*be* and *ce*). Calculate nodal displacement(s) and element forces if:

(a) $P_{xe} = 150$ kips.

(b) Element *ed* is fabricated 0.10 in. short.

(c) The roller support at *e* is removed and $V_e = -0.10$ in.

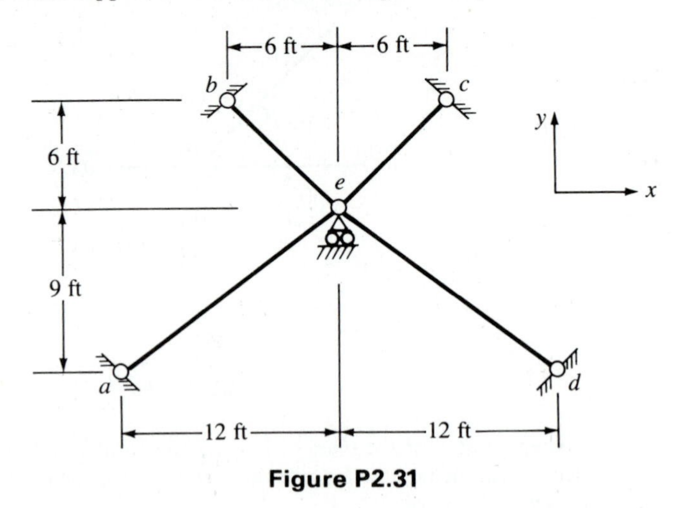

Figure P2.31

2.32 The truss shown in Fig. P2.32 is made from steel so that $E = 200$ GPa, $\alpha = 1.2 \times 10^{-5}$ mm/(mm · °C). Areas of all vertical elements are 200 mm^2; horizontal elements have areas of 300 mm^2; and diagonals have areas of 400 mm^2. Calculate nodal displacements and element forces if:

(a) Support *b* moves 10 mm to the right.

(b) Element *bd* experiences a temperature rise of 50°C above ambient.

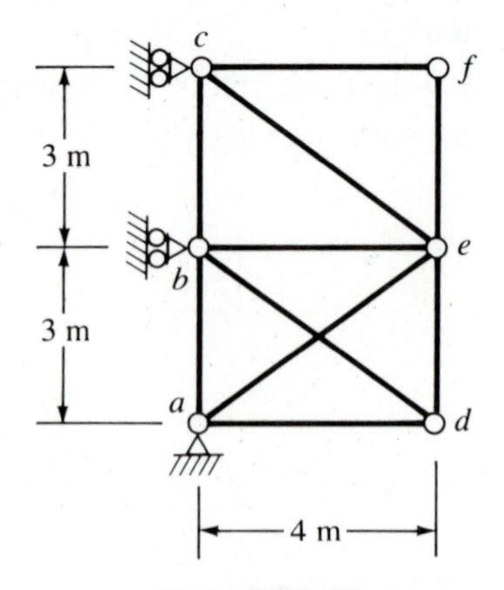

Figure P2.32

3 THE STIFFNESS METHOD USING VIRTUAL WORK

In analyzing any structure, three basic conditions must be satisfied. First, the structure and each of its components must be in equilibrium. In this book we consider only static *equilibrium*, but dynamic equilibrium must be investigated in some situations. Second, there can be no rips or tears in the structure; more precisely, the deformations must be continuous throughout the structure. This is the condition of *compatibility*. Third, the relationships between material stress and strain must be satisfied. These are the so-called *constitutive relationships*. In the previous chapter, we derived the equations of the stiffness method and applied them to trusses using the three basic equations. In some situations, it is extremely difficult to obtain a solution by applying the basic equations. Elements that are tapered from end to end and structures with unusual geometries (e.g., beams curved into a geometry not expressible by simple functions) are examples in which application of the basic equations is not tractable. For all structural solutions these three basic conditions must be satisfied, but this can be accomplished using an alternative approach employing *energy principles*.

There are a number of energy principles that can be used in lieu of the three basic equations. The computations and algebraic manipulations involved with energy principles are generally more straightforward than those encountered when invoking the basic conditions. There is an accompanying danger, in that the energy methods tend to mask the basic concepts so that the analyst can lose sight of the physical significance of the various parts of the solution procedure. Energy principles are extremely powerful for obtaining approximate solutions; thus tapered elements and arbitrarily curved beams can be investigated in a routine manner. Also, the calculations using energy methods are ideally cast for computer manipulation and solution of the equations. In Ch. 2 we applied the basic equations to formulate the stiffness method, but the results were rewritten in matrix form after development.

In this chapter we will investigate one of the energy methods known as the *principle of virtual work*. It has been used in its present form since the mid-1950s, but its origin can be traced back to Galileo (1564–1642) and his work on levers and pulleys. In his now-famous letter to Varignon, John Bernoulli (1667–1748) used virtual displacements. Energy principles can be broadly divided into two categories: *virtual work* and *complementary virtual work*. The former implicitly invokes equilibrium in constituting the system equations, whereas the latter uses compatibility to achieve the system relationships. In both of these two broad energy categories all three of the basic equations are satisfied, but this is done at different stages. The reader may recall using the slope-deflection equations (or moment distribution) and Castigliano's theorem (part I) in a first course in structural analysis. Both of these stem from virtual work. In contrast, the method of consistent displacements and Castigliano's theorem (part II) emanate from complementary virtual work. The principle of virtual work is the basis for the stiffness method, and this is the subject of investigation in this chapter.

For the reader starting the study of matrix structural analysis at this point in the text, we recommend a thorough study of all sections contained in this chapter, with the possible exception of Sec. 3.8, where a comparison of the basic equations and virtual work formulations is presented. For the student who has mastered the materials in Ch. 2, the contents of this chapter will expand and generalize your understanding. In this case we suggest either a complete study of all sections or a selective reading of Secs. 3.1, 3.2, 3.3, and 3.8.

3.1 THE PRINCIPLE OF VIRTUAL WORK

Virtual work is produced by moving a system slightly away from an equilibrium configuration. The system is perturbed by imposing small displacements that must be kinematically possible but need not necessarily correspond to the real displacements. The statement of virtual work implicitly ensures that equilibrium is satisfied. This broad energy principle will be examined in the subsequent sections for particles, rigid bodies, and deformable bodies.

3.1.1 VIRTUAL DISPLACEMENTS FOR PARTICLES AND RIGID BODIES The particle shown in Fig. 3.1 is in equilibrium under the system of forces $P_1, P_2, \ldots, P_i, \ldots, P_n$, which can be expressed as

$$\sum_{i=1}^{n} P_{xi} = \sum_{i=1}^{n} P_i \lambda_i = 0$$

and

$$\sum_{i=1}^{n} P_{yi} = \sum_{i=1}^{n} P_i \mu_i = 0 \tag{3.1}$$

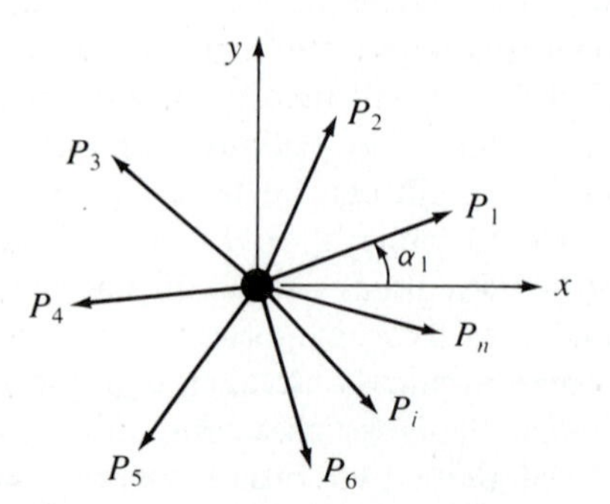

Figure 3.1 Forces on a particle

where λ_i and μ_i are the appropriate direction cosines that give the components of P_i in the x and y directions, respectively.

Suppose we subject the particle to a small *virtual displacement* (δu) in the x direction that is kinematically possible but not necessarily real. The corresponding *virtual work* (δW) results from the x components of the forces, P_{xi}, moving through δu; that is,

$$\delta W = \left(\sum_{i=1}^{n} P_{xi} \right) \delta u \tag{3.2}$$

but from equilibrium,

$$\sum_{i=1}^{n} P_{xi} = 0$$

Therefore $\delta W = 0$. Alternatively, if we first give the particle a virtual displacement δu followed by a virtual displacement δv in the y direction, the total virtual work is

$$\delta W = \left(\sum_{i=1}^{n} P_{xi} \right) \delta u + \left(\sum_{i=1}^{n} P_{yi} \right) \delta v = 0 \tag{3.3}$$

From equilibrium [Eq. (3.1)] we recognize that the two quantities in parentheses are zero, and $\delta W = 0$. This is the *principle of virtual work*, which can be stated as follows:

If a particle is in equilibrium under a system of forces, then for any virtual displacement the virtual work is zero.

The converse of this statement can also be established by investigating the behavior of the system under an arbitrary virtual displacement $\boldsymbol{\delta\Delta}$:

$$\boldsymbol{\delta\Delta} = \delta u \mathbf{i} + \delta v \mathbf{j} \tag{3.4}$$

where $\mathbf{i}$ and $\mathbf{j}$ are the unit vectors in the x and y directions, respectively; furthermore, δu and δv are independent and arbitrary virtual displacements in the x and y directions, respectively. Representing the forces acting on the particle as follows:

$$\mathbf{P} = \sum_{i=1}^{n} P_{xi} \mathbf{i} + \sum_{i=1}^{n} P_{yi} \mathbf{j} \tag{3.5}$$

The virtual work is the dot product of the vectors $\mathbf{P}$ and $\boldsymbol{\delta\Delta}$, which gives

$$\delta W = \mathbf{P} \cdot \boldsymbol{\delta\Delta} = \left(\sum_{i=1}^{n} P_{xi} \right) \delta u + \left(\sum_{i=1}^{n} P_{yi} \right) \delta v \tag{3.6}$$

Since $\boldsymbol{\delta\Delta}$ is arbitrary, the virtual displacements δu and δv must be *arbitrary* and *independent*; therefore, in order for δW to be zero the coefficients of δu and δv must each be zero. From Eq. (3.1) we observe that these two coefficients are the conditions of equilibrium. Thus, we can state the *converse of the statement of virtual work* as follows:

> **A particle subjected to a system of forces is in equilibrium if the virtual work is zero for every independent virtual displacement.**

These two statements also apply to a number of particles subjected to a system of forces in equilibrium, since we can envisage that the virtual work can be formulated for each individual particle and summed over the total number of particles. An aggregate of particles or a rigid body can be treated in a similar fashion.

Consider the rigid body in Fig. 3.2a with the illustrated loading and unyielding supports at points a and b. A free-body diagram of the system is shown in Fig. 3.2b. Since the body is rigid, a virtual displacement of the system must be of the form shown in Fig. 3.2c. Note that these are not the real displacements, which would require that v_a and v_b both be zero; rather the virtual displacements are kinematically correct since they introduce appropriate displacements during rigid-body motion. The virtual displacements are

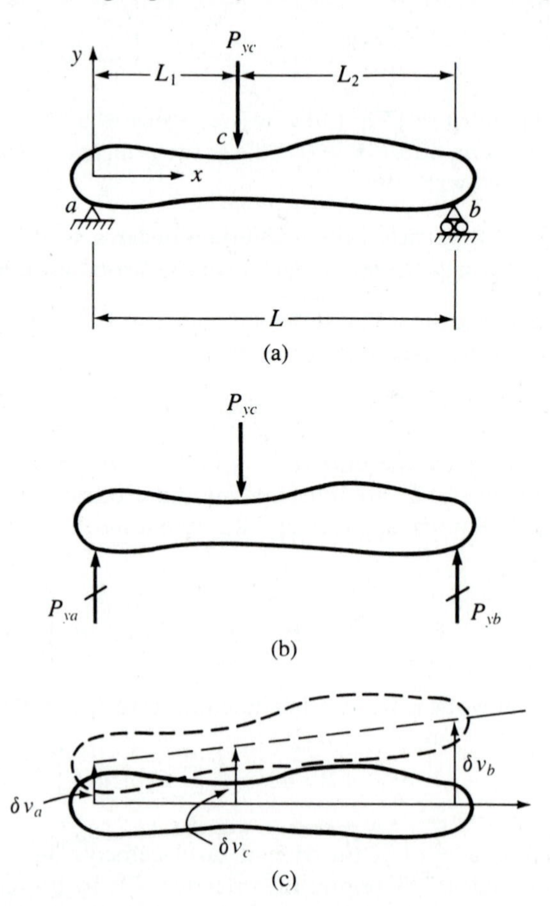

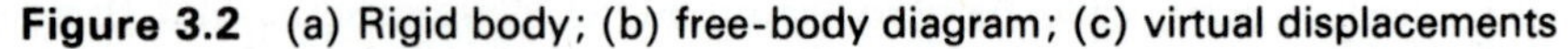

Figure 3.2 (a) Rigid body; (b) free-body diagram; (c) virtual displacements

expressed as follows:

$$\delta v = \left(1 - \frac{x}{L}\right)\delta v_a + \frac{x}{L}\,\delta v_b \tag{3.7}$$

The theorem of virtual work gives

$$\delta W = P_{ya}\,\delta v_a - P_{yc}\,\delta v_c + P_{yb}\,\delta v_b = 0 \tag{3.8}$$

Substituting Eq. (3.7) for δv_a into (3.8) yields

$$P_{ya}\,\delta v_a - P_{yc}\left(1 - \frac{L_1}{L}\right)\delta v_a - P_{yc}\,\frac{L_1}{L}\,\delta v_b + P_{yb}\,\delta v_b = 0 \tag{3.9}$$

or grouping terms gives

$$\left[P_{ya} - P_{yc}\left(1 - \frac{L_1}{L}\right)\right]\delta v_a + \left(-P_{yc}\,\frac{L_1}{L} + P_{yb}\right)\delta v_b = 0 \tag{3.10}$$

Since δv_a and δv_b are arbitrary and independent, their coefficients must each equal zero to satisfy Eq. (3.10), and this yields

$$\begin{aligned} P_{ya} &= P_{yc}\left(1 - \frac{L_1}{L}\right) \\ P_{yb} &= P_{yc}\,\frac{L_1}{L} \end{aligned} \tag{3.11}$$

Thus, by applying the theorem of virtual work we have obtained the equations of equilibrium for the rigid body. If the virtual displacement equation had contained only δv_a, for example, then only the first of Eqs. (3.11) would have been produced. Virtual work is applied to a planar rigid body with reactions and applied forces in two directions in Example 3.1.

EXAMPLE 3.1

Use the theorem of virtual work to calculate the reactions for the rigid body of Fig. E3.1a.

Solution

Using the virtual displacements shown in Fig. E3.1b, the virtual work done is

$$\begin{aligned} \delta W_{e1} = &\frac{4}{5}\,P_a\,\delta v_a + P_{yb}\,\delta v_b + P_{xb}\,\frac{\delta v_b - \delta v_a}{8}\,(6) \\ &-16\left[\delta v_a + \frac{5}{8}\,(\delta v_b - \delta v_a)\right] - 3\,\frac{\delta v_b - \delta v_a}{8}\,(2) \end{aligned}$$

Also, the virtual work associated with the virtual displacements of Fig. E3.1c is

$$\begin{aligned} \delta W_{e2} = &\frac{3}{5}\,P_a\,\delta u_a + P_{xb}\,\delta u_b - \frac{4}{5}\,P_a\,\frac{\delta u_b - \delta u_a}{6}\,(8) \\ &+16\,\frac{\delta u_b - \delta u_a}{6}\,(3) - 3\left[\delta u_a + \frac{\delta u_b - \delta u_a}{6}\,(2)\right] \end{aligned}$$

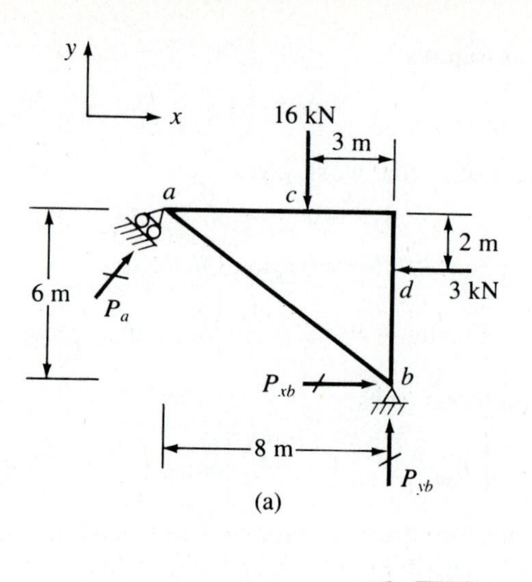

(a)

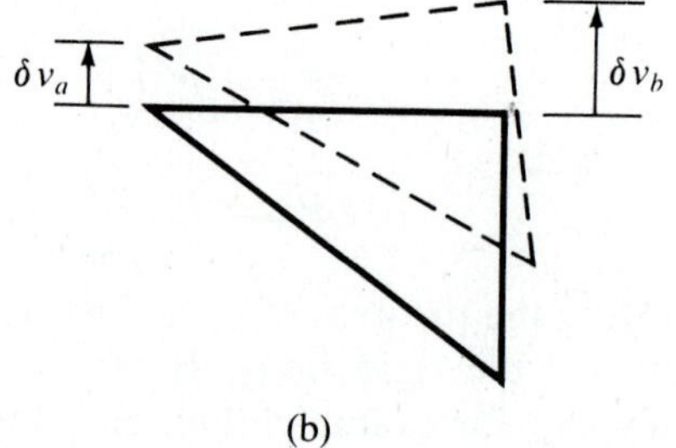

(b)

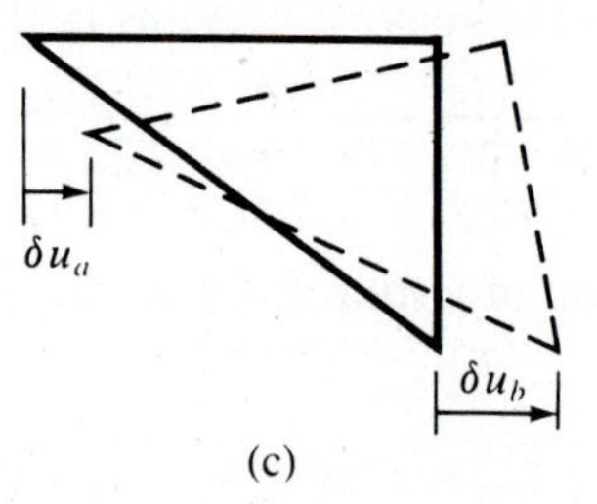

(c)

Figure E3.1 (a) Rigid body; (b) vertical virtual displacements; (c) horizontal virtual displacements

Applying the theorem of virtual work, i.e., $\delta W_e = \delta W_{e1} + \delta W_{e2} = 0$, and grouping terms gives

$$\left[\frac{3}{5}P_a + \frac{4}{3}\left(\frac{4}{5}P_a\right) - 8 - 2\right]\delta u_a + \left[P_{xb} - \frac{4}{3}\left(\frac{4}{5}P_a\right) + 8 - 1\right]\delta u_b$$
$$+\left(\frac{4}{5}P_a - \frac{3}{4}P_{xb} - 6 + \frac{3}{4}\right)\delta v_a + \left(P_{yb} + \frac{3}{4}P_{xb} - 10 - \frac{3}{4}\right)\delta v_b = 0$$

The virtual displacements will be kinematically correct if $\delta v_a = -(3/4)\delta u_a$. Substitution gives the following statement of virtual work:

$$\left(\frac{16}{15}P_a + \frac{9}{16}P_{xb} - \frac{97}{16}\right)\delta u_a + \left(P_{xb} - \frac{16}{15}P_a + 7\right)\delta u_b + \left(P_{yb} + \frac{3}{4}P_{xb} - \frac{43}{4}\right)\delta v_b = 0$$

Since δu_a, δu_b, and δv_b are arbitrary and independent, each of their coefficients in the above equation must be zero. This gives the three independent equations:

$$\frac{16}{15}P_a + \frac{9}{16}P_{xb} - \frac{97}{16} = 0$$

$$-\frac{16}{15}P_a + P_{xb} + 7 = 0$$

$$P_{yb} + \frac{3}{4}P_{xb} - \frac{43}{4} = 0$$

Upon solution we have $P_a = 6$ kN (↗); $P_{xb} = -0.60$ kN (↚); and $P_{yb} = 11.20$ kN(↕).

Discussion The principle of virtual work gives the three equations of static equilibrium for this rigid body with a statically determinate support system. Note in Fig. E3.1b and c that the translational displacements are accompanied in both cases by a rotation of the rigid body. The reaction at point *a* must act normal to the surface of the roller support; for purposes of writing the equation of virtual work, the reaction forces are considered to displace through the virtual displacements. At point *b* the vertical and horizontal displacements are independent, but at point *a* the rigid body must move parallel to the roller surface. Thus, the constraint equation $\delta v_a = -(3/4)\delta u_a$ was introduced.

3.1.2 VIRTUAL WORK FOR DEFORMABLE BODIES

The applied forces and corresponding displacements are functionally related for deformable bodies. Also, the element forces are related to deformations, which implies material properties relating stresses and strains. Throughout this text, unless otherwise noted, we will assume a linear relationship between associated quantities (i.e., force-displacement, force-deformation, and stress-strain).

Consider the prismatic linearly elastic axial force element in Fig. 3.3. For convenience the nodes at each end have been illustrated detached from the element. The forces applied on each of the three free-body diagrams are depicted as directed in the positive *x* coordinate direction. The subscripts on quantities refer to the node at which they are located: *P* denotes an applied

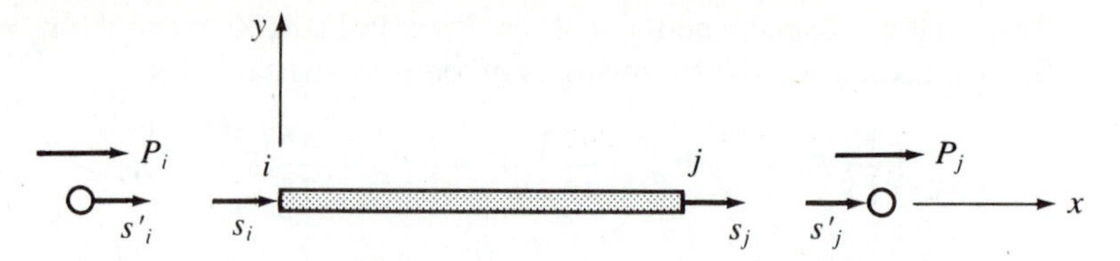

Figure 3.3 Single axial force element and end nodes

force; s indicates the force acting on the element; and s' designates the element force applied to the nodal point. Giving the nodal points the virtual displacements δu_i and δu_j, the virtual work for the two nodal points is

$$\delta W = (P_i + s'_i)\delta u_i + (P_j + s'_j)\delta u_j \tag{3.12}$$

Since these two nodal points are rigid bodies (i.e., they do not deform), $\delta W = 0$ from the theorem of virtual work for systems of particles, and since δu_i and δu_j are arbitrary and independent, the coefficients of these two quantities must be zero. This gives

$$\begin{aligned} &P_i + s'_i = 0 \quad \text{or} \quad P_i = -s'_i \\ \text{and} \quad &P_j + s'_j = 0 \quad \text{or} \quad P_j = -s'_j \end{aligned} \tag{3.13}$$

which are the equations of equilibrium for the two nodes. The forces interacting between the nodes and element are related using Newton's third law as follows:

$$s'_i = -s_i \quad \text{and} \quad s'_j = -s_j \tag{3.14}$$

Substituting Eqs. (3.14) into Eq. (3.12) yields

$$\begin{aligned} \delta W &= (P_i - s_i)\delta u_i + (P_j - s_j)\delta u_j \\ &= (P_i\,\delta u_i + P_j\,\delta u_j) - (s_i\,\delta u_i + s_j\,\delta u_j) \end{aligned} \tag{3.15}$$

But from equilibrium of the axial force element, $-s_i = s_j = s^{ij}$, where s^{ij} is the axial force in element ij; its sign is chosen so that a positive quantity denotes tension. Substituting into Eq. (3.15) gives

$$0 = (P_i\,\delta u_i + P_j\,\delta u_j) - [s^{ij}(\delta u_j - \delta u_i)] \tag{3.16}$$

The first quantity in parentheses represents the virtual work done by the applied external forces; it is denoted as δW_e. The graph of load versus displacement for node i is illustrated in Fig. 3.4, and the external virtual work done by P_i is denoted by the cross-hatched area. Observe from Fig. 3.4 that as the force is gradually increased the associated displacement increases proportionally. The point on the graph where P_i and u_i are attained is an equilibrium position, and the total virtual work done in increasing the displacement by δu_i (ΔW_e) is

$$\begin{aligned} \Delta W_e &= P_i\,\delta u_i + \tfrac{1}{2}\,\delta P_i\,\delta u_i \\ &= \delta W_e + \tfrac{1}{2}\,\delta^2 W_e \end{aligned} \tag{3.17}$$

where δW_e is the first variation of the external forces, which we shall refer to as the *external virtual work*, and $\delta^2 W_e$ is the second variation of the external

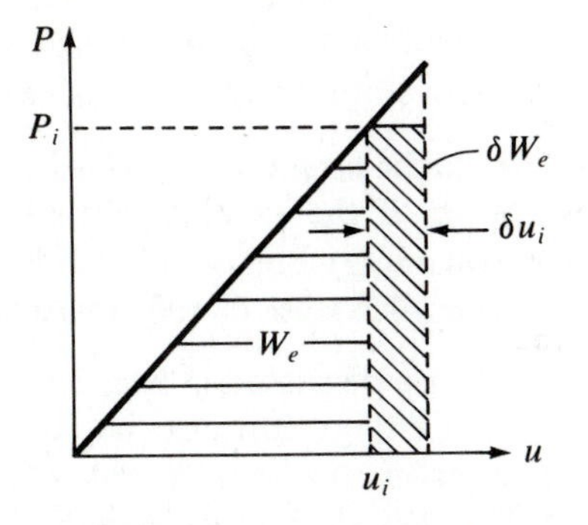

Figure 3.4 Force-displacement graph for a linear element

work. Considering δP_i and δu_i as small quantities, δW_e is the first-order approximation of the total increment in W_e. For nonlinear systems, the second variation must also be considered in ΔW_e. In general it is not necessary to include terms beyond the first variation for linearly elastic behavior, but higher order terms must be retained in describing such phenomena as plasticity and stability behavior.

The term $[s^{ij}(\delta u_j - \delta u_i)]$ in Eq. (3.16) can be understood with the help of Fig. 3.5. The deformation of the element (d^{ij}) equals $(u_j - u_i)$; therefore the virtual deformation (δd^{ij}) is

$$\delta d^{ij} = \delta u_j - \delta u_i \tag{3.18}$$

Note from Fig. 3.5 that the element force and deformation are linearly related and the state where s^{ij} and d^{ij} occur is an equilibrium position. Increasing the deformation by a virtual quantity δd^{ij} results in a first-order increase:

$$\delta W_i = s^{ij}\,\delta d^{ij} \tag{3.19}$$

plus a second-order change of $(\delta s^{ij}\,\delta d^{ij})/2$, which will be neglected for reasons similar to those used to ignore $\delta^2 W_e$ for the external work. The quantity δW_i is the *virtual work of the internal forces*, or the *virtual strain energy*. Hence Eq. (3.16) can be written as follows:

$$0 = \delta W_e - \delta W_i \tag{3.20}$$

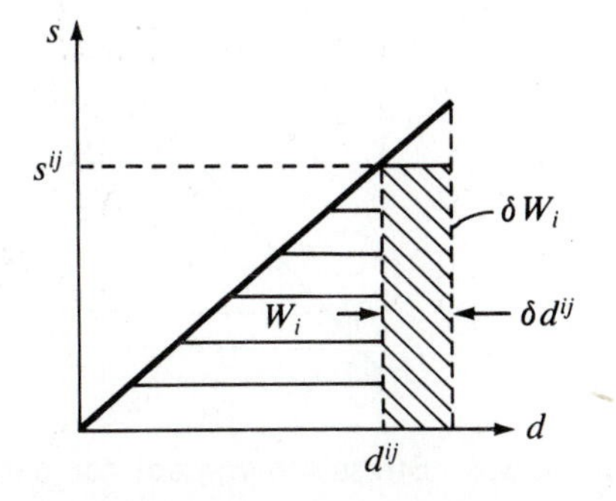

Figure 3.5 Force-deformation graph for a linear element

which is an equation of the theorem of virtual work for deformable bodies. From the above we can state the converse of the principle of virtual work:

> **A linearly elastic structure subjected to a system of forces is in equilibrium if the external virtual work is equal to the virtual strain energy for any virtual displacement state away from a compatible state of deformation.**

Virtual displacements are discussed in Sec. 3.1.3.

The virtual strain energy can also be computed directly from the stress-strain graph. For a prismatic linearly elastic axial force element with cross-sectional area A and length L, the element force and deformation can be expressed as follows:

$$s^{ij} = \sigma^{ij} A$$

and

$$d^{ij} = \varepsilon^{ij} L \tag{3.21}$$

where σ^{ij} and ε^{ij} are the axial stress and strain, respectively, in element ij. The virtual deformation associated with the virtual displacements is

$$\delta d^{ij} = \delta\varepsilon^{ij} L \tag{3.22}$$

where $\delta\varepsilon^{ij}$ is the *virtual strain*. Substituting Eqs. (3.21) and (3.22) into Eq. (3.19) yields

$$\delta W_i = \sigma^{ij}\, \delta\varepsilon^{ij} AL \tag{3.23}$$

Alternatively this can be written as

$$\delta W_i = \int_{\text{vol}} \sigma^{ij}\, \delta\varepsilon^{ij}\, dV \tag{3.24}$$

since the stress and strain are constant over the volume of a prismatic element and $\int_{\text{vol}} dV = AL$. The physical interpretation of Eq. (3.24) is depicted in Fig. 3.6. The element is loaded from a stress-free state and the stress and strain are linearly related up to the equilibrium point where the state $(\sigma^{ij}, \varepsilon^{ij})$ is attained. The area under the stress-strain graph is the *strain energy density* $\bar{W}_i$. The strain energy W_i is obtained by integrating the strain energy density over

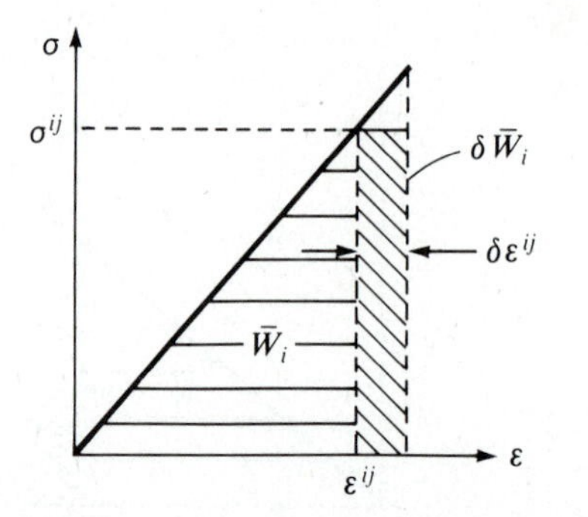

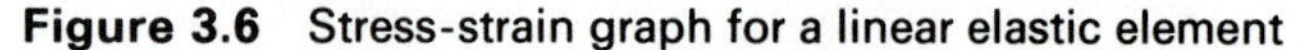

Figure 3.6 Stress-strain graph for a linear elastic element

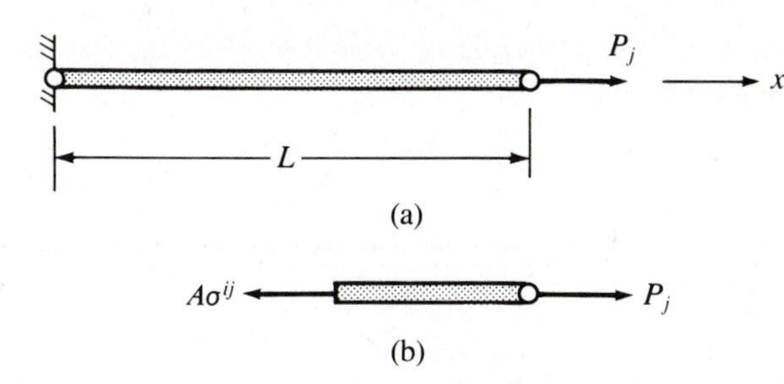

Figure 3.7 (a) Constrained axial force element; (b) free-body diagram of element segment

the volume of the element:

$$W_i = \int_{\text{vol}} \bar{W}_i \, dV \tag{3.25}$$

where

$$\bar{W}_i = \int_0^{\varepsilon^{ij}} \sigma \, d\varepsilon = \frac{1}{2} \sigma^{ij} \varepsilon^{ij} \tag{3.26}$$

Imposing a virtual strain $\delta\varepsilon^{ij}$ gives the *virtual strain energy density*

$$\delta\bar{W}_i = \sigma^{ij} \, \delta\varepsilon^{ij} \tag{3.27}$$

plus a second-order quantity $(\delta\sigma^{ij} \, \delta\varepsilon^{ij})/2$, which can be ignored for linear structural response.

Equation (3.20) is a mathematical expression of the principle of virtual work, but the statement made below the equation is considered to be the converse of the principle. To obtain an expression of the principle, consider the linearly elastic prismatic axial force element constrained at point i, as illustrated in Fig. 3.7a. Equilibrium of the free-body diagram of Fig. 3.7b requires that

$$P_j - \sigma^{ij} A = 0 \tag{3.28}$$

Imposing the virtual displacement δu_j on the element and multiplying the equation of equilibrium, Eq. (3.28), by this quantity yields

$$(P_j - \sigma^{ij} A)\delta u_j = 0$$

or

$$P_j \, \delta u_j - \sigma^{ij} A \, \delta u_j = 0 \tag{3.29}$$

Note that $P_j \, \delta u_j = \delta W_e$ and since $\delta u_j = \delta\varepsilon^{ij} L$, this gives

$$\sigma^{ij} A \, \delta u_j = \sigma^{ij} \, \delta\varepsilon^{ij} A L = \delta W_i$$

Thus, Eq. (3.29) is an expression of the *principle of virtual work*, $\delta W_e = \delta W_i$, and we can state that:

If a linearly elastic structure is in equilibrium under a system of forces, then for any state of virtual displacements away from a compatible state of deformation the external virtual work is equal to the virtual strain energy.

In Example 3.2 the principle of virtual work is used to investigate a statically indeterminate system.

EXAMPLE 3.2

Apply the principle of virtual work for deformable bodies to calculate the reaction forces for the two-element structure in Fig. E3.2. These are uniform axial force elements, and the slope of the force-deformation graph (e.g., see Fig. 3.5) is k_i. Observe that since $\sigma = E\varepsilon$, $\sigma = s/A$ and $\varepsilon = d/L$; therefore $k = AE/L$.

Solution

The free-body diagram of the structure with the reactions shown as applied forces is illustrated in Fig. E3.2b. Since there are three nodal points, an admissible virtual displacement state is shown in Fig. E3.2c. The quantities δu_a, δu_b, and δu_c are arbitrary and independent. In this case,

$$\delta W_e = P_a\,\delta u_a + 9\,\delta u_b + P_c\,\delta u_c$$

and

$$\delta W_i = k_1 u_b(\delta u_b - \delta u_a) - k_2\,u_b(\delta u_c - \delta u_b)$$

The principle of virtual work dictates that $\delta W_e = \delta W_i$; substituting these expressions for external virtual work and internal strain energy yields, with a rearrangement of terms,

$$(P_a + k_1 u_b)\delta u_a + (9 - k_1 u_b - k_2\,u_b)\delta u_b + (P_c + k_2\,u_b)\delta u_c = 0$$

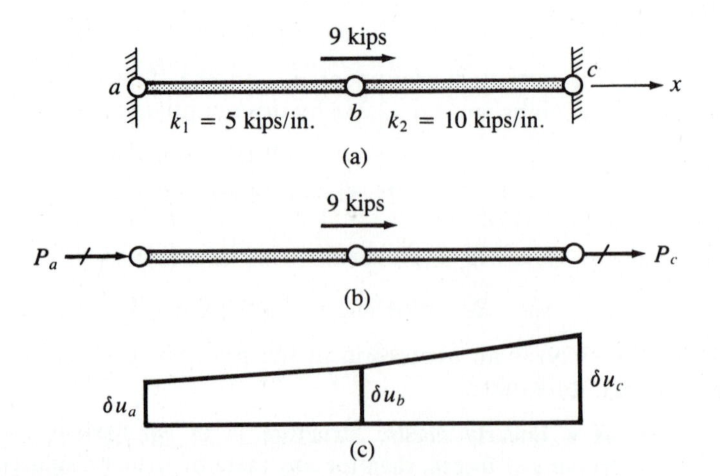

Figure E3.2 (a) Two-element assemblage; (b) free-body diagram; (c) virtual displacement state

Since δu_a, δu_b, and δu_c are arbitrary and independent, each of their coefficients must be zero for the above equation to be satisfied. These conditions give

$$P_a + k_1 u_b = 0$$
$$9 - k_1 u_b - k_2 u_b = 0$$
$$P_c + k_2 u_b = 0$$

The solution of these three equations gives $u_b = 0.60$ in. $(\rightarrow)$; $P_a = -3$ kips $(\nleftarrow)$; and $P_c = -6$ kips $(\nleftarrow)$.

Discussion By applying the principle of virtual work to this statically indeterminate structure both displacements and reaction forces are obtained. This occurs because the reaction forces are dependent upon the deformations of the structure, and the necessary number of equations for solving for all unknowns are formulated by giving each of the independent degrees of freedom of the structure a virtual change. Note that only reaction forces were obtained by applying the theorem of virtual work to the statically determinate structure in Example 3.1. In computing δW_i the real element forces result from the real displacements. Since $u_a = u_c = 0$, only u_b will produce forces within the elements. In contrast, for the virtual displacements selected and shown in Fig. E3.2c, δu_a, δu_b, and δu_c all participate in giving virtual deformations. See the next section for a detailed description of these terms.

3.1.3 REAL, ACTUAL, AND ADMISSIBLE DISPLACEMENTS The principle of virtual work for deformable bodies involves the virtual work of the external forces (δW_e) and the internal virtual work or virtual strain energy (δW_i). The physical meaning of the former scalar quantity is illustrated in Fig. 3.4, and the latter is shown in Fig. 3.5. The virtual strain energy density ($\delta \overline{W}_i$) is portrayed in Fig. 3.6. In each case, the virtual quantity is the result of a virtual displacement. That is, the virtual displacement yields virtual deformations of the element (δd) and corresponding virtual strains ($\delta \varepsilon$). The virtual displacements need not be those experienced by the structure, but they must be *kinematically admissible.* For example, the single axial force element in Fig. 3.8a is constrained at node a, but by envisaging the reaction force at a as another applied force we can impose an admissible displacement state as shown. Thus, even though u_a must be zero, the virtual displacement at a need not duplicate the support conditions of the structure. The admissible displacement state in Fig. 3.8c is continuous between nodes a and b, and does not introduce any rips or tears in the element. In contrast, the displacements in Fig. 3.8d would not be admissible because there is a discontinuity (a tear) at point c, and this violates the kinematic conditions of the element.

The element in Fig. 3.8a is uniformly tapered from end to end, with the cross-sectional area decreasing linearly. By making an imaginary cut through the element and envisaging a free-body diagram of everything to the right of the cut, we deduce that the element will have an axial force of P throughout its

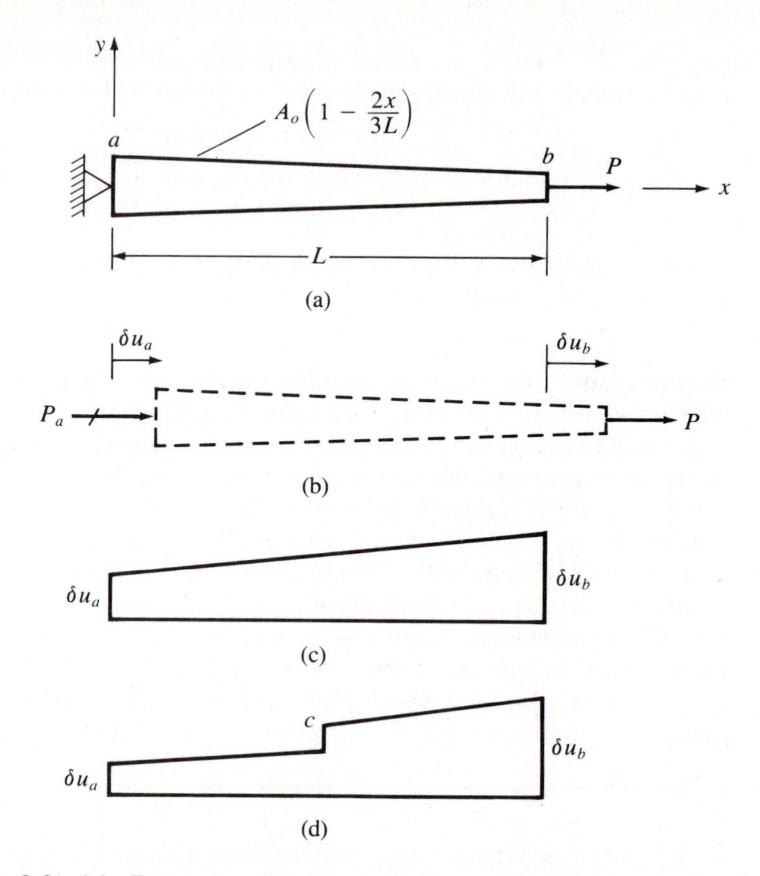

Figure 3.8 (a) Tapered axial force element; (b) the free-body diagram; (c) admissible and (d) nonadmissible virtual displacement states

length. Consequently, the actual stress in the element varies from end to end as follows:

$$\sigma = \frac{P}{A_o(1 - 2x/3L)}$$

Since $\varepsilon = \sigma/E$ and $du/dx = \varepsilon$ [see Eq. (2.2)], therefore

$$\frac{du}{dx} = \frac{P}{A_o E} \frac{1}{1 - 2x/3L}$$

Integration gives

$$u = -\frac{3PL}{2A_o E} \ln\left(1 - \frac{2x}{3L}\right)$$

The constant of integration is zero since $u_a = 0$.

Therefore, to solve this problem exactly we must express the displacements experienced by the element as a natural log function. If the exact solution were obtained we would have the *actual displacements*. The displacements corresponding to an approximate solution are the *real displacements*.

For structures with complex geometries and/or applied forces, it is frequently preferable to obtain an approximate rather than an exact solution; the principle of virtual work is a powerful method to generate approximate solutions. In the subsequent text the displacements experienced by the structure or element will usually be referred to as real.

3.1.4 DISCUSSION The principle of virtual work results from perturbing the displacements (and corresponding strains) of the structure from an equilibrium position. The resulting first-order change in internal strain energy is the virtual strain energy. The *virtual strain energy density* is the change in area below the stress-strain curve resulting from the virtual strain (see Fig. 3.6). The *virtual strain* corresponds to *admissible displacements*, which need not correspond to the real displacements but must simply satisfy kinematic constraints of the structure. The *virtual strain energy* is obtained by integrating the strain energy density over the volume of the structure. The analysis in the foregoing sections has been limited to the axial force element, but we will see in subsequent chapters that the strain energy for any structure is obtained by integrating the product of the real stress and virtual strains over the volume of the structure.

External virtual work is the first-order change in the external work resulting from perturbing the displacements of a structure from an equilibrium position. That is, it is the change in area below the force-displacement diagram (see Fig. 3.4). For point forces it is the product of the real force and the corresponding virtual displacement. The virtual work of a distributed axial force $q(x)$ (see Fig. 3.9) applied to the element between points a and b is

$$\delta W_e = \int_a^b q(x)\, \delta u(x)\, dx \tag{3.30}$$

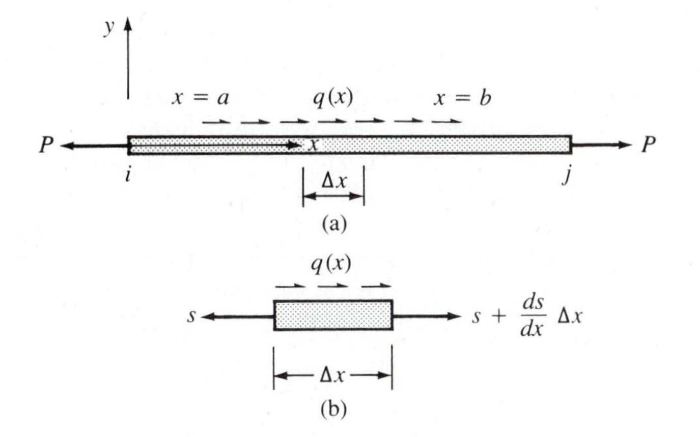

Figure 3.9 (a) Axial force element with distributed load; (b) free-body diagram of a segment of the element

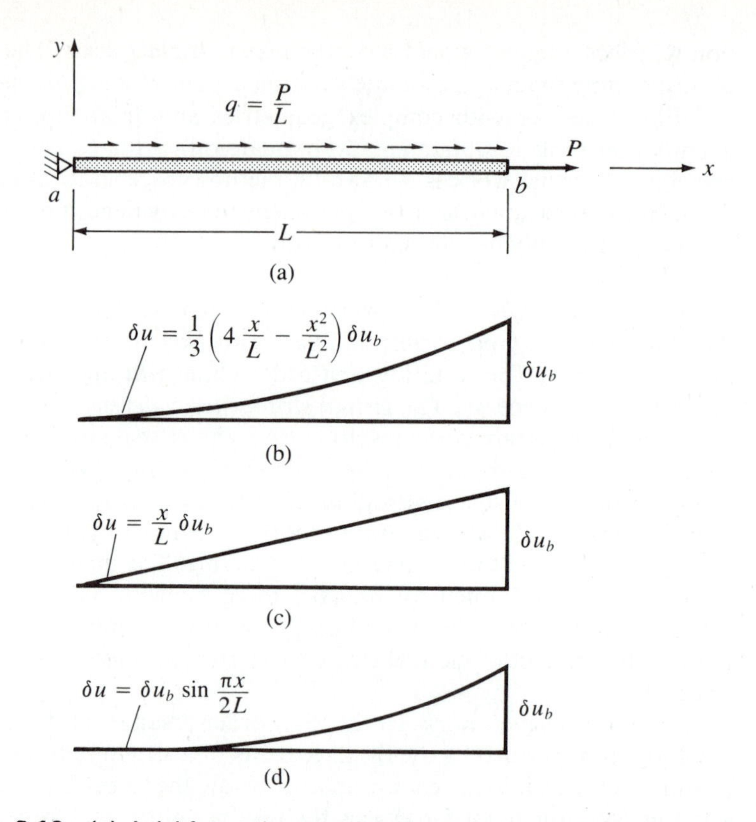

Figure 3.10 (a) Axial force element; (b) quadratic virtual displacement function; (c) linear virtual displacement function; (d) sinusoidal virtual displacement function

where $\delta u(x)$ is a description of the virtual displacements along the element [$q(x)$ is sometimes referred to as a surface traction]. The virtual displacements need only be admissible. That is, they may correspond to the real displacements for computational convenience, but they can also be chosen so that they only satisfy the kinematic conditions of the structure.

We will study the effect of selecting various displacement functions by analyzing an axial force element with applied point and distributed axial loading as illustrated in Fig.3.10a. We will obtain the exact solution using the basic governing equations so that the accuracy of the energy solutions can be evaluated. The distributed axial force of intensity q (also referred to as a shear flow) has units of force per length. The small segment Δx has the applied axial forces shown in Fig. 3.9b, and equilibrium requires that

$$s + \frac{ds}{dx}\,\Delta x - s + q(x)\,\Delta x = 0$$

or

$$\left[\frac{ds}{dx} + q(x)\right]\Delta x = 0$$

Since $\Delta x \neq 0$, the quantity in brackets must be zero; this gives the differential equation of equilibrium as

$$\frac{ds}{dx} + q(x) = 0 \tag{3.31}$$

The distributed axial force q is a constant for the linearly elastic prismatic axial force member in Fig. 3.10; therefore Eq. (3.31) mandates that

$$s = -\frac{P}{L} x + C_1$$

where C_1 is a constant of integration. Since $s = P$ at $x = L$, therefore $C_1 = 2P$. The fact that the cross-sectional area is constant implies that the axial stress is a linear function of x; that is,

$$\sigma = \frac{s}{A} = \frac{1}{A}\left(-P\frac{x}{L} + 2P\right) = \frac{P}{A}\left(2 - \frac{x}{L}\right)$$

Using $\varepsilon = \sigma/E$, plus the strain-displacement equation $\varepsilon = du/dx$, and integrating the result we have

$$u = \frac{P}{2AE}\left(4x - \frac{x^2}{L}\right) + C_2$$

The constant of integration C_2 is zero since the member is constrained at $x = 0$. The displacement at b $(x = L)$ is

$$u_b = \frac{3PL}{2AE}$$

Thus,

$$P = \frac{2AE}{3L} u_b$$

and

$$u = \frac{1}{3}\left(4\frac{x}{L} - \frac{x^2}{L^2}\right)u_b \tag{3.32}$$

This structural member can also be examined using the principle of virtual work. The external virtual work is

$$\delta W_e = P\,\delta u_b + \int_0^L \frac{P}{L}\,\delta u\,dx \tag{3.33}$$

The internal virtual strain energy is

$$\delta W_i = \int_{\text{vol}} \sigma\,\delta\varepsilon\,dV \tag{3.34}$$

Using Eqs. (3.33) and (3.34) we will examine the effect of using various real and virtual displacement functions in conjunction with the principle of virtual work.

CASE 1 Use the actual displacement function for both real and virtual displacements, as shown in Fig. 3.10b; that is,

$$u = \frac{1}{3}\left(4\frac{x}{L} - \frac{x^2}{L^2}\right)u_b$$

and

$$\delta u = \frac{1}{3}\left(4\frac{x}{L} - \frac{x^2}{L^2}\right)\delta u_b$$

Substituting the above in Eqs. (3.33) and (3.34) gives

$$\delta W_e = \frac{14}{9} P\,\delta u_b$$

and

$$\delta W_i = \frac{28AE}{27L} u_b\,\delta u_b$$

Thus $\delta W_i = \delta W_e$ gives $u_b = 3PL/2AE$.

CASE 2 Use the actual displacement function for u (i.e., the same used in case 1) and let $\delta u = (x/L)\,\delta u_b$ (see Fig. 3.10c). Note that the virtual displacement function satisfies the kinematic boundary conditions [i.e., $\delta u(x = 0) = 0$ and $u(x = L) = \delta u_b$]; therefore, it satisfies the conditions of admissibility. Substituting into Eqs. (3.33) and (3.34) gives

$$\delta W_e = \frac{3}{2} P\,\delta u_b$$

and

$$\delta W_i = \frac{AE}{L} u_b\,\delta u_b$$

In this situation $\delta W_i = \delta W_e$ also gives the exact answer $u_b = 3PL/2AE$. This case illustrates that the solution is independent of the choice of the virtual displacements as long as they are admissible.

CASE 3 As further evidence of the freedom of choice for virtual displacements we will use the exact form of u and let $\delta u = \delta u_b \sin(\pi x/2L)$, as shown in Fig. 3.10d. Note that the virtual displacements are admissible. Substituting these displacement functions into Eqs. (3.33) and (3.34) yields

$$\delta W_e = 1.6366P\,\delta u_b$$

and

$$\delta W_i = 1.6366\,\frac{2AE}{3L} u_b\,\delta u_b$$

Again $\delta W_i = \delta W_e$ gives $u_b = 3PL/2AE$, which indicates that the solution using the principle of virtual work is not dependent upon the choice of the virtual displacements chosen.

Much of the appeal of the principle of virtual work is in obtaining approximate solutions. Assuming that the exact solution to the above problem is not known, we can choose any displacement function that satisfies the kinematic boundary conditions. It is generally convenient to select the real and virtual displacement functions as the same; therefore, in this case assume that

$$u = u_b \sin \frac{\pi x}{2L}$$

and

$$\delta u = \delta u_b \sin \frac{\pi x}{2L}$$

Substituting into Eqs. (3.33) and (3.34) gives

$$\delta W_e = 1.6366P \, \delta u_b$$

and

$$\delta W_i = 1.2337 \frac{AE}{L} u_b \, \delta u_b$$

Applying the principle of virtual work yields $u_b = 1.3266PL/AE$, which is slightly less than the exact solution, but this stems from the approximate nature of the analysis. It can be shown that an appropriate but approximate displacement function gives deflections that are equal to or less than the exact ones. This assertion will not be proven, but we can envisage that in choosing an approximate displacement function the internal strain energy is generally overestimated. That is, more internal strain energy is required to deform the element into a higher-order displacement shape than that associated with the actual deformed configuration. In the above example, the unknown deflection is obtained by dividing δW_e by a coefficient derived from δW_i; therefore, the approximated deflection should be less than or equal to the actual value.

In the above calculations we have been using the physical coordinates in describing the displacements, but it is possible to use parameters that do not refer to specific displacements. For example, assume that

$$u = a_0 + a_1 x + a_2 x^2$$

where the a_i are the *generalized displacements* (or *generalized coordinates*). The displacement function for the member in Fig. 3.10a must be admissible so $a_0 = 0$ since $u(0) = 0$. Using the same form to describe the virtual displacements we have

$$\delta u = \delta a_1 x + \delta a_2 x^2$$

where the δa_i are arbitrary and independent generalized displacements. With these displacement functions the real stresses are

$$\sigma = E(a_1 + 2a_2 x)$$

and the virtual strains are

$$\delta \varepsilon = \delta a_1 + 2 \, \delta a_2 x$$

Substituting these into Eqs. (3.33) and (3.34) yields

$$\delta W_e = P(\tfrac{3}{2}L\ \delta a_1 + \tfrac{4}{3}L^2\ \delta a_2)$$

and

$$\delta W_i = AE[(a_1 L + a_2 L^2)\ \delta a_1 + (a_1 L^2 + \tfrac{4}{3}a_2 L^3)\delta a_2]$$

Equating δW_e and δW_i and grouping terms gives

$$[(AE(a_1 L + a_2 L^2) - \tfrac{3}{2}PL]\ \delta a_1 + [AE(a_1 L^2 + \tfrac{4}{3}a_2 L^3) - \tfrac{4}{3}PL^2]\ \delta a_2 = 0$$

Since δa_1 and δa_2 are arbitrary and independent, the above equation can only be satisfied if each of their coefficients is zero, thus giving the following equations:

$$a_1 + a_2 L = \frac{3}{2}\frac{P}{AE}$$

$$a_1 + \frac{4}{3}a_2 L = \frac{4}{3}\frac{P}{AE}$$

Solution gives

$$a_1 = \frac{2P}{AE} \qquad \text{and} \qquad a_2 = -\frac{P}{2AEL}$$

Thus,

$$u = \frac{2P}{AE}x - \frac{P}{2AEL}x^2 = \frac{PL}{AE}\left(2\frac{x}{L} - \frac{1}{2}\frac{x^2}{L^2}\right)$$

Evaluating gives

$$u(x = L) = u_b = \frac{3PL}{2AE}$$

which is the actual displacement as obtained previously. Since u was assumed to be quadratic, it is reasonable to assume that this displacement function would give the exact solution.

Trigonometric functions can also be assumed using generalized displacements. For example, with

$$u = a_3 \sin\frac{\pi x}{2L} \qquad \text{and} \qquad \delta u = \delta a_3 \sin\frac{\pi x}{2L}$$

we obtain from the principle of virtual work

$$a_3 = 1.3266\frac{PL}{AE}$$

Since $u(x = L) = a_3$, this is the same value obtained previously with the displacements in the form of a sine function. More terms can be used to improve the accuracy of the solution. Using

$$u = a_4 \sin\frac{\pi x}{2L} + a_5 \sin\frac{3\pi x}{2L}$$

and applying the principle of virtual work yields

$$a_4 = 1.3266 \qquad \text{and} \qquad a_5 = -0.0710$$

or

$$u = \frac{PL}{AE}\left(1.3266 \sin\frac{\pi x}{2L} - 0.0710 \sin\frac{3\pi x}{2L}\right)$$

With this approximation $u_b = 1.3976(PL/AE)$, which is a slight improvement in the solution obtained using the one-term sine approximation.

3.2 THE ELEMENT STIFFNESS MATRIX

For some structural elements it is difficult to use the basic equations to obtain the exact element stiffness relationships; the principle of virtual work is a convenient alternative for investigating the behavior of tapered elements and elements with distributed forces. The stress-strain relationships, as well as the compatibility equations, are imposed when applying the principle of virtual work, and equilibrium is implicit in the principle. Typically, an approximate solution is formulated and used to obtain the corresponding stiffness matrix; this requires a description of the deflected shape of the element under loading. Polynomials are generally chosen, but trigonometric and other functions may also be used. The displacement functions typically characterize displacement continuity between elements, but in some cases (e.g., beams) continuity of slopes at element nodes may also be mandated. Usually, the continuous element displacements are described in terms of the nodal values (e.g., displacements and/or slopes); these relationships are contained in the shape functions.

3.2.1 SHAPE FUNCTIONS The interpolation functions used to describe the displacements throughout an element are called shape functions; these can be formulated using polynomials, trigonometric, logarithmic, or other functions. Generally shape functions are expressed using *complete polynomials* (e.g., an nth degree polynomial in one variable must contain all terms of order n and lower). These functions must generally be integrated and/or differentiated to give properties such as element stiffnesses and stresses; therefore, polynomials are more convenient for performing these manipulations. The displacements for the axial force element in Fig. 3.11 can be described as follows:

$$u = a_0 + a_1 x \tag{3.35}$$

or in matrix form

$$u = [1 \quad x]\begin{bmatrix} a_0 \\ a_1 \end{bmatrix}$$

$$u = \quad \boldsymbol{x} \quad \boldsymbol{a} \tag{3.36}$$

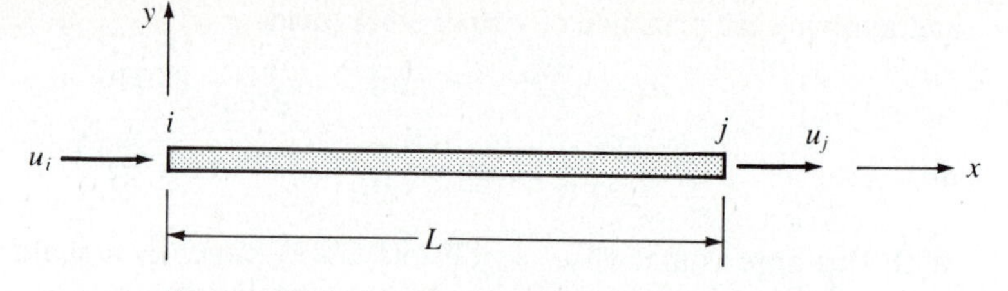

Figure 3.11 Unconstrained axial force element

At the nodal points $u(0) = u_i$ and $u(L) = u_j$; substituting into Eq. (3.36) gives

$$\begin{bmatrix} u_i \\ u_j \end{bmatrix} = \begin{bmatrix} 1 & 0 \\ 1 & L \end{bmatrix} \begin{bmatrix} a_0 \\ a_1 \end{bmatrix}$$

$$\mathbf{u} = \mathbf{L} \quad \mathbf{a} \tag{3.37}$$

Thus

$$\mathbf{a} = \mathbf{L}^{-1}\,\mathbf{u} \tag{3.38}$$

where

$$\mathbf{L}^{-1} = \frac{1}{L}\begin{bmatrix} L & 0 \\ -1 & 1 \end{bmatrix}$$

Substituting Eq. (3.38) into Eq. (3.36) yields

$$u = [1 \quad x]\,\frac{1}{L}\begin{bmatrix} L & 0 \\ -1 & 1 \end{bmatrix}\begin{bmatrix} u_i \\ u_j \end{bmatrix}$$

$$= \frac{1}{L}\,[L - x \quad x]\begin{bmatrix} u_i \\ u_j \end{bmatrix} = \begin{bmatrix} 1 - \dfrac{x}{L} & \dfrac{x}{L} \end{bmatrix}\begin{bmatrix} u_i \\ u_j \end{bmatrix}$$

$$u = \mathbf{N}\mathbf{u} \tag{3.39}$$

where $\mathbf{N}$ is the matrix of shape functions with $N_1 = 1 - x/L$ and $N_2 = x/L$. These two shape functions are displayed in Fig. 3.12. Note that N_i is unity at $x = x_i$ and zero at $x = x_j$; furthermore, N_j is zero at $x = x_i$ and unity at $x = x_j$.

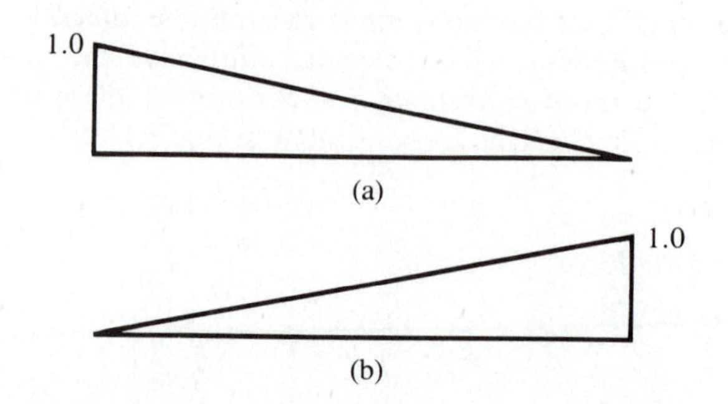

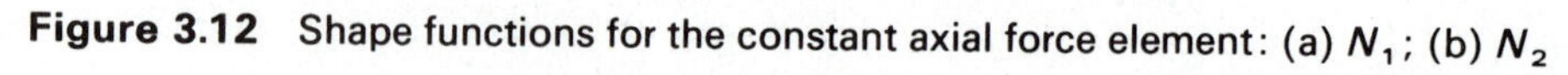
Figure 3.12 Shape functions for the constant axial force element: (a) N_1; (b) N_2

It is frequently useful to think of the shape functions as nondimensional since they are the multipliers of the nodal displacements to yield u. By using the nondimensional coordinate $r = 2x/L - 1$ for the axial force element with linear displacements the element extends from -1 to $+1$. Substituting into Eq. (3.39) the shape functions become

$$\mathbf{N} = [\tfrac{1}{2}(1 - r) \ \tfrac{1}{2}(1 + r)]$$

The shape functions must contain both the rigid-body displacements and elastic deformations. In the case of the axial force element discussed above, a_0 describes the rigid-body translation of the element, and a_1 characterizes the element deformations since the strain is the first derivative of u. In summary, in the above discussion we obtained the shape functions by: starting with a polynomial such as that shown in Eq. (3.35); methodically satisfying the displacement boundary conditions $u(0) = u_i$ and $u(L) = u_j$; and calculating $\mathbf{N} = \mathbf{x}\mathbf{L}^{-1}$. Alternatively, the shape interpolation functions can be envisaged as polynomials of the appropriate order and satisfying the associated displacement boundary conditions. This latter approach can be extremely useful for more complex elements, but for this discussion the more systematic formulation will be followed to avoid introducing unecessarily complicating aspects.

Shape functions for other elements can be obtained in a fashion similar to that used for the linear displacement axial force element. Example 3.3 illustrates the use of polynomials for a three-node element.

EXAMPLE 3.3

Obtain the shape functions for the three-node axial force element in Fig. E3.3a.

Solution

Since there are three nodal points and three known displacements, u_i, u_j, and u_k, the following second-order polynomial will be used:

$$u = a_0 + a_1 x + a_2 x^2$$

or

$$u = [1 \quad x \quad x^2] \begin{bmatrix} a_0 \\ a_1 \\ a_2 \end{bmatrix}$$

$$\begin{bmatrix} u_i \\ u_j \\ u_k \end{bmatrix} = \begin{bmatrix} 1 & 0 & 0 \\ 1 & L & L^2 \\ 1 & L/2 & L^2/4 \end{bmatrix} \begin{bmatrix} a_0 \\ a_1 \\ a_2 \end{bmatrix}$$

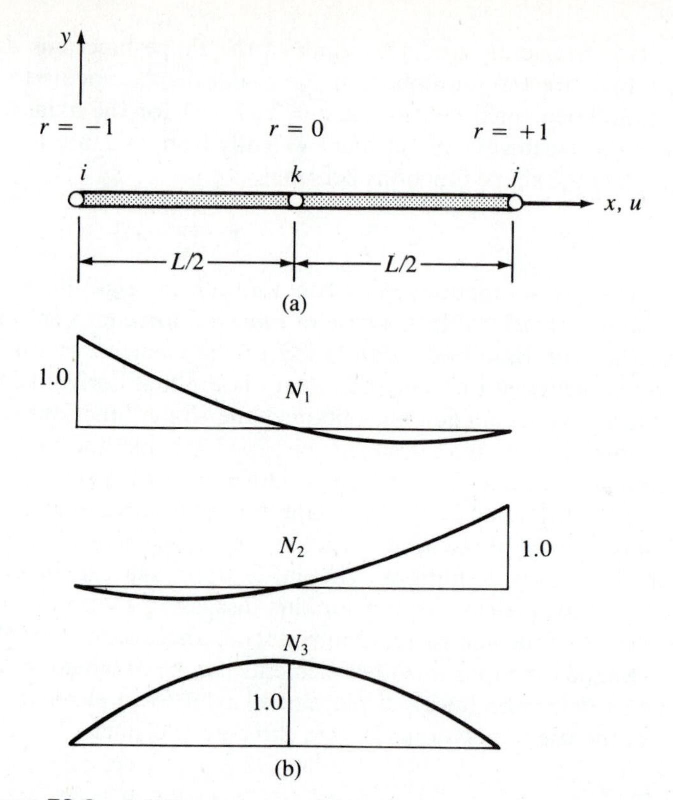

Figure E3.3 (a) Three-node axial force element; (b) shape functions

Thus

$$\mathbf{L}^{-1} = \begin{bmatrix} 1 & 0 & 0 \\ -\dfrac{3}{L} & -\dfrac{1}{L} & \dfrac{4}{L} \\ \dfrac{2}{L^2} & \dfrac{2}{L^2} & -\dfrac{4}{L^2} \end{bmatrix}$$

and the shape functions are

$$\begin{aligned} \mathbf{N} &= \mathbf{x}\mathbf{L}^{-1} \\ &= \left[\left(1 - 3\frac{x}{L} + 2\frac{x^2}{L^2}\right)\left(-\frac{x}{L} + 2\frac{x^2}{L^2}\right)\left(4\frac{x}{L} - 4\frac{x^2}{L^2}\right)\right] \end{aligned}$$

....

Discussion The shape functions are illustrated in Fig. E3.3b. Since

$$u = N_1 u_i + N_2 u_j + N_3 u_k$$

each N_i must pass through unity at the node for which it serves as a multiplier in this displacement equation; furthermore, it must pass through zero at the other two nodal points. Using the nondimensional coordinate

$$r = 2\frac{x}{L} - 1 \qquad (-1 \le r \le +1)$$

the shape functions become

$$N_1 = \frac{r}{2}(r-1) \qquad N_2 = \frac{r}{2}(r+1) \qquad N_3 = 1 - r^2$$

These representations make it easier to visualize the shape functions, and they can also be convenient when analyzing the element properties. This formulation is frequently used in the finite element method, but in this book nondimensionalized coordinates will not be used other than as a visual aid. The three-node element is useful for tapered members and other applications where the stress is not constant along the length of the element.

3.2.2 THE AXIAL FORCE ELEMENT Upon knowing the displacements, the strains and stresses within an element can be obtained by applying the strain-displacement equation, plus the material behavior relationship. Using the principle of virtual work, we can derive the equations relating forces and displacements at the ends of an element; this gives the element stiffness matrix.

The procedure will be demonstrated for the prismatic element with a cross-sectional area A and subjected to forces that are colinear with its centroidal axis (see Fig. 3.13). Under loading the element remains straight, and plane sections normal to the element axis remain plane. We will assume that the element undergoes sufficiently small displacements so that linear strain-displacement relations are valid and the equilibrium equations apply to the undeformed state. Also, the element experiences small strains and is made from a homogeneous, orthotropic, and linearly elastic material so that Hooke's law is applicable. The only strain in this case, ε_x, is obtained from the displacements using Eq. (2.2); that is,

$$\varepsilon = \frac{\partial u}{\partial x}$$

The displacement along the element is expressed in terms of the end displacements by the shape functions; substituting Eq. (3.39) into the strain-

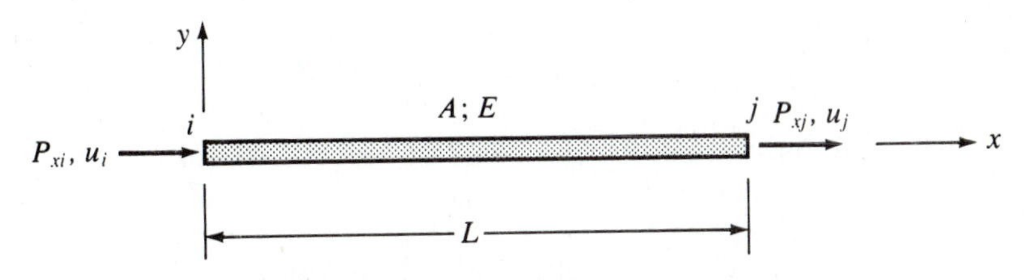

Figure 3.13 The axial force element

displacement equations yields

$$\varepsilon = \frac{\partial}{\partial x}\,\mathbf{Nu} = \mathbf{Bu} \tag{3.40}$$

where

$$\mathbf{B} = \frac{\partial}{\partial x}\,\mathbf{N} \tag{3.41}$$

In the case of the prismatic axial force element

$$\mathbf{B} = \frac{1}{L}[-1 \quad 1] \tag{3.42}$$

Since the strains are the product of **Bu**, we note from Eq. (3.42) that this element has constant strains throughout its length.

Substituting Eq. (3.40) into Hooke's law [i.e., Eq. (2.3)] gives

$$\sigma = E\mathbf{Bu} \tag{3.43}$$

For the prismatic axial force element the stress is

$$\sigma = \frac{E}{L}[-1 \quad 1]\begin{bmatrix} u_i \\ u_j \end{bmatrix} = \frac{E}{L}(u_j - u_i) \tag{3.44}$$

That is, $(u_j - u_i)/L$ is the strain, and the product of the modulus of elasticity, E, and the strain gives the stress. Note that this element has constant stress as one would anticipate. This is an element that would be used to represent a truss member, and a prismatic element with forces applied only at the ends must display a state of constant stress along its length.

The internal virtual work is obtained using the actual stresses and the virtual strains, which result from virtual displacements. The virtual strains are described in a form similar to that of the actual strains in Eq. (3.40):

$$\delta\varepsilon = \mathbf{B}\ \boldsymbol{\delta}\mathbf{u} \tag{3.45}$$

The definition of internal virtual work (virtual strain energy) in Eq. (3.24) presumes that the stress and strain are defined as scalars, but we can envisage the more general case where each of the quantities is defined as a column matrix; in this case

$$\delta W_i = \int_{\text{vol}} \boldsymbol{\delta\varepsilon}^T \boldsymbol{\sigma}\, dV \tag{3.46}$$

Substituting Eqs. (3.43) and (3.45) into Eq. (3.46) gives

$$\delta W_i = \int_{\text{vol}} \boldsymbol{\delta}\mathbf{u}^T \mathbf{B}^T E \mathbf{Bu}\, dV$$

The real and virtual displacements are constants and can be taken outside of the integral; this yields

$$\delta W_i = \boldsymbol{\delta}\mathbf{u}^T \int_{\text{vol}} \mathbf{B}^T E \mathbf{B}\, dV\ \mathbf{u} \tag{3.47}$$

Multiple forces and the associated displacements must be represented as column matrices; the external virtual work is formulated by generalizing Eq. (3.17) as follows:

$$\delta W_e = \boldsymbol{\delta}\mathbf{u}^T\mathbf{p} \tag{3.48}$$

Substituting Eqs. (3.47) and (3.48) into the theorem of virtual work [Eq. (3.20)] we have

$$\boldsymbol{\delta}\mathbf{u}^T\mathbf{p} = \boldsymbol{\delta}\mathbf{u}^T \int_{\text{vol}} \mathbf{B}^T E\mathbf{B}\, dV\, \mathbf{u}$$

or

$$\boldsymbol{\delta}\mathbf{u}^T\left(\mathbf{p} - \int_{\text{vol}} \mathbf{B}^T E\mathbf{B}\, dV\, \mathbf{u}\right) = 0$$

Since $\boldsymbol{\delta}\mathbf{u}$ is arbitrary and nonzero, the quantity in parentheses must be zero; that is,

$$\mathbf{p} = \int_{\text{vol}} \mathbf{B}^T E\mathbf{B}\, dV\, \mathbf{u} = \mathbf{k}\mathbf{u} \tag{3.49}$$

where

$$\mathbf{k} = \int_{\text{vol}} \mathbf{B}^T E\mathbf{B}\, dV \tag{3.50}$$

This is the general formulation for calculating an element stiffness matrix. After choosing an appropriate set of shape functions, they are operated on in accordance with Eq. (3.41) to give the matrix **B**. From Eq. (3.50) we observed that **k** is a symmetric matrix. This can be verified by recalling that for the product of three matrices **a**, **b**, and **c**,

$$(\mathbf{abc})^T = \mathbf{c}^T\mathbf{b}^T\mathbf{a}^T$$

Therefore,

$$\mathbf{k}^T = \int_{\text{vol}} (\mathbf{B}^T E\mathbf{B})^T\, dV = \int_{\text{vol}} \mathbf{B}^T E\mathbf{B}\, dV = \mathbf{k}$$

That is, since $\mathbf{k} = \mathbf{k}^T$ we conclude that the element stiffness matrix is symmetric.

Using Eq. (3.50) and substituting **B** as shown in Eq. (3.42) for the axial force element of Fig. 3.12 gives

$$\mathbf{k} = \int_{\text{vol}} \frac{1}{L}\begin{bmatrix} -1 \\ 1 \end{bmatrix} E \frac{1}{L}[-1 \quad 1]\, dV$$

$$= \frac{E}{L^2}\begin{bmatrix} 1 & -1 \\ -1 & 1 \end{bmatrix} \int_{\text{vol}} dV$$

But $\int_{\text{vol}} dV = AL$, which gives

$$\mathbf{k} = \frac{AE}{L}\begin{bmatrix} 1 & -1 \\ -1 & 1 \end{bmatrix} \tag{3.51}$$

The stiffness matrix for the three-node element of Example 3.3 is derived in Example 3.4.

EXAMPLE 3.4

Obtain the stiffness matrix for the prismatic three-node element of Example 3.3; use the theorem of virtual work in the form of Eq. (3.50).

Solution

Operating on the matrix **N** from Example 3.3 using Eq. (3.41) gives the following:

$$\mathbf{B} = \frac{1}{L}\left[\left(-3 + 4\frac{x}{L}\right)\left(-1 + 4\frac{x}{L}\right)\left(4 - 8\frac{x}{L}\right)\right]$$

Since the element is prismatic, Eq. (3.50) reduces to the form

$$\mathbf{k} = AE\int_0^L \mathbf{B}^T\mathbf{B}\,dx = \frac{AE}{L^2}\int_0^L \mathbf{j}\,dx$$

where

$$\begin{aligned}
j(1, 1) &= 9 - 24x/L + 16x^2/L^2\\
j(1, 2) &= 3 - 16x/L + 16x^2/L^2\\
j(1, 3) &= -12 + 40x/L - 32x^2/L^2\\
j(2, 1) &= 3 - 16x/L + 16x^2/L^2\\
j(2, 2) &= 1 - 8x/L + 16x^2/L^2\\
j(2, 3) &= -4 + 24x/L - 32x^2/L^2\\
j(3, 1) &= -12 + 40x/L - 32x^2/L^2\\
j(3, 2) &= -4 + 24x/L - 32x^2/L^2\\
j(3, 3) &= 16 - 64x/L + 64x^2/L^2
\end{aligned}$$

Integrating gives the following stiffness matrix for the three-node prismatic axial force element:

$$\mathbf{k} = \frac{AE}{3L}\begin{bmatrix} 7 & 1 & -8\\ 1 & 7 & -8\\ -8 & -8 & 16 \end{bmatrix}$$

3.3 COORDINATE TRANSFORMATIONS

The axial force elements in the previous section were analyzed using a coordinate system with the x axis extending along the length of the element. When investigating a complete truss composed of many individual elements, it is necessary to use one coordinate system for the entire structure; generally, the nodal forces and displacements are described in terms of a single orthogonal coordinate system. The coordinates associated with each individual element comprise the *local coordinate system*, and all the nodal forces and displacements of the total structure are characterized using just one *global coordinate system*. Quantities in local coordinates are designated with an overbar (e.g., $\bar{u}_i$), and the global coordinate values have no overbar. Lower-case symbols are associated with the element, and upper-case letters denote displacements, forces, etc., that are referenced to the assembled structure. A third system, the *nodal coordinate system*, is described in Sec. 6.2; this type is required to describe forces and displacements related to supports that are not aligned with the global coordinates. Orthogonal coordinate transformations are discussed in Sec. 2.5; these were derived using simple geometrical arguments. The subsequent analysis will invoke the principle of virtual work, and more general transformations will be described.

The nodal displacements for an element such as the axial force element in Fig. 3.14 can be transformed from the global to the local coordinate system using the transformation matrix **T** as follows:

$$\bar{\mathbf{u}} = \mathbf{T}\mathbf{u} \tag{3.52}$$

where $\bar{\mathbf{u}}$ and $\mathbf{u}$ are column matrices containing the nodal displacement components in the local and global coordinate systems, respectively. The *degrees of freedom* of an element (or total structure) is the number of independent *generalized displacements* (or *generalized coordinates*) required to describe its displaced configuration relative to an initial state under zero loading.

An element (or structure) with n degrees of freedom and displacements $\mathbf{u}$ is given a set of small virtual displacements $\delta\mathbf{u}$. The virtual work done by all of

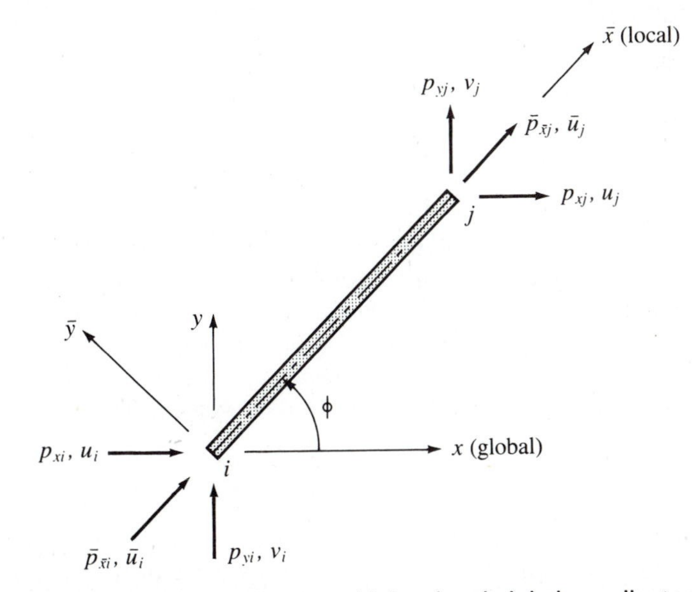

Figure 3.14 Axial force element with local and global coordinate systems

the forces acting on the system during these changes is

$$\begin{aligned}\delta W_e &= p_1\,\delta u_1 + p_2\,\delta u_2 + p_3\,\delta u_3 + \cdots + p_n\,\delta u_n \\ &= \mathbf{p}^T\,\boldsymbol{\delta}\mathbf{u} \\ &= \boldsymbol{\delta}\mathbf{u}^T\mathbf{p}\end{aligned} \tag{3.53}$$

The p_i are the *generalized forces.* Note that the component of generalized force p_i produces the virtual work $p_i\,\delta u_i$, but no virtual work is done with any other component of $\mathbf{u}$. That is, the nodal displacement and force components are coupled in the energy sense, and the force and displacement vectors are called *conjugate vectors.* The term *vector* is used for convenience as a mathematical generalization to denote a one-dimensional array and should not be confused with a physical vector representing a force or displacement. Hence, a mathematical vector may contain components from several physical vectors. If force and virtual displacement are conjugate vector sets expressed in both local and global coordinates, the virtual work done by the nodal forces during linear elastic response in each is

$$\delta W_e = \boldsymbol{\delta}\bar{\mathbf{u}}^T\bar{\mathbf{p}} \tag{3.54}$$

and

$$\delta W_e = \boldsymbol{\delta}\mathbf{u}^T\mathbf{p} \tag{3.55}$$

Since virtual work is a scalar, it is invariant under a rotation of coordinates, that is,

$$\boldsymbol{\delta}\bar{\mathbf{u}}^T\bar{\mathbf{p}} = \boldsymbol{\delta}\mathbf{u}^T\mathbf{p} \tag{3.56}$$

From Eq. (3.52) we obtain

$$\boldsymbol{\delta}\bar{\mathbf{u}} = \mathbf{T}\,\boldsymbol{\delta}\mathbf{u} \tag{3.57}$$

Substituting Eq. (3.57) into Eq. (3.56) gives

$$\boldsymbol{\delta}\mathbf{u}^T\mathbf{T}^T\bar{\mathbf{p}} = \boldsymbol{\delta}\mathbf{u}^T\mathbf{p} \tag{3.58}$$

Thus

$$\mathbf{p} = \mathbf{T}^T\bar{\mathbf{p}} \tag{3.59}$$

That is, we have shown that the transformation of displacements in Eq. (3.52) implies the corresponding transformation of forces in Eq. (3.59); these are called *contragredient transformations.* Note that the displacement transformation can be deduced from the force transformation and vice versa.

The transformation of the stiffness matrix from local to global coordinates can also be accomplished using the argument that the virtual work is invariant with respect to the coordinate system. Thus, for a linearly elastic element

$$\delta W_e = \boldsymbol{\delta}\bar{\mathbf{u}}^T\bar{\mathbf{p}} \tag{3.60}$$

In developing Eq. (3.49) there was no need to make a distinction between local and global coordinates. We can envisage that in that derivation these two coordinate systems were coincidental. Now we must note that Eq. (3.49) actually refers to the local coordinate system, that is

$$\bar{\mathbf{p}} = \bar{\mathbf{k}}\bar{\mathbf{u}} \tag{3.61}$$

Substituting Eq. (3.61) into Eq. (3.60) and using the transformation of Eq. (3.52) yields

$$\delta W_e = \boldsymbol{\delta}\bar{\mathbf{u}}^T\bar{\mathbf{k}}\bar{\mathbf{u}} = \boldsymbol{\delta}\mathbf{u}^T\mathbf{T}^T\bar{\mathbf{k}}\mathbf{T}\mathbf{u} \tag{3.62}$$

The virtual work can also be described using all quantities in global coordinates as follows:

$$\delta W_e = \boldsymbol{\delta}\mathbf{u}^T\mathbf{p} \tag{3.63}$$

If the element force-displacement relations are described in global coordinates, we have

$$\mathbf{p} = \mathbf{k}\mathbf{u} \tag{3.64}$$

where $\mathbf{k}$ is the element stiffness matrix expressed in the global coordinates. Substituting Eq. (3.64) into Eq. (3.63) gives the virtual work in global coordinates:

$$\delta W_e = \boldsymbol{\delta}\mathbf{u}^T\mathbf{k}\mathbf{u} \tag{3.65}$$

Equating Eq. (3.62) and Eq. (3.65) gives

$$\boldsymbol{\delta}\mathbf{u}^T\mathbf{T}^T\bar{\mathbf{k}}\mathbf{T}\mathbf{u} = \boldsymbol{\delta}\mathbf{u}^T\mathbf{k}\mathbf{u} \tag{3.66}$$

Since $\boldsymbol{\delta}\mathbf{u}$ is arbitrary and nonzero,

$$\mathbf{k} = \mathbf{T}^T\bar{\mathbf{k}}\mathbf{T} \tag{3.67}$$

This is called a symmetric bilinear form since $\bar{\mathbf{k}}$ is symmetric. We observe from Eq. (3.67) that if the element stiffness matrix is prescribed in local coordinates, it can be obtained in global coordinates using Eq. (3.67). In the case of the axial force element in Fig. 3.14,

$$\bar{u}_i = u_i \cos\phi + v_i \sin\phi \tag{3.68a}$$

and

$$\bar{u}_j = u_j \cos\phi + v_j \sin\phi \tag{3.68b}$$

Thus,

$$\underset{\bar{\mathbf{u}}}{\begin{bmatrix} \bar{u}_i \\ \bar{u}_j \end{bmatrix}} = \underset{\mathbf{T}}{\begin{bmatrix} \cos\phi & \sin\phi & 0 & 0 \\ 0 & 0 & \cos\phi & \sin\phi \end{bmatrix}} \underset{\mathbf{u}}{\begin{bmatrix} u_i \\ v_i \\ u_j \\ v_j \end{bmatrix}} \tag{3.69}$$

The contragredient transformation for the forces is

$$\begin{bmatrix} p_{xi} \\ p_{yi} \\ p_{xj} \\ p_{yj} \end{bmatrix} = \begin{bmatrix} \cos\phi & 0 \\ \sin\phi & 0 \\ 0 & \cos\phi \\ 0 & \sin\phi \end{bmatrix} \begin{bmatrix} \bar{p}_{\bar{x}i} \\ \bar{p}_{\bar{x}j} \end{bmatrix} \tag{3.70}$$

Substituting the matrix from Eq. (3.69) and the matrix $\bar{\mathbf{k}}$ as given in Eq. (3.51) into the transformation of Eq. (3.67) gives the stiffness matrix **k** for the axial force element in global coordinates as follows:

$$\mathbf{k} = \mathbf{T}^T\bar{\mathbf{k}}\mathbf{T}$$

$$= \frac{AE}{L}\begin{bmatrix} \cos^2\phi & \sin\phi\cos\phi & -\cos^2\phi & -\sin\phi\cos\phi \\ \sin\phi\cos\phi & \sin^2\phi & -\sin\phi\cos\phi & -\sin^2\phi \\ -\cos^2\phi & -\sin\phi\cos\phi & \cos^2\phi & \sin\phi\cos\phi \\ -\sin\phi\cos\phi & -\sin^2\phi & \sin\phi\cos\phi & \sin^2\phi \end{bmatrix} \tag{3.71}$$

This is the same element stiffness matrix obtained using the governing equations and shown in Eq. (2.22).

Alternatively, the displacements and forces could be described using the following contragredient transformations:

$$\mathbf{u} = \mathbf{T}^T\bar{\mathbf{u}} \tag{3.72}$$

and

$$\bar{\mathbf{p}} = \mathbf{T}\mathbf{p} \tag{3.73}$$

In this case the transformation of the stiffness matrix in global coordinates can be transformed to the local coordinates as follows:

$$\bar{\mathbf{k}} = \mathbf{T}\mathbf{k}\mathbf{T}^T \tag{3.74}$$

The reader is urged to verify these transformations and demonstrate them for the axial force element.

These same transformation equations were derived in Sec. 2.5, but in that case **T** was an orthogonal transformation, that is,

$$\mathbf{T}^T = \mathbf{T}^{-1} \quad \text{or} \quad \mathbf{T}^T\mathbf{T} = \mathbf{I}$$

where

$$\mathbf{T} = \begin{bmatrix} \cos\phi & \sin\phi & 0 & 0 \\ -\sin\phi & \cos\phi & 0 & 0 \\ 0 & 0 & \cos\phi & \sin\phi \\ 0 & 0 & -\sin\phi & \cos\phi \end{bmatrix}$$

The stiffness transformations similar to Eqs. (3.67) and (3.74) were derived using the orthogonality property. Note that an orthogonal transformation is contragredient to itself.

For production-level computer programs it is probably most efficient to use the element stiffness matrix expressed in global coordinates as in Eq. (3.71). To analyze a complete truss it is necessary to use the transformations developed herein for setting up the nodal force-displacement equations and obtaining the element forces.

3.4 THE STRUCTURAL STIFFNESS MATRIX

Equation (3.50) indicates how to calculate the stiffness matrix for an element, and this matrix can be transformed into global coordinates using Eq. (3.67). To analyze a complete truss the individual element stiffness matrices **k** must be combined to give the *structural stiffness matrix* **K** of the entire assemblage of elements. The principle of virtual work can be used to obtain the relationship between the nodal forces **P** and the corresponding displacements **U** for the total structure.

For the mth element of a structure, we know from Eq. (3.47) that the virtual work in local coordinates is

$$\delta W_i^m = \delta \bar{\mathbf{u}}^{m^T} \int_{V^m} \mathbf{B}^{m^T} E \mathbf{B}^m \, dV \, \bar{\mathbf{u}}^m \tag{3.75}$$

For the total truss the principle of virtual work gives

$$\sum \delta u_i P_i = \boldsymbol{\delta}\mathbf{U}^T \mathbf{P} = \sum_m \left(\delta \bar{\mathbf{u}}^{m^T} \int_{V^m} \mathbf{B}^{m^T} E \mathbf{B}^m \, dV \, \bar{\mathbf{u}}^m \right) \tag{3.76}$$

where $\boldsymbol{\delta}\mathbf{U}$ is the matrix of all nodal virtual displacements for the structure and **P** contains all nodal forces corresponding to the δU_i. Note that the summation on the right-hand side of Eq. (3.76) is taken over all elements in the structure, while that on the left extends over all nodes. Rewriting Eq. (3.76) yields

$$\boldsymbol{\delta}\mathbf{U}^T \mathbf{P} = \boldsymbol{\delta}\mathbf{U}^T \left(\sum_m \mathbf{T}^{m^T} \int_{V^m} \mathbf{B}^{m^T} E \mathbf{B}^m \, dV \, \mathbf{T}^m \right) \mathbf{U} \tag{3.77}$$

$$= \boldsymbol{\delta}\mathbf{U}^T \sum_m \mathbf{K}^m \mathbf{U} = \boldsymbol{\delta}\mathbf{U}^T \, \mathbf{K}\mathbf{U} \tag{3.78}$$

where

$$\mathbf{K}^m = \mathbf{T}^{m^T} \int_{V^m} \mathbf{B}^{m^T} E \mathbf{B}^m \, dV \, \mathbf{T}^m$$

$$= \mathbf{T}^{m^T} \bar{\mathbf{k}}^m \mathbf{T}^m \tag{3.79}$$

and

$$\mathbf{K} = \sum_m \mathbf{K}^m \tag{3.80}$$

The matrix $\mathbf{K}^m$ is the stiffness matrix of the mth element expressed in global coordinates. Since $\boldsymbol{\delta}\mathbf{U}$ is arbitrary and nonzero, from Eq. (3.78) we have

$$\mathbf{P} = \mathbf{K}\mathbf{U} \tag{3.81}$$

Enforcing the principle of virtual work ensures that the system is in equilibrium; therefore Eq. (3.81) is the statement of nodal equilibrium for the total structure.

Summing the element volume integrals in Eq. (3.79) expresses the direct addition of the element stiffness matrices $\mathbf{K}^m$ to obtain the stiffness matrix for the total element assemblage. For conformability of matrix addition the element stiffness matrices $\mathbf{K}^m$ must be the same order as the structural stiffness matrix **K**. Considering the makeup of the $\mathbf{K}^m$, nonzero elements are only in those rows and columns that correspond to degrees of freedom associated with the mth element. Therefore we only need to store the compacted element stiffness matrix that has dimensions equal to the number of element degrees of freedom (four in the case of the axial force element). Techniques for efficiently assembling **K** are described in Sec. 2.3.

Example 3.5 illustrates the above derivation of the nodal equilibrium equations for a simple assemblage.

EXAMPLE 3.5

Calculate the structural stiffness matrix for the truss of Fig. E3.5a using the principle of virtual work. $A_{ab} = 1.2$ in.^{2}, $A_{ac} = 1.0$ in.^{2}, $A_{ad} = 3.6$ in.^{2}, and $E = 29 \times 10^3$ kips/in.^{2}.

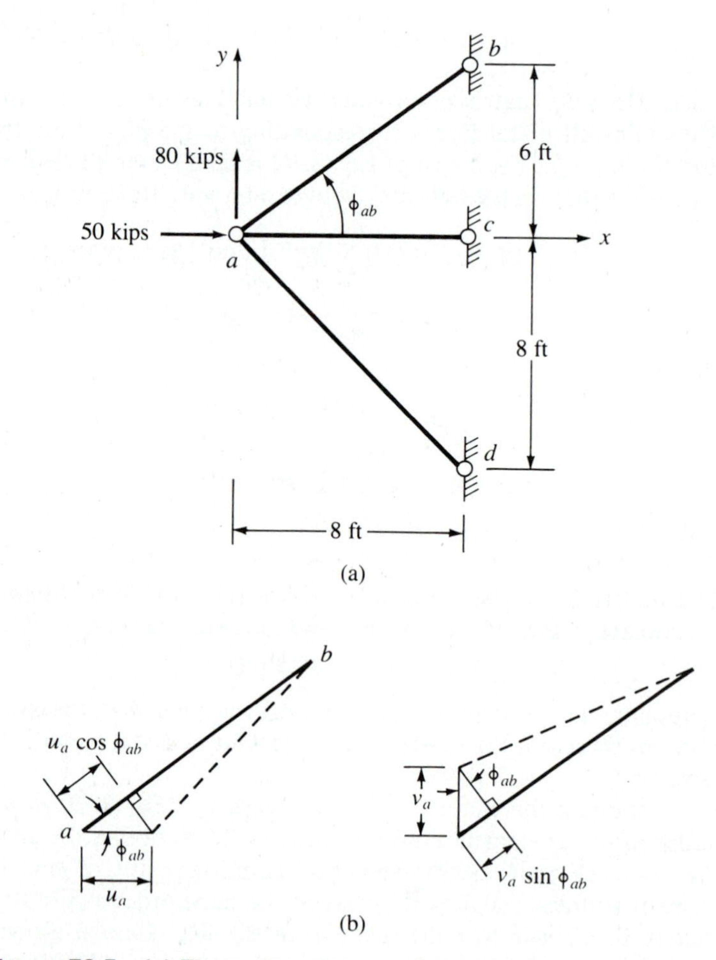

Figure E3.5 (a) Three-element truss; (b) deformations of element *ab*

Solution

For element ab ($\phi_{ab} = 36.87°$; see Fig. E3.5b),

$$\varepsilon^{ab} = -\frac{u_a \cos\phi_{ab}}{L_{ab}} - \frac{v_a \sin\phi_{ab}}{L_{ab}}$$

For element ac ($\phi_{ac} = 0°$),

$$\varepsilon^{ac} = -\frac{u_a}{L_{ac}}$$

For element ad ($\phi_{ad} = 315°$),

$$\varepsilon^{ad} = -\frac{u_a \cos\phi_{ad}}{L_{ad}} - \frac{v_a \sin\phi_{ad}}{L_{ad}}$$

The internal virtual work is

$$\begin{aligned}\delta W_i &= \sum_{(m)} \int_{V(m)} \delta\varepsilon^m \sigma^m \, dV = \sum_m \delta\varepsilon^m \sigma^m A_m L_m \\ &= \sum_m \delta\varepsilon^m \varepsilon^m E A_m L_m\end{aligned}$$

Choosing the virtual strains with the same form as the real strains gives

$$\begin{aligned}\delta W_i = &\left[u_a\, \delta u_a \left(\frac{AE}{L}\right)_{ab} \cos^2\phi_{ab} + u_a\, \delta v_a \left(\frac{AE}{L}\right)_{ab} \sin\phi_{ab}\cos\phi_{ab} \right. \\ &\left. + v_a\, \delta u_a \left(\frac{AE}{L}\right)_{ab} \sin\phi_{ab}\cos\phi_{ab} + v_a\, \delta v_a \left(\frac{AE}{L}\right)_{ab} \sin^2\phi_{ab} \right] \\ &+ u_a\, \delta u_a \left(\frac{AE}{L}\right)_{ac} \cos^2\phi_{ac} \\ &+ \left[u_a\, \delta u_a \left(\frac{AE}{L}\right)_{ad} \cos^2\phi_{ad} + u_a\, \delta v_a \left(\frac{AE}{L}\right)_{ad} \sin\phi_{ad}\cos\phi_{ad} \right. \\ &\left. + v_a\, \delta u_a \left(\frac{AE}{L}\right)_{ad} \sin\phi_{ad}\cos\phi_{ad} + v_a\, \delta v_a \left(\frac{AE}{L}\right)_{ad} \sin^2\phi_{ad} \right]\end{aligned}$$

In units of kips per inch: $(AE/L)_{ab} = 290$; $(AE/L)_{ac} = 302$; $(AE/L)_{ad} = 769$. Also, $\cos\phi_{ab} = 0.8$; $\sin\phi_{ab} = 0.6$; $\cos\phi_{ac} = 1.0$; $\cos\phi_{ad} = 0.707$; $\sin\phi_{ad} = -0.707$. Thus,

$$\begin{aligned}\delta W_i = &(185.6u_a + 139.2v_a + 302.0u_a + 384.5u_a - 384.5v_a)\delta u_a \\ &+ (139.2u_a + 104.4v_a - 384.5u_a + 384.5u_a)\delta v_a\end{aligned}$$

The external virtual work done by the applied loads is

$$\delta W_e = 50\, \delta u_a + 80\, \delta v_a$$

From the principle of virtual work, $\delta W_e = \delta W_i$. Since δu_a and δv_a are arbitrary, independent, and nonzero, we can equate the coefficients of these two values. Thus,

$$872.1u_a - 245.3v_a = 50$$
$$-245.3u_a + 488.9v_a = 80$$

Solution gives $u_a = 0.120$ in. and $v_a = 0.224$ in.

Discussion Alternatively, we may wish to follow the approach implied by Eqs. (3.79) through (3.81). In this case, the matrices $\mathbf{k}^m$ are given by Eq. (3.51), $\mathbf{T}$ is shown in Eq. (3.69), and the nonzero portions of $\mathbf{K}^m$ can be calculated using Eq. (3.71). To carry out the summation of Eq. (3.80), each of the $\mathbf{K}^m$ must be augmented with zeros to be an 8 by 8 matrix. For example, the stiffness matrix for element ab in this form is

$$290\begin{bmatrix} 0.64 & 0.48 & -0.64 & -0.48 & 0.0 & 0.0 & 0.0 & 0.0 \\ 0.48 & 0.36 & -0.48 & -0.36 & 0.0 & 0.0 & 0.0 & 0.0 \\ -0.64 & -0.48 & 0.64 & 0.48 & 0.0 & 0.0 & 0.0 & 0.0 \\ -0.48 & -0.36 & 0.48 & 0.36 & 0.0 & 0.0 & 0.0 & 0.0 \\ 0.0 & \cdots & \cdots & \cdots & \cdots & \cdots & \cdots & 0.0 \\ \vdots & & & & & & & \vdots \\ 0.0 & \cdots & \cdots & \cdots & \cdots & \cdots & \cdots & 0.0 \end{bmatrix}$$

Adding these three 8 by 8 matrices will give an 8 by 8 structural stiffness matrix. This differs from the final 2 by 2 matrix obtained in the solution above. The 8 by 8 matrix of course does not acknowledge the existence of supports at nodes b, c, and d; when these boundary conditions are imposed, $\mathbf{K}$ reduces to a 2 by 2 matrix. Imposing boundary conditions is discussed in the next section (it is also described in Sec. 2.4).

3.5 NODAL DISPLACEMENTS

The equations of global nodal equilibrium can be partitioned as follows:

$$\begin{bmatrix} \mathbf{P}_f \\ \hline \mathbf{P}_s \end{bmatrix} = \left[\begin{array}{c|c} \mathbf{K}_{ff} & \mathbf{K}_{fs} \\ \hline \mathbf{K}_{sf} & \mathbf{K}_{ss} \end{array}\right] \begin{bmatrix} \mathbf{U}_f \\ \hline \mathbf{U}_s \end{bmatrix} \tag{3.82}$$

where the subscript f refers to the free or unrestrained nodes and the subscript s designates a supported or constrained node. Hence, $\mathbf{P}_f$ contains applied forces and $\mathbf{U}_f$ is the column matrix of corresponding unknown displacements, while $\mathbf{U}_s$ represents prescribed displacements and $\mathbf{P}_s$ contains the associated unknown forces. Multiplying the partitioned matrices in Eq. (3.82) gives the two matrix equations

$$\mathbf{P}_f = \mathbf{K}_{ff}\mathbf{U}_f + \mathbf{K}_{fs}\mathbf{U}_s$$

and

$$\mathbf{P}_s = \mathbf{K}_{sf}\mathbf{U}_f + \mathbf{K}_{ss}\mathbf{U}_s \tag{3.83}$$

For the case in which $\mathbf{U}_s$ is null (unyielding supports),

$$\mathbf{P}_f = \mathbf{K}_{ff}\mathbf{U}_f \tag{3.84}$$

and

$$\mathbf{P}_s = \mathbf{K}_{sf}\mathbf{U}_f \tag{3.85}$$

In this case $\mathbf{P}_s$ contains the reaction forces. It is usually neither necessary nor useful to find the reaction forces using Eq. (3.85); therefore, it is sufficient to work only with Eq. (3.84). This topic is discussed in more detail in Sec. 2.4.

3.6 ELEMENT FORCES

The forces in each individual element are calculated subsequent to computing the nodal displacements. Using the strain-displacement equation of Eq. (3.40) and the stress-strain relationship [Eq. (2.3)], the stress in element ij is

$$\sigma^{ij} = E\mathbf{B}\bar{\mathbf{u}}^{ij} \tag{3.86}$$

Note that this equation refers to the element displacements expressed in local coordinates. Equation (3.52) transforms displacements from global to local coordinates; substituting this relationship into Eq. (3.86) yields

$$\sigma^{ij} = E\mathbf{B}\mathbf{T}\mathbf{u}^{ij} \tag{3.87}$$

For a prismatic element with uniform cross-sectional area A, the axial element force (s^{ij}) is

$$\begin{aligned} s^{ij} &= A\sigma^{ij} = AE\mathbf{B}\mathbf{T}\mathbf{u}^{ij} \\ &= \mathbf{e}\mathbf{u}^{ij} \end{aligned} \tag{3.88}$$

where the force-transformation matrix $\mathbf{e}$ is

$$\mathbf{e} = AE\mathbf{B}\mathbf{T} \tag{3.89}$$

For the prismatic axial force element

$$\begin{aligned} \mathbf{e} &= AE\frac{1}{L}[-1 \quad 1]\begin{bmatrix} \cos\phi & \sin\phi & 0 & 0 \\ 0 & 0 & \cos\phi & \sin\phi \end{bmatrix} \\ &= \frac{AE}{L}[-\cos\phi \quad -\sin\phi \quad \cos\phi \quad \sin\phi] \end{aligned} \tag{3.90}$$

Equation (3.88) is formulated so that a positive sign for s^{ij} indicates a tensile axial force element and a negative result implies a compression in the element.

For example, the axial force in element ab of Example 3.5 is calculated as follows:

$$\begin{aligned} s^{ab} &= 290[-0.8 \quad -0.6 \quad 0.8 \quad 0.6]\begin{bmatrix} 0.120 \\ 0.224 \\ 0.000 \\ 0.000 \end{bmatrix} \\ &= -66.8 \text{ kips (c)} \end{aligned}$$

The same force-transformation matrix (e) was obtained in Sec. 2.6 by applying the basic equations. Additional examples illustrating numerical calculation of element forces are presented in that previous section.

3.7 INITIAL AND THERMAL STRAINS

Temperature changes, fabrication errors, prestressing, and other phenomena all cause structural response that can be analyzed in a similar fashion. Even though all these effects may not occur at the inception of a structure, for convenience we will refer to them collectively in the subsequent text as initial strain effects. The real stress in an element is related to the strains ε and initial strains ε^o as follows:

$$\begin{aligned}\sigma &= E(\varepsilon - \varepsilon^o) + \sigma^o \\ &= E\mathbf{B}\bar{\mathbf{u}} - E\varepsilon^o + \sigma^o\end{aligned} \tag{3.91}$$

Applying the principle of virtual work ($\delta W_e = \delta W_i$), with the real stresses from Eq. (3.91) and the virtual strains $\delta\varepsilon = \mathbf{B}\ \delta\bar{\mathbf{u}}$, yields

$$\boldsymbol{\delta}\bar{\mathbf{u}}^T\bar{\mathbf{p}} = \boldsymbol{\delta}\bar{\mathbf{u}}^T\left(\int_{\text{vol}} \mathbf{B}^T E\mathbf{B}\ dV\ \bar{\mathbf{u}} - \int_{\text{vol}} \mathbf{B}^T E\varepsilon^o\ dV + \int_{\text{vol}} \mathbf{B}^T\sigma^o\ dV\right) \tag{3.92}$$

From Eq. (3.50) we note that the first integral on the right-hand side of the equation is $\bar{\mathbf{k}}$. Using the fact that $\boldsymbol{\delta}\mathbf{u}$ is arbitrary and nonzero, Eq. (3.92) reduces to

$$\bar{\mathbf{p}} = \bar{\mathbf{k}}\bar{\mathbf{u}} + \bar{\mathbf{p}}^o \tag{3.93}$$

where the element initial force matrix, $\bar{\mathbf{p}}^o$, expressed in local coordinates is

$$\bar{\mathbf{p}}^o = -\int_{\text{vol}} \mathbf{B}^T E\varepsilon^o\ dV + \int_{\text{vol}} \mathbf{B}^T\sigma^o\ dV \tag{3.94}$$

For a temperature increase ΔT, $\varepsilon^o = \alpha\ \Delta T$. If the element is constrained at the ends $\varepsilon^o = 0$ and $\sigma^o = -E\alpha\ \Delta T$; thus the initial force matrix for a prismatic axial force element with cross-sectional area A is

$$\bar{\mathbf{p}}^o = AE\alpha\ \Delta T\{1 \quad -1\} \tag{3.95}$$

If a prismatic axial force element is fabricated ΔL too long, $\varepsilon^o = \Delta L/L$. If the element is constrained at the ends $\varepsilon^o = 0$, $\sigma^o = -E\ \Delta L/L$, and the initial force matrix is

$$\bar{\mathbf{p}}^o = \frac{AE\ \Delta L}{L}\{1 \quad -1\} \tag{3.96}$$

These same two results are shown as Eqs. (2.63) and (2.64), respectively, but in those previous cases the $\bar{\mathbf{p}}^o$ are shown as 4 by 1 matrices. This was necessary because the orthogonal transformation matrix was used with those earlier results. The initial forces can be transformed from local to global coordinates using Eq. (3.59), that is,

$$\mathbf{p}^o = \mathbf{T}^T \bar{\mathbf{p}}^o \tag{3.97}$$

In analyzing a complete truss the element initial forces must be combined at the nodal points when equilibrium of the assemblage is enforced. This fact can be demonstrated using the theorem of virtual work for the assemblage. For the mth element

$$\delta W_i^m = \int_{V(m)} \boldsymbol{\delta\varepsilon}^{m^T} \boldsymbol{\sigma}^m \, dV$$

Substituting Eqs. (3.45) and (3.91) for $\boldsymbol{\delta\varepsilon}^m$ and $\boldsymbol{\sigma}^m$, respectively, gives the following:

$$\delta W_i^m = \boldsymbol{\delta}\bar{\mathbf{u}}^{m^T}\left(\int_{V(m)} \mathbf{B}^{m^T} E \mathbf{B}^m \, dV \, \bar{\mathbf{u}}^m - \int_{V(m)} \mathbf{B}^{m^T} E \boldsymbol{\varepsilon}^{o(m)} \, dV + \int_{V(m)} \mathbf{B}^{m^T} \sigma^{o(m)} \, dV\right) \tag{3.98}$$

The principle of virtual work for the entire truss yields

$$\begin{aligned} \boldsymbol{\delta}\mathbf{U}^T\mathbf{P} &= \boldsymbol{\delta}\mathbf{U}^T\left(\sum_m \mathbf{T}^{m^T} \int_{V(m)} \mathbf{B}^{m^T} E \mathbf{B}^m \, dV \, \mathbf{T}\right)\mathbf{U} - \mathbf{T}^{m^T} \int_{V(m)} \mathbf{B}^{m^T} E \boldsymbol{\varepsilon}^{o(m)} \, dV \\ &\quad + \mathbf{T}^{m^T} \int_{V(m)} \mathbf{B}^{m^T} \boldsymbol{\sigma}^{o(m)} \, dV \qquad (3.99) \\ &= \boldsymbol{\delta}\mathbf{U}^T \sum_m \mathbf{K}^m \mathbf{U} + \boldsymbol{\delta}\mathbf{U}^T \sum_m \mathbf{P}^{o(m)} \qquad (3.100) \end{aligned}$$

where

$$\mathbf{P}^{o(m)} = -\int_{V(m)} \mathbf{B}^{m^T} E \boldsymbol{\varepsilon}^{o(m)} \, dV + \int_{V(m)} \mathbf{B}^{m^T} \boldsymbol{\sigma}^{o(m)} \, dV \tag{3.101}$$

$\mathbf{P}^{o(m)}$ is the matrix of initial forces for the mth element expressed in global coordinates. Let

$$\mathbf{P}^o = \sum_m \mathbf{P}^{o(m)} \tag{3.102}$$

Since $\boldsymbol{\delta}\mathbf{U}$ is arbitrary and nonzero, from Eq. (3.100) we have

$$\mathbf{P} = \mathbf{K}\mathbf{U} + \mathbf{P}^o \tag{3.103}$$

This is the equation of nodal equilibrium when initial forces are imposed on the structure. This same equation was obtained in Sec. 2.7 by enforcing the basic equations [see Eq. (2.65)]. The equations of overall structural response must also be partitioned in a fashion similar to that of Eq. (3.82) in which initial forces are not present.

The element forces are computed after the nodal displacements have been determined. The force in a prismatic axial force element with cross-sectional area A is obtained using Eq. (3.91) and the fact that

$$\begin{aligned} s^{ij} &= A\sigma^{ij} \\ &= AE\mathbf{B}\bar{\mathbf{u}}^{ij} - AE\varepsilon^{o} + A\sigma^{o} \\ &= \frac{AE}{L}[-1 \quad 1]\bar{\mathbf{u}}^{ij} - AE\varepsilon^{o} + A\sigma^{o} \\ &= \mathbf{e}\mathbf{u}^{ij} + \bar{s}^{o} \end{aligned} \tag{3.104}$$

where

$$\bar{s}^{o} = \begin{cases} -AE\alpha\,\Delta T & \text{for a temperature increase } \Delta T \\ -\dfrac{AE\,\Delta L}{L} & \text{for a fabrication error } \Delta L \end{cases} \tag{3.105}$$

Examples 2.12 and 2.13 illustrate the analysis for fabrication errors and thermal forces, respectively.

3.8 COMPARISON OF DIRECT AND VIRTUAL WORK METHODS

The principle of virtual work is used in this chapter to derive the stiffness method of matrix structural analysis. Virtual work is classified as an equilibrium method, whereas complementary virtual work is a dual energy method invoking compatibility. The latter method is used in Ch. 7 to derive the flexibility method.

The flowchart for the stiffness method is shown in Fig. 2.13; this was initially obtained by invoking the basic equations of equilibrium, compatibility, and stress-strain. The appropriate equation numbers are referenced beside each process. The reader is urged to review that flowchart and note the relevant equation numbers from that chapter for each step.

In deriving the basic matrices for an element using virtual work, we ensure compatibility by choosing shape functions that are continuous over the element [see Eqs. (3.39) and (3.40)]. The stress-strain relationships are explicitly included in the virtual work formulation [see Eq. (3.43)]. The element stiffness matrix obtained using the principle of virtual work [i.e., Eq. (3.50)] is in equilibrium since this basic equation is implicit to the principle. The governing matrix equation for the assemblage [i.e., Eqs. (3.81) and (3.103)] obtained from virtual work ensures both nodal equilibrium and compatibility. It follows that the computations for element forces shown in Eq. (3.88) imply that equilibrium, compatibility, and the stress-strain equations are satisfied, since an approach similar to that used for the element stiffness matrix is used.

A primary advantage of using the theorem of virtual work is that it is possible to obtain *approximate solutions* for elements with complex shape or geometry by selecting the appropriate interpolation functions to describe the element displacements. Another important benefit of virtual work is that the equations of the stiffness method can be formulated ab initio using matrices.

3.9 Problems

3.1–3.4 Calculate the reactions for the beams shown in Figs. P3.1 to P3.4 using the principle of virtual work.

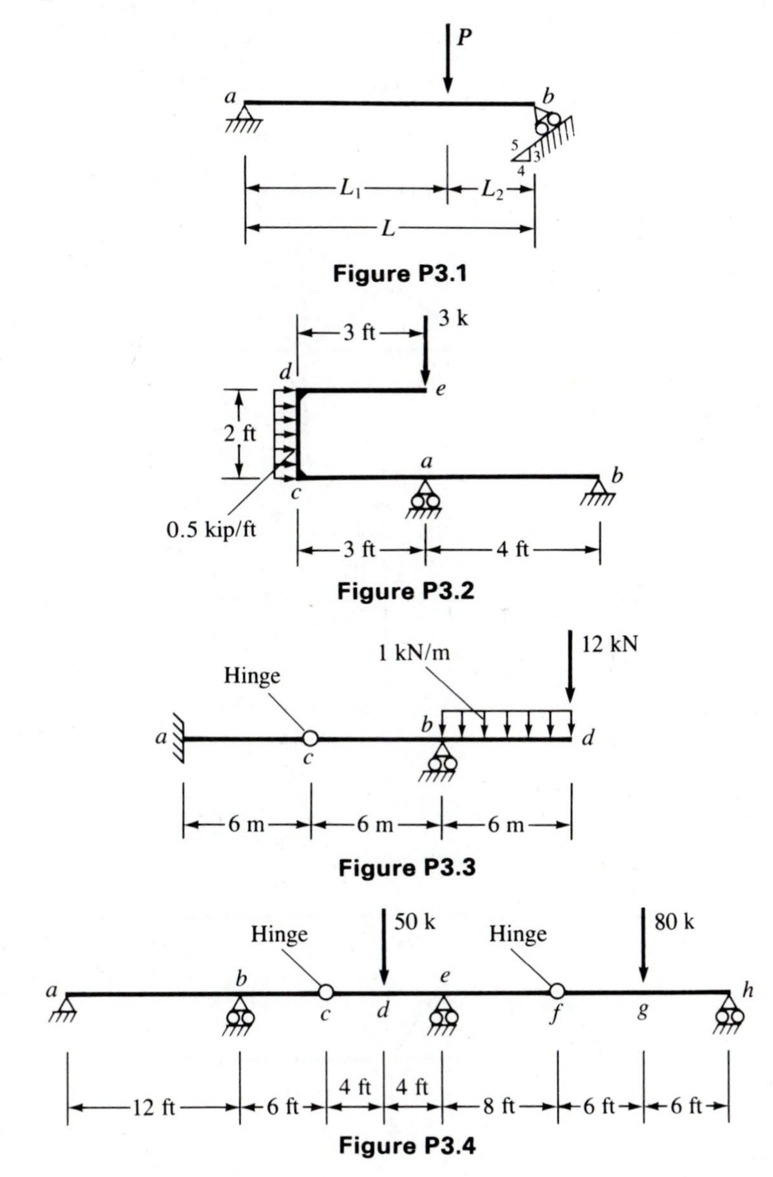

Figure P3.1

Figure P3.2

Figure P3.3

Figure P3.4

3.5–3.8 Calculate the force in element *bd* for the trusses shown in Figs. P3.5 to P3.8 using the principle of virtual work.

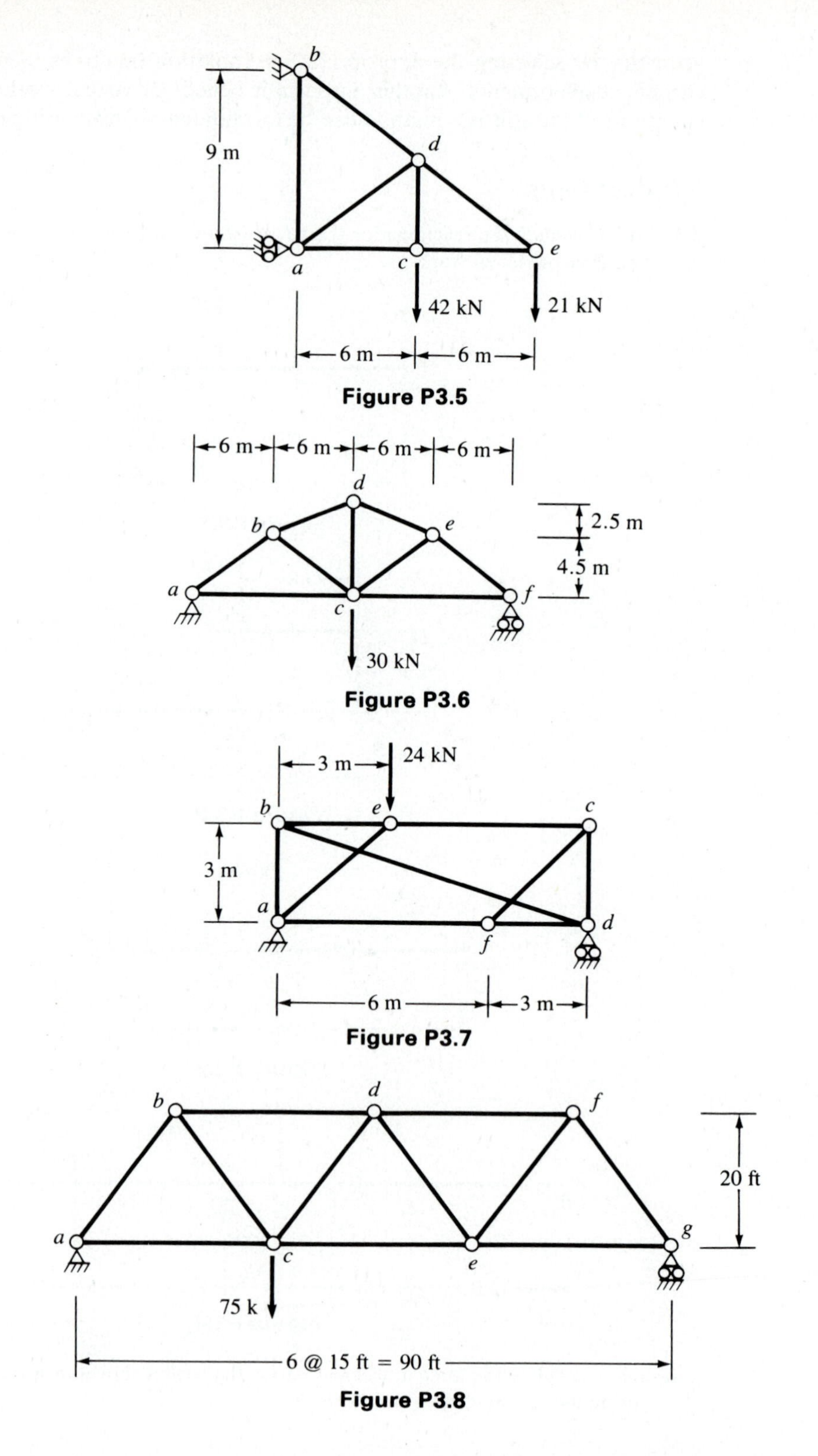

Figure P3.5

Figure P3.6

Figure P3.7

Figure P3.8

3.9 Obtain the element stiffness matrix for the axial force member in Fig. P3.9 with linearly varying area by: **(a)** combining **k** for two constant stress axial force elements, each of length $L/2$; **(b)** using a single linear displacement function; **(c)** using a quadratic displacement function; and **(d)** using the exact displacement function. Comment on the strengths and weaknesses of each **k**.

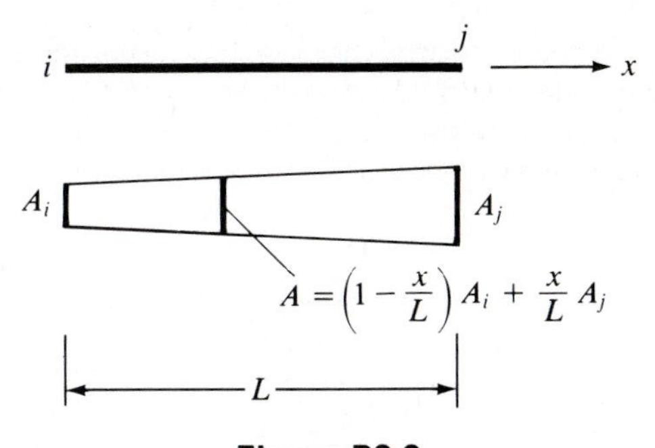

Figure P3.9

3.10 Obtain the element stiffness matrix for the tapered axial force element shown in Fig. P3.10 by: **(a)** deriving the exact expression for displacement and using the corresponding shape functions and **(b)** using a linear displacement function. Comment on the two formulations.

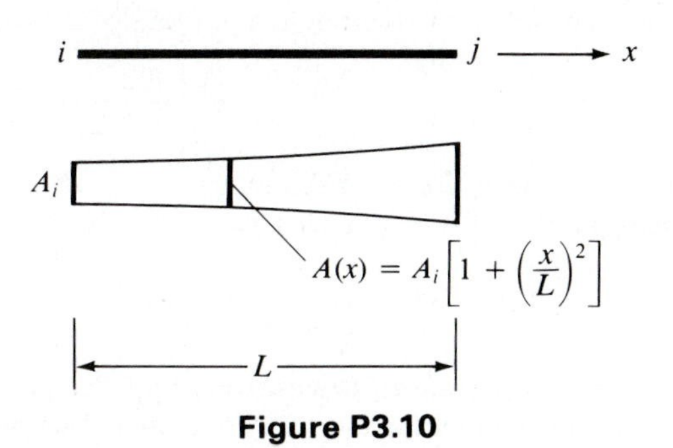

Figure P3.10

3.11 Obtain the element stiffness matrix (using the principle of virtual work) for the three-node axial force element in Fig. P3.11; AE is constant. Use the shape functions in Example 3.3 for the actual displacements and the following for the virtual displacements:

$$N_i = \frac{1}{2}\left(\cos\frac{\pi x}{2L} + \cos\frac{3\pi x}{2L}\right)$$

$$N_k = \frac{1}{\sqrt{2}}\left(\cos\frac{\pi x}{2L} - \cos\frac{3\pi x}{2L}\right)$$

$$N_j = \frac{1}{2}\cos\left(\frac{\pi x}{2L} + \cos\frac{3\pi x}{2L}\right) - \cos\frac{\pi x}{L}$$

Contrast your result with the stiffness of Example 3.4.

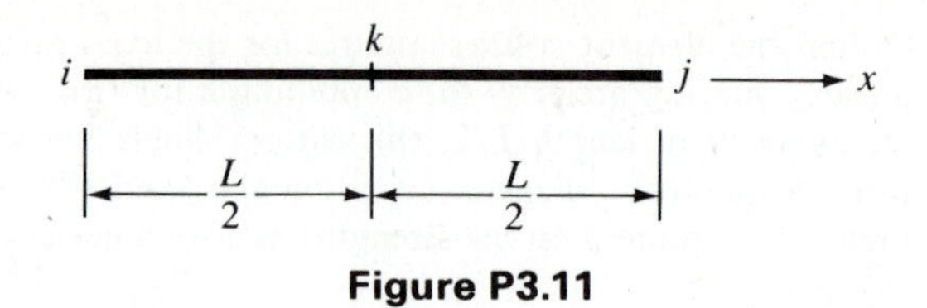

Figure P3.11

3.12 Obtain the element stiffness matrix (using the principle of virtual work) for the three-node axial force element in Fig. P3.12 with linearly varying area. **(a)** Use the shape functions in Example 3.3 for both actual and virtual displacements. **(b)** Use the shape functions from Prob. 3.11. Comment on the two formulations.

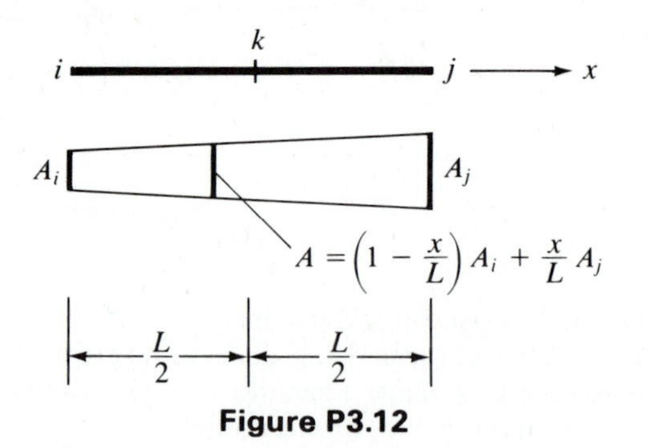

Figure P3.12

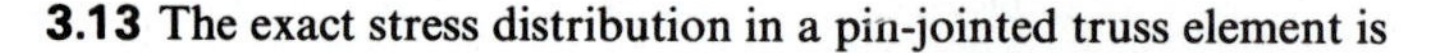

3.13 The exact stress distribution in a pin-jointed truss element is

$$\sigma = \frac{E(u_j - u_i)}{L}$$

where u_j and u_i are the displacements of the two ends, E is the modulus of elasticity, and L is the length. Using a compatible displacement distribution, that is,

$$u = u_i + \frac{x}{L}(5u_i - u_j) + \frac{x^2}{L^2}(-6u_i + 2u_j)$$

and the exact stresses, demonstrate that the principle of virtual work gives the correct expression for the stiffness matrix. Identify the shape functions and plot them.

3.14 The following displacement function is to be used for the four-node axial force element in Fig. P3.14 with AE constant:

$$u = u_i + \left(-\frac{11}{2}u_i + u_j + 9u_k - \frac{9}{2}u_l\right)\frac{x}{L}$$
$$+ \left(9u_i - \frac{9}{2}u_j - \frac{45}{2}u_k + 18u_l\right)\frac{x^2}{L^2}$$
$$+ \left(-\frac{9}{2}u_i + \frac{9}{2}u_j + \frac{27}{2}u_k - \frac{27}{2}u_l\right)\frac{x^3}{L^3}$$

Identify the shape functions and plot them. Is this a compatible displacement distribution? Demonstrate how the principle of virtual work gives the expression for one element of the stiffness matrix. Use the same shape function for the exact stress distribution that you use for the strain distribution. Will the stiffness matrix for this axial force element be exact?

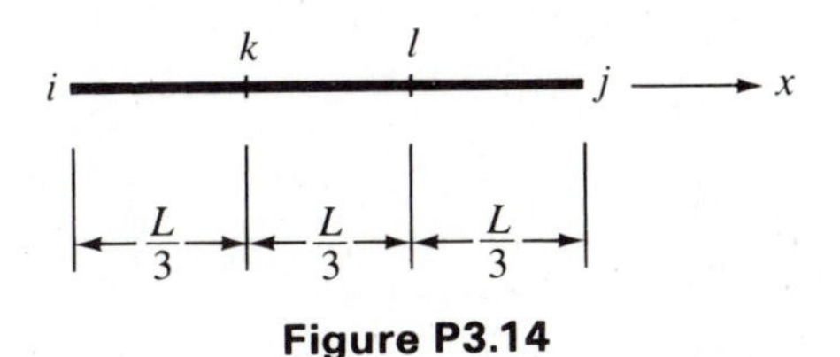

Figure P3.14

3.15 Use the principle of virtual work to develop the method for calculating consistent loads; that is,

$$\mathbf{p} = \int_{\text{sur}} \mathbf{N}^T \mathbf{q}\, dS + \mathbf{N}^T \Big|_{x_i} P$$

Use this result to obtain the consistent nodal loads for the axial force elements shown in Fig. P3.15. Assume that $N_1 = 1 - x/L$ and $N_2 = x/L$.

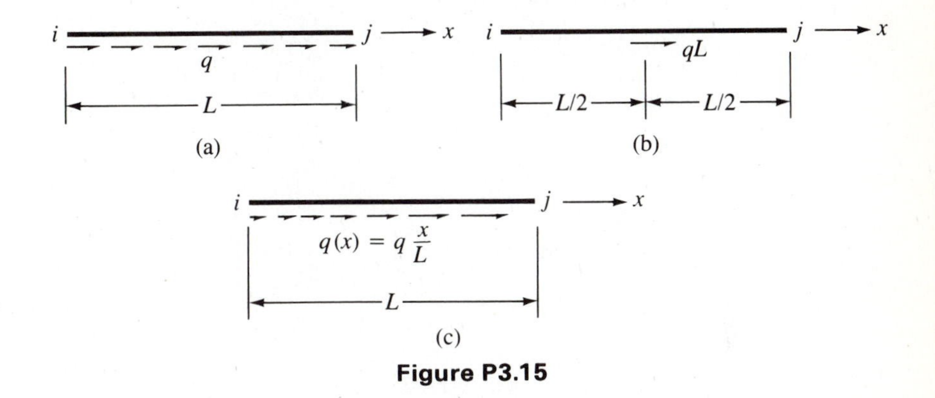

Figure P3.15

3.16 Calculate nodal displacements for the truss in Fig. P3.16 using the principle of virtual work. $E = 30 \times 10^3$ ksi; $A_{ab} = A_{ad} = A\sqrt{2}$ and $A_{ac} = A$.

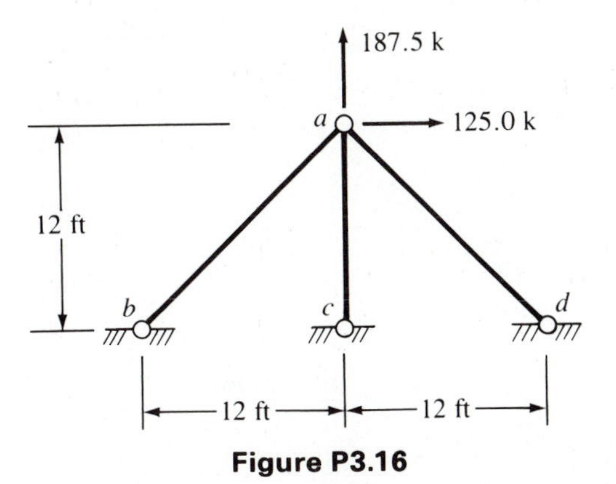

Figure P3.16

In Probs. 3.17 to 3.27 use the principle of virtual work to calculate nodal displacements and element forces for the original problems.

3.17 Prob. 2.9.

3.18 Prob. 2.18.

3.19 Prob. 2.19.

3.20 Prob. 2.21.

3.21 Prob. 2.23.

3.22 Prob. 2.14.

3.23 Prob. 2.17.

3.24 Prob. 2.22.

3.25 Prob. 2.28 (parts a, c, and d).

3.26 Prob. 2.30 (parts a and b).

3.27 Prob. 2.31 (parts a and b).

4 THE STIFFNESS METHOD FOR BEAMS AND PLANAR FRAMES

The precepts of the stiffness method were described for trusses in Chs. 2 and 3. The same broad concepts can be used to analyze structures experiencing axial and flexural deformations. The analysis of beams and rigid frames poses problems that differ somewhat from those encountered for trusses. The individual elements of a truss are two-force members that resist only tension and compression. In contrast, beams withstand both bending moments and loads applied normal to their longitudinal axis; the cardinal internal effects are flexure and shear. The members of planar frames with rigid joints are subjected to axial forces, in addition to flexure and shear. In this chapter we will focus on structures composed of linear elastic beam elements.

The general solution algorithm for the stiffness method as shown in Fig. 2.13 can be extended to the analysis of structures composed of flexural members if the specific equation references in that flowchart are ignored. The basic governing equations are satisfied for beams and rigid-frame structures in a manner similar to that for trusses. The material constitutive relations are fulfilled within the individual elements, as are the equilibrium and compatibility equations. Assembling individual beam element stiffness matrices into the structural stiffness matrix assures that nodal equilibrium is satisfied. Alternatively, using energy methods to derive the relationships, virtual work yields the pertinent element matrices, and the nodal equilibrium equations for the assemblage are embodied in the statement of virtual work for the total structure.

The generalized displacements for beams consist of linear deflections and rotations; these must be functionally related to the corresponding generalized forces through the element stiffness matrix. Also, the element force-transformation matrix, relating nodal displacements to internal element forces, must be derived for beams. It is not necessary to transform quantities from local to global coordinates (or vice versa) for beams since the element and structural coordinates can be made to coincide. This is not the case for planar frames; thus the appropriate coordinate transformation matrices must be used to relate quantities in the two fundamental coordinate systems.

The nodal points of a truss are uniquely defined as the physical panel points; this is not the case for beams and planar frames. In analyzing structures composed of beam elements, it is necessary for the engineer to decide where the nodal points are located. This must be done by comparing the properties of the beam element with the anticipated behavior of the total structure. This process, called *idealization*, can only be done by someone well versed in the stiffness method and having a good intuitive understanding of structural behavior. In this chapter special attention is devoted to structural idealization after the beam element is fully described.

Beams and planar frames subject to initial strains induced by support settlement, fabrication errors, and temperature can be analyzed in a three-step procedure similar to that outlined for trusses. Unique initial force matrices for beam elements are derived for these effects.

Analyses throughout this chapter employ both the basic governing equations and the principle of virtual work, where appropriate. The reader is urged to study both methods to understand the interrelationships of these two fundamental approaches of structural analysis.

4.1 THE PRISMATIC BEAM ELEMENT

A beam is a slender member more than four to five times as long as either of its cross-sectional dimensions; it is typically subjected to transverse loads. Beams can experience axial, flexural, shear, and torsional deformations. The beam element investigated in this section is prismatic (i.e., it has a uniform cross section along its entire length) and is constructed from material that is homogeneous, isotropic, and linearly elastic. Displacements are assumed to be sufficiently small so that engineering beam theory is applicable, and geometric nonlinearities do not exist. The element is assumed to be axially rigid; the beam element with axial deformations is examined in Sec. 4.5.1. The prismatic beam element with length L and moment of inertia I is shown in Fig. 4.1. The generalized displacements for the element consist of the end displacements ($\bar{v}_i$ and $\bar{v}_j$) and the end rotations ($\bar{\theta}_i$ and $\bar{\theta}_j$). The corresponding generalized forces are $\bar{p}_{\bar{y}i}$, $\bar{p}_{\bar{y}j}$, $\bar{m}_i$, and $\bar{m}_j$. Note that the rotations and moments are about the $\bar{z}$ axis. Throughout this book all planar elements are considered to be in the $\bar{x}$-$\bar{y}$ plane; the subscript $\bar{z}$ on the moments is implied but not stated. The notation for beams and planar frames is similar to that used for trusses in that the forces and displacements associated with the element are denoted by lowercase

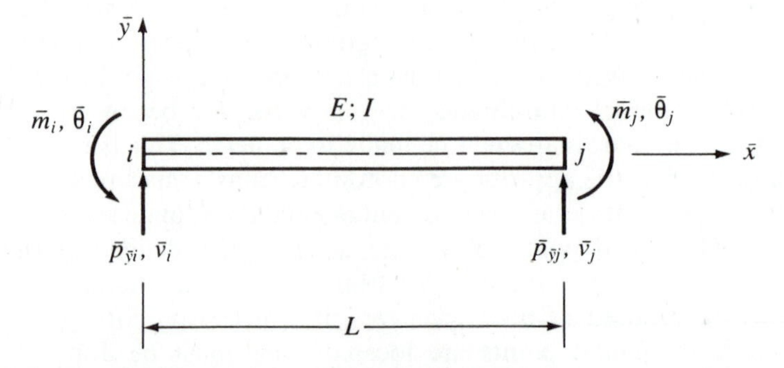

Figure 4.1 The beam element with generalized displacements and forces in local coordinates

letters, and all quantities referred to the local coordinate system are distinguished by a bar above the symbol.

The stiffness matrix relating nodal forces and displacements is obtained by three different approaches. In Sec. 4.1.1 this is accomplished by solving the basic governing equations. Two different energy principles are used to obtain the same element stiffness matrix in Secs. 4.1.2 and 4.1.3. The relationships between element forces (i.e., shear and moments) and nodal displacements is described in Sec. 4.1.4.

4.1.1 THE BASIC EQUATIONS AND $\bar{\mathbf{k}}$ The moment-curvature relation for a linearly elastic beam element symmetrical about the y axis is

$$\frac{d^2\bar{v}}{d\bar{x}^2} = \frac{M}{EI} \tag{4.1}$$

where $\bar{v}$ is the displacement of the beam in the $\bar{y}$ direction; V and M are the beam shear and moment, respectively; E is the modulus of elasticity; and I is the moment of inertia of the beam (i.e., $I = \int y^2\, dA$). The differential equations of equilibrium for a beam with no concentrated forces or moments are

$$\frac{dM}{d\bar{x}} = V \tag{4.2}$$

and

$$\frac{dV}{d\bar{x}} = q \tag{4.3}$$

Substituting Eq. (4.1) into Eqs. (4.2) and (4.3) yields

$$\frac{d^3\bar{v}}{d\bar{x}^3} = \frac{V}{EI} \tag{4.4}$$

and

$$\frac{d^4\bar{v}}{d\bar{x}^4} = \frac{q}{EI} \tag{4.5}$$

For the beam element under consideration $q = 0$; substituting into Eq. (4.5) and integrating once gives

$$\frac{d^3\bar{v}}{d\bar{x}^3} = \frac{V}{EI} = C_1 \tag{4.6}$$

where C_1 is a constant of integration. Subsequent integrations yield

$$\frac{d^2\bar{v}}{d\bar{x}^2} = \frac{M}{EI} = C_1\bar{x} + C_2 \tag{4.7}$$

$$\frac{d\bar{v}}{d\bar{x}} = \bar{\theta} = \tfrac{1}{2}C_1\bar{x}^2 + C_2\bar{x} + C_3 \tag{4.8}$$

and

$$\bar{v} = \tfrac{1}{6}C_1\bar{x}^3 + \tfrac{1}{2}C_2\bar{x}^2 + C_3\bar{x} + C_4 \tag{4.9}$$

Using the fact that

$$\bar{v}(0) = \bar{v}_i, \qquad \bar{v}(L) = \bar{v}_j, \qquad \frac{d\bar{v}}{d\bar{x}}(0) = \bar{\theta}_i, \qquad \text{and} \qquad \frac{d\bar{v}}{d\bar{x}}(L) = \bar{\theta}_j$$

yields

$$\begin{aligned} C_1 &= \frac{12}{L^3}(\bar{v}_i - \bar{v}_j) + \frac{6}{L^2}(\bar{\theta}_i + \bar{\theta}_j) \\ C_2 &= \frac{6}{L^2}(-\bar{v}_i + \bar{v}_j) - \frac{1}{L}(4\bar{\theta}_i + 2\bar{\theta}_j) \\ C_3 &= \bar{\theta}_i \\ C_4 &= \bar{v}_i \end{aligned} \tag{4.10}$$

Substituting these constants of integration in Eqs. (4.6) and (4.7) we have

$$V = \frac{12EI}{L^3}(\bar{v}_i - \bar{v}_j) + \frac{6EI}{L^2}(\bar{\theta}_i + \bar{\theta}_j) \tag{4.11}$$

$$M = \frac{6EI}{L^3}(2\bar{x} - L)(\bar{v}_i - \bar{v}_j) + \frac{6EI\bar{x}}{L^2}(\bar{\theta}_i + \bar{\theta}_j) - \frac{2EI}{L}(2\bar{\theta}_i + \tilde{\theta}_j) \tag{4.12}$$

Thus

$$V(0) = V(L) = \frac{12EI}{L^3}(\bar{v}_i - \bar{v}_j) + \frac{6EI}{L^2}(\bar{\theta}_i + \bar{\theta}_j) \tag{4.13}$$

$$M(0) = \frac{6EI}{L^2}(-\bar{v}_i + \bar{v}_j) - \frac{2EI}{L}(2\tilde{\theta}_i + \bar{\theta}_j) \tag{4.14}$$

$$M(L) = \frac{6EI}{L^2}(\bar{v}_i - \bar{v}_j) + \frac{2EI}{L}(\bar{\theta}_i + 2\bar{\theta}_j) \tag{4.15}$$

The moments and shear in Eqs. (4.13) through (4.15) are expressed with a strength-of-materials sign convention. That is, positive shear forms a clockwise couple with the resultant force acting on the element; moments are positive if they deform the element into the shape associated with a positive derivative. Since the forces on the end of the beam element of Fig. 4.1 are expressed in terms of the coordinates, the deformation signs associated with Eqs. (4.13) through (4.15) are transformed as follows:

$$\bar{v}_i = -\bar{v}_j = V(0) \tag{4.16}$$

$$\bar{m}_i = -M(0) \tag{4.17}$$

$$\bar{m}_j = M(L) \tag{4.18}$$

Equations (4.13) through (4.18) give the following matrix equation expressing the relationships between generalized element displacements and forces:

$$\begin{bmatrix} \bar{p}_{\bar{y}i} \\ \bar{m}_i \\ \bar{p}_{\bar{y}j} \\ \bar{m}_j \end{bmatrix} = \frac{EI}{L} \begin{bmatrix} \frac{12}{L^2} & \frac{6}{L} & -\frac{12}{L^2} & \frac{6}{L} \\ \frac{6}{L} & 4 & -\frac{6}{L} & 2 \\ -\frac{12}{L^2} & -\frac{6}{L} & \frac{12}{L^2} & -\frac{6}{L} \\ \frac{6}{L} & 2 & -\frac{6}{L} & 4 \end{bmatrix} \begin{bmatrix} \bar{v}_i \\ \bar{\theta}_i \\ \bar{v}_j \\ \bar{\theta}_j \end{bmatrix} \tag{4.19}$$

$$\bar{\mathbf{p}} = \bar{\mathbf{k}} \, \bar{\mathbf{u}}$$

where $\bar{\mathbf{p}}$ and $\bar{\mathbf{u}}$ are column matrices of generalized nodal forces and displacements, respectively, and $\bar{\mathbf{k}}$ is the element stiffness matrix (all elements of the three matrices are expressed in element coordinates).

Note that the element stiffness matrix in Eq. (4.19) is symmetric, which was also true for the element stiffness matrix for the axial force element [Eq. (2.22)]. This property can be explained using the theorem of virtual work (see Sec. 3.2). The element stiffness matrix is also singular with respect to solu-

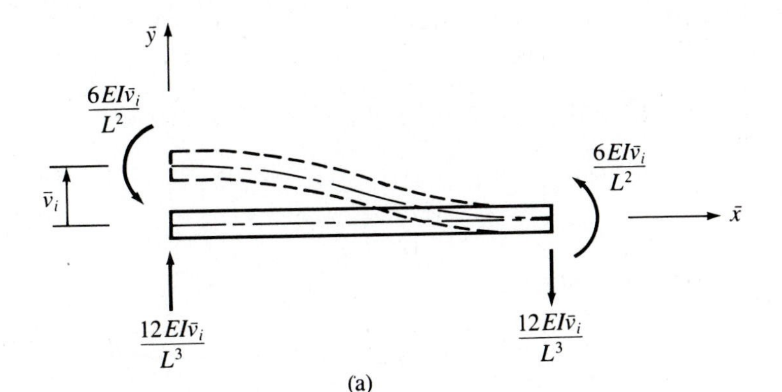

(a)

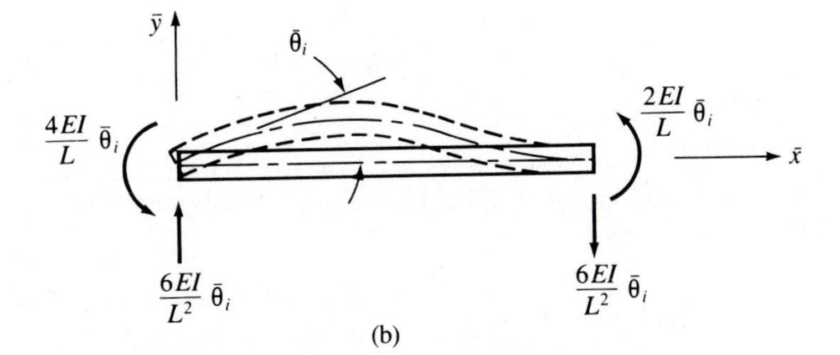

(b)

Figure 4.2 Generalized forces imposed on the deformed beam element: (a) $\bar{v}_i \neq 0$ and $\bar{\theta}_i = \bar{v}_j = \bar{\theta}_j = 0$; $\bar{\theta}_i \neq 0$ and $\bar{v}_i = \bar{v}_j = \bar{\theta}_j = 0$

tion (i.e., it has no unique inverse). This stems from the fact that since the beam element is not supported, it will exhibit rigid-body motion; therefore, the deflections resulting from an arbitrary set of applied nodal forces will be undetermined. This problem does not arise when the element is used, alone or in combination with others, to describe an appropriately restrained structure.

A given column of the element stiffness matrix can be envisaged as the forces required to impose a unit value of the corresponding displacement. For example, the elements in the first column of $\bar{\mathbf{k}}$ are the forces associated with a deformed shape, as shown in Fig. 4.2a. Any of the classical methods of structural analysis can be used to verify these results; note that this deformed shape corresponds to that used in the moment-distribution method to obtain the fixed-end moments for an element subjected to joint translation. Similarly, the second column of $\bar{\mathbf{k}}$ is composed of the forces required to maintain the deformed shape of Fig. 4.2b, which corresponds to a beam supported on a roller at node i and fixed at node j. Again these forces can be found using the methods of classical structural analysis. This deformed state corresponds to the configuration investigated in the moment-distribution method to obtain the rotational stiffness and the carryover factor.

EXAMPLE 4.1

Calculate the rotations and reactions at points a and b for the uniform beam in Fig. E4.1 with $I = 450$ in.^{4} and $E = 29 \times 10^3$ kips/in.^{2}.

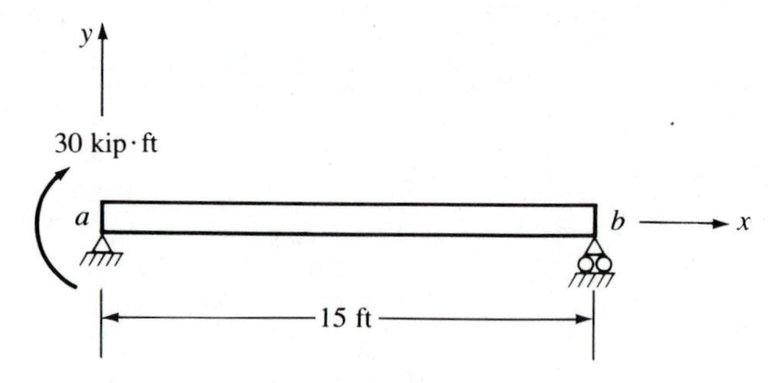

Figure E4.1

Solution

$EI/L = 72{,}500$ kip·in.^{2}. The global and local coordinates coincide. From Eq. (4.19),

$$\begin{bmatrix} p_{ya} \\ m_a \\ p_{yb} \\ m_b \end{bmatrix} = 7.25 \begin{bmatrix} 3.7037 & 333.33 & -3.7037 & 333.33 \\ 333.33 & 40000. & -333.33 & 20000. \\ -3.7037 & -333.33 & 3.7037 & -333.33 \\ 333.33 & 20000. & -333.33 & 40000. \end{bmatrix} \begin{bmatrix} v_a \\ \theta_a \\ v_b \\ \theta_b \end{bmatrix}$$

Note that all quantities are expressed in terms of kips and in.; $v_a = v_b = 0$ and $m_a = -360$ kip · in. From the second and fourth equations

$$-360 = 10^4(29.0\theta_a + 14.5\theta_b)$$
$$0 = 10^4(14.5\theta_a + 29.0\theta_b)$$

Solving gives $\theta_a = -16.56 \times 10^{-3}$ rad and $\theta_b = 8.28 \times 10^{-3}$ rad. The first and third equations give the reactions, that is,

$$p_{ya} = 7.25(333.33\theta_a + 333.33\theta_b) = -2 \text{ kips}$$
$$p_{yb} = 7.25(-333.33\theta_a - 333.33\theta_b) = 2 \text{ kips}$$

These results can be checked using the equations of statics and a classical approach such as the moment area method.

4.1.2 THE PRINCIPLE OF VIRTUAL WORK AND $\bar{\mathbf{k}}$

The stiffness matrix for a beam element as shown in Fig. 4.1 can also be obtained from the principle of virtual work using the general procedure outlined in Sec. 3.2. The displacements of the beam in the direction normal to its longitudinal axis are described by the matrix equation

$$\bar{v} = \mathbf{N}\bar{\mathbf{u}} \tag{4.20}$$

where $\mathbf{N}$ is the matrix of shape functions and $\bar{\mathbf{u}}$ contains the generalized displacements. Using Eqs. (4.9) and (4.10) we have

$$\bar{v} = \left[2\left(\frac{\bar{x}}{L}\right)^3 - 3\left(\frac{\bar{x}}{L}\right)^2 + 1\right]\bar{v}_i + \left(\frac{\bar{x}^3}{L^2} - 2\frac{\bar{x}^2}{L} + \bar{x}\right)\bar{\theta}_i$$
$$+ \left[-2\left(\frac{\bar{x}}{L}\right)^3 + 3\left(\frac{\bar{x}}{L}\right)^2\right]\bar{v}_j + \left(\frac{\bar{x}^3}{L^2} - \frac{\bar{x}^2}{L}\right)\bar{\theta}_j \tag{4.21}$$

or

$$\bar{v} = N_1\bar{v}_i + N_2\bar{\theta}_i + N_3\bar{v}_j + N_4\bar{\theta}_j \tag{4.22}$$

where

$$\begin{aligned} N_1 &= 2\left(\frac{\bar{x}}{L}\right)^3 - 3\left(\frac{\bar{x}}{L}\right)^2 + 1 \\ N_2 &= \frac{\bar{x}^3}{L^2} - 2\frac{\bar{x}^2}{L} + \bar{x} \\ N_3 &= -2\left(\frac{\bar{x}}{L}\right)^3 + 3\left(\frac{\bar{x}}{L}\right)^2 \\ N_4 &= \frac{\bar{x}^3}{L^2} - \frac{\bar{x}^2}{L} \end{aligned} \tag{4.23}$$

These shape functions are illustrated in Fig. 4.3. The longitudinal stress in terms of the applied bending moment is

$$\sigma = -\frac{M\bar{y}}{I} \tag{4.24}$$

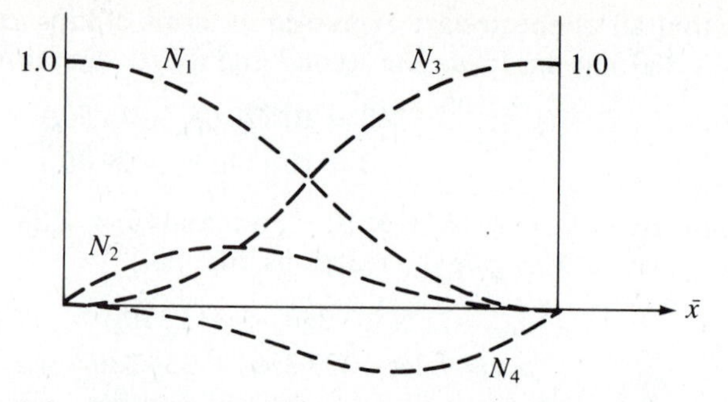

Figure 4.3 Shape functions for flexure of a uniform beam element

Substituting the moment-curvature relation [Eq. (4.1)] yields

$$\sigma = -\frac{d^2\bar{v}}{d\bar{x}^2} E\bar{y} \tag{4.25}$$

Substituting Eqs. (4.22) and (4.23) into Eq. (4.25) we have

$$\sigma = -E\bar{y}\left[\left(12\frac{\bar{x}}{L^3} - 6\frac{1}{L^2}\right) \left(6\frac{\bar{x}}{L^2} - 4\frac{1}{L}\right) \left(-12\frac{\bar{x}}{L^3} - 6\frac{1}{L^2}\right) \left(6\frac{\bar{x}}{L^2} - 2\frac{1}{L}\right)\right]\begin{bmatrix}\bar{v}_i \\ \bar{\theta}_i \\ \bar{v}_j \\ \bar{\theta}_j\end{bmatrix}$$

$$\sigma = E \qquad \mathbf{B} \qquad \mathbf{u} \tag{4.26}$$

where

$$\mathbf{B} = -\bar{y}\frac{d^2}{d\bar{x}^2}\mathbf{N} \tag{4.27}$$

Equation (3.40) defines **B** as the matrix relating strains and nodal displacements; combining Hooke's law with Eq. (4.26) we note that **B** has the same definition for a beam. That is, the longitudinal strain is

$$\varepsilon = \frac{\sigma}{E} = \mathbf{Bu} \tag{4.28}$$

This is the same definition of **B** given in Eq. (3.40), but the matrix **B** takes a different form for the beam element than it has for the axial force element. Note that for small displacements, the beam curvature, κ, is approximated as

$$\kappa \approx \frac{d^2\bar{v}}{d\bar{x}^2} = \frac{1}{\bar{y}}\mathbf{Bu} \tag{4.29}$$

The virtual longitudinal strain is given in a form similar to the real strain, that is,

$$\delta\varepsilon = \mathbf{B}\,\boldsymbol{\delta}\bar{\mathbf{u}} \tag{4.30}$$

The matrix of real generalized forces $\bar{\mathbf{p}} = \{\bar{p}_{\bar{y}i} \quad \bar{m}_i \quad \bar{p}_{\bar{y}j} \quad \bar{m}_j\}$ and the virtual generalized displacements $\boldsymbol{\delta}\bar{\mathbf{u}} = \{\delta\bar{v}_i \quad \delta\bar{\theta}_i \quad \delta\bar{v}_j \quad \delta\bar{v}_j\}$. Using the real stress from Eq. (4.26), the virtual strain according to Eq. (4.30), and the theorem of virtual work [Eq. (3.20)] for deformable bodies, we obtain

$$\begin{aligned}\boldsymbol{\delta}\bar{u}^T\bar{\mathbf{p}} &= \int_{\text{vol}} \boldsymbol{\delta\varepsilon}^T\boldsymbol{\sigma}\, dV = \int_{\text{vol}} \boldsymbol{\delta}\bar{\mathbf{u}}^T\mathbf{B}^T E\mathbf{B}\bar{\mathbf{u}}\, dV \\ &= \boldsymbol{\delta}\bar{\mathbf{u}}^T \int_{\text{vol}} \mathbf{B}^T E\mathbf{B}\, dV\bar{\mathbf{u}} = \boldsymbol{\delta}\bar{\mathbf{u}}^T\bar{\mathbf{k}}\bar{\mathbf{u}}\end{aligned} \tag{4.31}$$

where

$$\bar{\mathbf{k}} = \int_{\text{vol}} \mathbf{B}^T E\mathbf{B}\, dV \tag{4.32}$$

The elements of **B** are functions of $\bar{x}$; the product of the scalar multiplier of **B** and $\mathbf{B}^T$, y, when integrated over the cross section yields I. Substituting the matrices from Eq. (4.26) in Eq. (4.32),

$$\bar{\mathbf{k}} = EI \int_0^L \begin{bmatrix} \frac{6}{L^2}\left(-2\frac{\bar{x}}{L}+1\right) \\ \frac{2}{L}\left(-3\frac{\bar{x}}{L}+2\right) \\ \frac{6}{L^2}\left(2\frac{\bar{x}}{L}-1\right) \\ \frac{2}{L}\left(-3\frac{\bar{x}}{L}+1\right) \end{bmatrix} \tag{4.33}$$

$$\times\left[\frac{6}{L^2}\left(-2\frac{\bar{x}}{L}+1\right) \quad \frac{2}{L}\left(-3\frac{\bar{x}}{L}+2\right) \quad \frac{6}{L^2}\left(2\frac{\bar{x}}{L}-1\right) \quad \frac{2}{L}\left(-3\frac{\bar{x}}{L}+1\right)\right] dx$$

$$\bar{\mathbf{k}} = \frac{EI}{L}\begin{bmatrix} 12/L^2 & 6/L & -12/L^2 & 6/L \\ & 4 & -6/L & 2 \\ \text{symmetric} & & 12/L^2 & -6/L \\ & & & 4 \end{bmatrix} \tag{4.34}$$

This is the same result obtained in Sec. 4.1.1 by solving the basic governing equations [see Eq. (4.19)].

4.1.3 CASTIGLIANO'S THEOREM (PART I) AND $\bar{\mathbf{k}}$ Alternatively, we can calculate the element stiffness matrix for the beam of Fig. 4.1 using Castigliano's theorem (part I), which is based on the principle of virtual work. For a linearly elastic structure with a given system of loads and temperature distribution, the first partial derivative of the strain energy with respect to a particular displacement, u_r, is equal to the corresponding force, p_r; that is,

$$\frac{\partial W_i}{\partial \bar{u}_r} = \bar{p}_r \tag{4.35}$$

In general

$$\bar{\mathbf{p}} = \bar{\mathbf{k}}\bar{\mathbf{u}} \tag{4.36}$$

Therefore, Castigliano's theorem (part I) gives

$$\frac{\partial W_i}{\partial \bar{u}_r} = \sum_s \bar{k}_{rs}\bar{u}_s \tag{4.37}$$

Furthermore, the second derivative of the strain energy yields

$$\frac{\partial^2 W_i}{\partial \bar{u}_r \, \partial \bar{u}_s} = \bar{k}_{rs} \tag{4.38}$$

Since W_i is a continuously differentiable function

$$\frac{\partial^2 W_i}{\partial \bar{u}_r \, \partial \bar{u}_s} = \frac{\partial^2 W_i}{\partial \bar{u}_s \, \partial \bar{u}_r} \tag{4.39}$$

Thus

$$\bar{k}_{rs} = \bar{k}_{sr} \tag{4.40}$$

That is, the element stiffness matrix is symmetric.

The stiffness elements can also be obtained by observing the general form of the strain energy. Using Eqs. (3.25) and (3.26) and referring to Fig. 3.6, we note that if an element jk is loaded from a stress-free state and subjected to the stress σ^{jk} with the corresponding linearly-related strain ε^{jk}, the strain energy is

$$W_i = \frac{1}{2}\int_{\text{vol}} \sigma^{jk}\varepsilon^{jk}\, dV \tag{4.41}$$

From Eqs. (4.26) and (4.28) for a beam we have $\varepsilon^{jk} = \varepsilon = \mathbf{B}\bar{\mathbf{u}}$ and $\sigma^{jk} = \sigma = E\mathbf{B}\bar{\mathbf{u}}$; therefore the strain energy is

$$W_i = \frac{1}{2}\int_{\text{vol}} \boldsymbol{\sigma}^T\boldsymbol{\varepsilon}\, dV = \frac{1}{2}\bar{\mathbf{u}}^T \int_{\text{vol}} \mathbf{B}^T E\mathbf{B}\, dV\, \bar{\mathbf{u}} \tag{4.42}$$

Note that the matrix product under the integral is symmetric. Recognizing that for a linearly elastic structure $W_i = W_e$ yields

$$W_i = \tfrac{1}{2}\bar{\mathbf{u}}^T\bar{\mathbf{k}}\bar{\mathbf{u}} \tag{4.43}$$

we deduce that the integrated product in Eq. (4.42) is the element stiffness matrix. Expressing Eq. (4.43) in summation form we have

$$W_i = \frac{1}{2}\sum_m \sum_n \bar{u}_m \bar{k}_{mn} \bar{u}_n \tag{4.44}$$

Operating on Eq. (4.44) gives

$$\frac{\partial W_i}{\partial \bar{u}_r} = \frac{1}{2}\sum_n \bar{k}_{rn}\bar{u}_n + \frac{1}{2}\sum_m \bar{u}_m \bar{k}_{mr} \tag{4.45}$$

but since $\bar{k}_{ij} = \bar{k}_{ji}$ we have

$$\frac{\partial W_i}{\partial \bar{u}_r} = \sum_n \bar{k}_{rn}\bar{u}_n \tag{4.46}$$

Thus,

$$\frac{\partial^2 W_i}{\partial \bar{u}_r \, \partial \bar{u}_s} = \bar{k}_{rs}$$

which is the same result obtained in Eq. (4.38).

Obtaining the element stiffness matrix using Castigliano's theorem (part I) is an alternate way to envisage the process. Since this theorem is based upon the principle of virtual work, the equations (and resulting stiffness matrix) are the same as those obtained in the previous section. That is, using the definition of strain energy given in Eq. (4.42), and operating upon it according to Eq. (4.38), yields the stiffness matrix in precisely the same form produced by applying the principle of virtual work [see Eqs. (4.32) through (4.34)].

4.1.4 ELEMENT FORCES The shear and bending moments in the beam element are required for design purposes and can be calculated using an element force-transformation matrix of the form in Eq. (2.57a), that is,

$$\mathbf{s}^{ij} = \bar{\mathbf{e}}\bar{\mathbf{u}}^{ij} \tag{4.47}$$

The shear force in the beam is constant, and the bending moment varies linearly along the length of the element [see Eqs. (4.11) and (4.12)]. Therefore, we need calculate only one value of shear plus the bending moments at two points, and all internal element forces can be obtained from these quantities. By selecting the first, second, and fourth equations from the force-displacement equations [i.e., Eq. (4.19) or Eq. (4.34)] we obtain the following matrix equations for calculating the internal generalized forces in each individual beam element:

$$\underset{\mathbf{s}^{ij}}{\begin{bmatrix} V \\ M_i \\ M_j \end{bmatrix}} = \underset{\bar{\mathbf{e}}}{\frac{EI}{L}\begin{bmatrix} \frac{12}{L^2} & \frac{6}{L} & -\frac{12}{L^2} & \frac{6}{L} \\ \frac{6}{L} & 4 & -\frac{6}{L} & 2 \\ \frac{6}{L} & 2 & -\frac{6}{L} & 4 \end{bmatrix}} \underset{\bar{\mathbf{u}}^{ij}}{\begin{bmatrix} \bar{v}_i \\ \bar{\theta}_i \\ \bar{v}_j \\ \bar{\theta}_j \end{bmatrix}} \tag{4.48}$$

where the forces for the beam element between nodes i and j are described by the column matrix $\mathbf{s}^{ij}$ and $\bar{\mathbf{e}}$ is the element force-transformation matrix. The element forces obtained from Eq. (4.48) are specified with respect to the local coordinates as shown in Fig. 4.1, and strength-of-materials (deformation) sign convention must be used for plotting shear and moment diagrams.

4.2 NODAL EQUILIBRIUM, NODAL DISPLACEMENTS, AND STRUCTURAL IDEALIZATION

The interaction of individual elements in a beam structure is achieved by imposing nodal equilibrium as we did for trusses in Ch. 2. In the case of beam structures it is necessary to prescribe equilibrium for both moments and direct forces. The element stiffness matrices, $\bar{\mathbf{k}}$, are combined to give the stiffness matrix for the total structure, $\mathbf{K}$. For beams the global x axis is usually aligned with the beam axis so both local and global axes coincide. Therefore, the coordinate-transformation matrix ($\mathbf{T}$) similar to that used for trusses need not be used since the element stiffness matrices in local and global coordinates are identical ($\bar{\mathbf{k}} = \mathbf{k}$).

Consider the beam with ten nodes in Fig. 4.4a. Since the beam element has two degrees of freedom per node, the structural stiffness matrix will be 20 by 20. Consider the elements connected to node r in Fig. 4.4b. The force-displacement equations for each of the two beam elements attached to this node are expressed using Eq. (4.19) as follows:

$$\begin{bmatrix} p_{yq}^{qr} \\ m_q^{qr} \\ p_{yr}^{qr} \\ m_r^{qr} \end{bmatrix} = \begin{bmatrix} k_{11}^{qr} & k_{12}^{qr} & k_{13}^{qr} & k_{14}^{qr} \\ k_{21}^{qr} & k_{22}^{qr} & k_{23}^{qr} & k_{24}^{qr} \\ k_{31}^{qr} & k_{32}^{qr} & k_{33}^{qr} & k_{34}^{qr} \\ k_{41}^{qr} & k_{42}^{qr} & k_{43}^{qr} & k_{44}^{qr} \end{bmatrix} \begin{bmatrix} V_q \\ \Theta_q \\ V_r \\ \Theta_r \end{bmatrix} \tag{4.49a}$$

$$\begin{bmatrix} p_{yr}^{rs} \\ m_r^{rs} \\ p_{ys}^{rs} \\ m_s^{rs} \end{bmatrix} = \begin{bmatrix} k_{11}^{rs} & k_{12}^{rs} & k_{13}^{rs} & k_{14}^{rs} \\ k_{21}^{rs} & k_{22}^{rs} & k_{23}^{rs} & k_{24}^{rs} \\ k_{31}^{rs} & k_{32}^{rs} & k_{33}^{rs} & k_{34}^{rs} \\ k_{41}^{rs} & k_{42}^{rs} & k_{43}^{rs} & k_{44}^{rs} \end{bmatrix} \begin{bmatrix} V_r \\ \Theta_r \\ V_s \\ \Theta_s \end{bmatrix} \tag{4.49b}$$

The subscripts denote the position of the element in the matrix, and the superscripts designate the element. No superscripts are required for the displacements since for compatibility all elements must be connected to a common node; that is,

$$v_r^{qr} = v_r^{rs} = V_r$$

$$\theta_r^{qr} = \theta_r^{rs} = \Theta_r$$

and so forth.

The free-body diagrams for the elements connected to node r are shown in Fig. 4.4b. The generalized forces applied externally to the node are P_{yr} and M_r. Nodal equilibrium in the y direction dictates that

$$P_{yr} = p_{yr}^{qr} + p_{yr}^{rs} \tag{4.50}$$

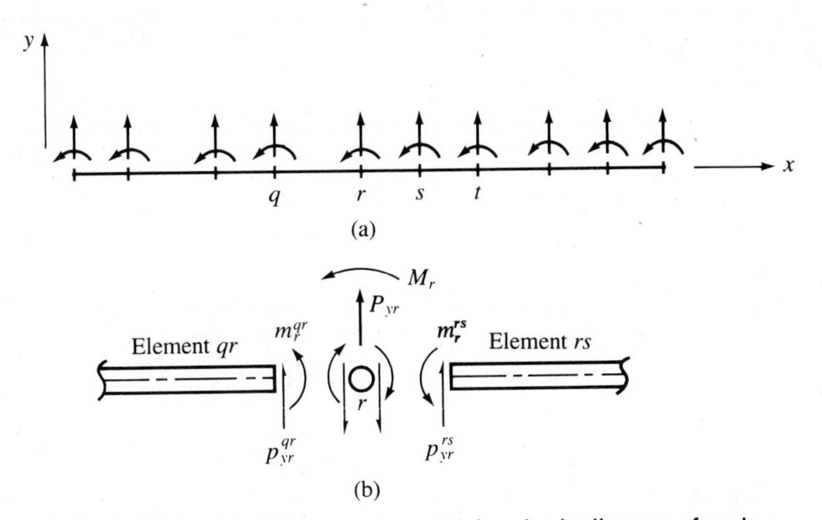

Figure 4.4 (a) Beam structure; (b) free-body diagram of node *r*

Substituting Eqs. (4.49a) and (4.49b), the equation of equilibrium becomes

$$P_{yr} = (k_{33}^{qr} + k_{11}^{rs})V_r + (k_{34}^{qr} + k_{12}^{rs})\Theta_r$$
$$+ k_{31}^{qr}V_q + k_{32}^{qr}\Theta_q + k_{13}^{rs}V_s + k_{14}^{rs}\Theta_s$$

or

$$P_{yr} = K_{jh}V_q + K_{ji}\Theta_q + K_{jj}V_r + K_{jk}\Theta_r + K_{jl}V_s + K_{jm}\Theta_s \tag{4.51}$$

where

$$K_{jj} = k_{33}^{qr} + k_{11}^{rs} \qquad \text{and} \qquad K_{jk} = k_{34}^{qr} + k_{12}^{rs} \tag{4.52}$$

and so forth. That is, the displacement in the y direction at node $r(V_r)$ is the jth degree of freedom for the beam, while the slope at node $r(\theta_r)$ is the kth degree of freedom, etc. The stiffnesses contributed to a given degree of freedom by the various structural elements must be added into the appropriate location in the structural stiffness matrix (**K**). This process is identical to that described in Sec. 2.3. A similar equation of equilibrium for the moments at node r leads to the following equation:

$$M_r = k_{41}^{qr}V_q + k_{42}^{qr}\Theta_q + (k_{43}^{qr} + k_{21}^{rs})V_r$$
$$+ (k_{44}^{qr} + k_{22}^{rs})\Theta_r + k_{23}^{rs}V_s + k_{24}^{rs}\Theta_r$$

or

$$M_r = K_{kh}V_q + K_{ki}\Theta_q + K_{kj}V_r + K_{kk}\Theta_r + K_{kl}V_s + K_{km}\Theta_s \tag{4.53}$$

where

$$K_{kj} = k_{43}^{qr} + k_{21}^{rs} \qquad \text{and} \qquad K_{kk} = k_{44}^{qr} + k_{22}^{rs} \tag{4.54}$$

and so forth. The procedure of constructing **K** is illustrated for the two-element beam in Example 4.2.

EXAMPLE 4.2

Obtain the structural stiffness matrix for the two-element beam structure in Fig. E4.2 with $I_{ab} = 240 \times 10^{-6}$ m^4, $I_{bc} = 150 \times 10^{-6}$ m^4, and $E = 200$ GPa.

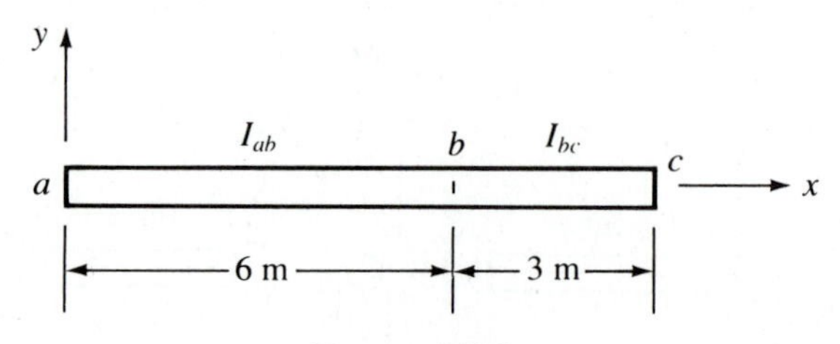

Figure E4.2

Solution

For both elements $\bar{\mathbf{k}} = \mathbf{k}$. From Eq. (4.19),

$$\begin{bmatrix} p_{ya} \\ m_a \\ p_{yb} \\ m_b \end{bmatrix}^{ab} = 8000 \begin{bmatrix} 0.3333 & 1.0000 & -0.3333 & 1.0000 \\ 1.0000 & 4.0000 & -1.0000 & 2.0000 \\ -0.3333 & -1.0000 & 0.3333 & -1.0000 \\ 1.0000 & 2.0000 & -1.0000 & 4.0000 \end{bmatrix} \begin{bmatrix} v_a \\ \theta_a \\ v_b \\ \theta_b \end{bmatrix}$$

$$\begin{bmatrix} p_{yb} \\ m_b \\ p_{yc} \\ m_c \end{bmatrix}^{bc} = 10{,}000 \begin{bmatrix} 1.3333 & 2.0000 & -1.3333 & 2.0000 \\ 2.0000 & 4.0000 & -2.0000 & 2.0000 \\ -1.3333 & -2.0000 & 1.3333 & -2.0000 \\ 2.0000 & 2.0000 & -2.0000 & 4.0000 \end{bmatrix} \begin{bmatrix} v_b \\ \theta_b \\ v_c \\ \theta_c \end{bmatrix}$$

Enforcing nodal equilibrium yields

$$\underset{\mathbf{P}}{\begin{bmatrix} P_{ya} \\ M_a \\ P_{yb} \\ M_b \\ P_{yc} \\ M_c \end{bmatrix}} = 10^3 \underset{\mathbf{K}}{\begin{bmatrix} 2.6666 & 8.0000 & -2.6666 & 8.0000 & 0.0000 & 0.0000 \\ 8.0000 & 32.0000 & -8.0000 & 16.0000 & 0.0000 & 0.0000 \\ -2.6666 & -8.0000 & 16.0000 & 12.0000 & -13.3333 & 20.0000 \\ 8.0000 & 16.0000 & 12.0000 & 72.0000 & -20.0000 & 20.0000 \\ 0.0000 & 0.0000 & -13.3333 & -20.0000 & 13.3333 & -20.0000 \\ 0.0000 & 0.0000 & 20.0000 & 20.0000 & -20.0000 & 40.0000 \end{bmatrix}} \underset{\mathbf{U}}{\begin{bmatrix} V_a \\ \Theta_a \\ V_b \\ \Theta_b \\ V_c \\ \Theta_c \end{bmatrix}}$$

Discussion The fact that **K** is a 6 by 6 matrix is a consequence of the fact that there are three nodes and two degrees of freedom per node. Note that **K** is symmetrical.

The force-displacement equations for an assemblage of beam elements can also be obtained using the theorem of virtual work. The internal virtual work for element m is given by Eq. (4.31) as follows:

$$\delta W_i^m = \boldsymbol{\delta}\mathbf{u}^{m^T}\mathbf{k}^m\mathbf{u}^m \tag{4.55}$$

where the superscript denotes quantities associated with element m. All quantities are expressed in global coordinates since local and global axes coincide for beams. The total internal virtual work for an assemblage of N elements is

$$\delta W_i = \sum_{m=1}^{N} \boldsymbol{\delta}\mathbf{u}^{m^T}\mathbf{k}^m\mathbf{u}^m = \boldsymbol{\delta}\mathbf{u}^{m^T} \sum_{m=1}^{N} \mathbf{k}^m\mathbf{u}^m \tag{4.56}$$

The displacement matrix $\boldsymbol{\delta}\mathbf{u}^m$ is 4 by 1 and contains only those degrees of freedom associated with the element m. Similarly, $\mathbf{k}^m$ is a 4 by 4 matrix with the elements of Eq. (4.19). Expanding each of the matrices to include all degrees of freedom for the problem gives

$$\delta W_i = \boldsymbol{\delta}\mathbf{U}^T \sum_{m=1}^{N} \mathbf{K}^m\mathbf{U} \tag{4.57}$$

For the beam structure of Fig. 4.4, $\mathbf{K}^m$ is a 20 by 20 matrix for each individual element. Invoking the principle of virtual work we have

$$\delta W_e = \boldsymbol{\delta}\mathbf{U}^T\mathbf{P} = \delta W_i = \boldsymbol{\delta}\mathbf{U}^T \sum_{m=1}^{N} \mathbf{K}^m\mathbf{U} \tag{4.58}$$

Since $\boldsymbol{\delta}\mathbf{U}$ is arbitrary and nonzero

$$\mathbf{P} = \mathbf{K}\mathbf{U} \tag{4.59}$$

where

$$\mathbf{K} = \sum_{m=1}^{N} \mathbf{K}^m \tag{4.60}$$

It is implicit in the principle of virtual work that the system is in equilibrium; therefore, Eq. (4.59) implies nodal equilibrium for the structural assemblage.

This is the same result and method used for trusses in Sec. 3.4, but for beams, $\mathbf{U}$ contains the rotational degrees of freedom. Furthermore, invoking Eq. (4.59) implies moment equilibrium, as well as direct force equilibrium. The summation in Eq. (4.60) expresses direct addition of the element stiffness matrices $\mathbf{K}^m$ to obtain the stiffness matrix for the total assemblage. Nonzero elements in $\mathbf{K}^m$ exist only in those rows and columns corresponding to degrees of freedom associated with element m. Thus, we need only store the compact (4 by 4) element stiffness matrix $\mathbf{k}^m$. Techniques discussed in Sec. 2.3 for efficient assembly of $\mathbf{K}$ for trusses are applicable to beams as well, but rotational degrees of freedom are involved for the latter type of structure. Example 4.3 demonstrates how the force-displacement equations are obtained for an assemblage of beams.

EXAMPLE 4.3

Obtain the structural stiffness matrix for the three-element structure in Fig. E4.3 with $I = 1200$ in.^{4}, $L = 14.5$ ft, and $E = 29 \times 10^3$ kips/in.^{2}.

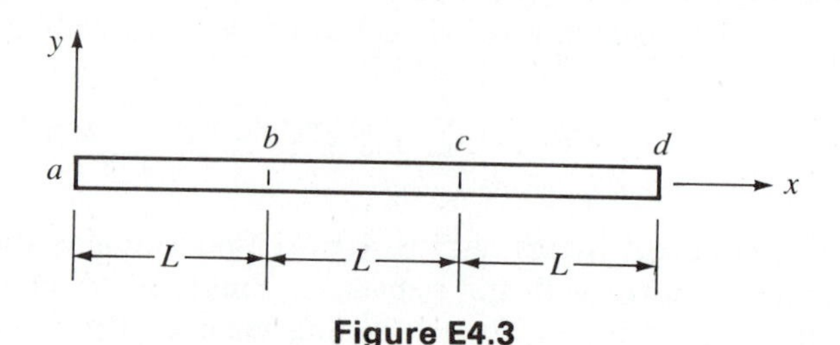

Figure E4.3

Solution

Since the global and local axes are aligned, $\bar{\mathbf{k}} = \mathbf{k}$. The elements are identical so all $\mathbf{k}$ matrices are equal:

$$\mathbf{k} = 10^2 \begin{bmatrix} 0.79271 & 68.966 & -0.79271 & 68.966 \\ 68.966 & 8000.0 & -68.966 & 4000.0 \\ -0.79271 & -68.966 & 0.79271 & -68.966 \\ 68.966 & 4000.0 & -68.966 & 8000.0 \end{bmatrix}$$

$$\mathbf{K} = 10^2 \begin{bmatrix} 0.79271 & 68.966 & -0.79271 & 68.966 & 0000.0 & 0000.0 & 0000.0 & 0000.0 \\ 68.966 & 8000.0 & -68.966 & 4000.0 & 0000.0 & 0000.0 & 0000.0 & 0000.0 \\ -0.79271 & -68.966 & 1.58542 & 0000.0 & -0.79271 & 68.966 & 0000.0 & 0000.0 \\ 68.966 & 4000.0 & 0000.0 & 16000.0 & -68.966 & 4000.0 & 0000.0 & 0000.0 \\ 0000.0 & 0000.0 & -0.79271 & -68.966 & 1.58542 & 0000.0 & -0.79271 & 68.966 \\ 0000.0 & 0000.0 & 68.966 & 4000.0 & 0000.0 & 16000.0 & -68.966 & 4000.0 \\ 0000.0 & 0000.0 & 0000.0 & 0000.0 & -0.79271 & -68.966 & 0.79271 & -68.966 \\ 0000.0 & 0000.0 & 0000.0 & 0000.0 & 68.966 & 4000.0 & -68.966 & 8000.0 \end{bmatrix}$$

$\mathbf{P} = \{P_{ya} \quad M_a \quad P_{yb} \quad M_b \quad P_{yc} \quad M_c \quad P_{yd} \quad M_d\}$ and $\mathbf{U} = \{V_a \quad \Theta_a \quad V_b \quad \Theta_b \quad V_c \quad \Theta_c \quad V_d \quad \Theta_d\}$

The stiffness matrix for the prismatic beam element has the characteristics that the shear is constant and the bending moment varies linearly between nodal points. These two properties are the consequences of using the moment-curvature relationship in conjunction with the differential equilibrium equations and the assumption that there are no concentrated forces or moments along the element length [see Eqs. (4.11) and (4.12)].

We must specify nodal points and individual elements for a beam structure since neither are defined by the construction features. This is in contrast with truss analysis where the physical panel points and the nodes generally coincide. The first step in analyzing a structure is *idealization*; it is the process of discretizing the structure into a set of individual elements so that the model

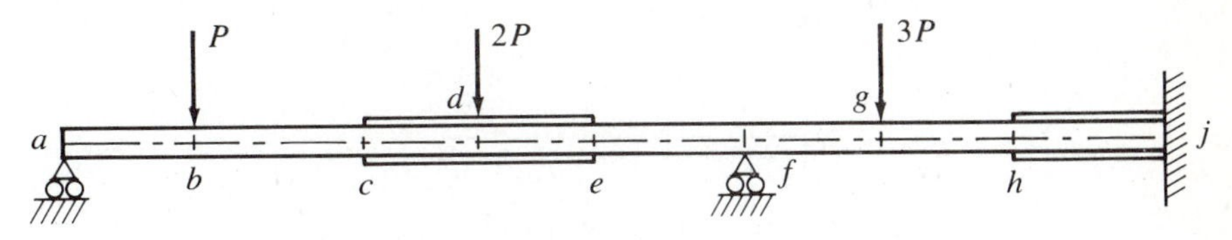

Figure 4.5 Idealization of a continuous beam

accurately characterizes the actual behavior of the structure. Since the beam element discussed in this chapter is uniform, with constant shear and linearly varying moment, each element of the structure should also have these characteristics. The continuous steel beam in Fig. 4.5 has been reinforced with cover plates between points c and e and also between h and j. Thus, it is apparent that a single uniform beam element cannot be used over the entire structure. There is another primary factor that prevents using one element: the presence of concentrated loads at b, d, and g, plus the support reaction forces at a, f, and j. That is, the shear and moment vary along the length of the structure. Using a single element between points of applied loads (or reactions) is acceptable since, in such a segment, constant shear is accompanied by a linearly varying moment; these are the assumptions incorporated into the beam element. Thus, we observe that by placing a node at each of the labeled points, the structure is accurately idealized since each element is uniform, has constant shear, and possesses moments that vary linearly.

Nodal displacements are computed subsequent to idealization, assembling the structural force-displacement equations [Eq. (4.59)] and imposing the restraint conditions of the structure. The problem solution can be carried out using logic identical to that employed for the analysis of trusses. In Sec. 2.4 (and also in Sec. 3.5) we noted that the structural force-displacement equations are partitioned so that known forces ($\mathbf{P}_f$) are grouped with unknown displacements ($\mathbf{U}_f$) and unknown forces ($\mathbf{P}_s$) are associated with known displacements ($\mathbf{U}_s$). This gives the following two matrix equations:

$$\mathbf{P}_f = \mathbf{K}_{ff}\mathbf{U}_f + \mathbf{K}_{fs}\mathbf{U}_s$$

and

$$\mathbf{P}_s = \mathbf{K}_{sf}\mathbf{U}_f + \mathbf{K}_{sf}\mathbf{U}_s$$

which can be solved for the unknown displacements and forces.

Refer back to the general flowchart for the stiffness method in Fig. 2.13 and note that we now have all the basic matrices for beam structures. Thus we are prepared to solve beam structures with applied loads. Examples 4.4 through 4.7 illustrate how the stiffness method is used for calculating deflections and element forces for various beam structures. These examples use the structural stiffness matrices obtained in Examples 4.2 and 4.3.

EXAMPLE 4.4

Calculate the nodal displacements and element forces for the beam of example 4.2 if it is supported as shown in Fig. E4.4.

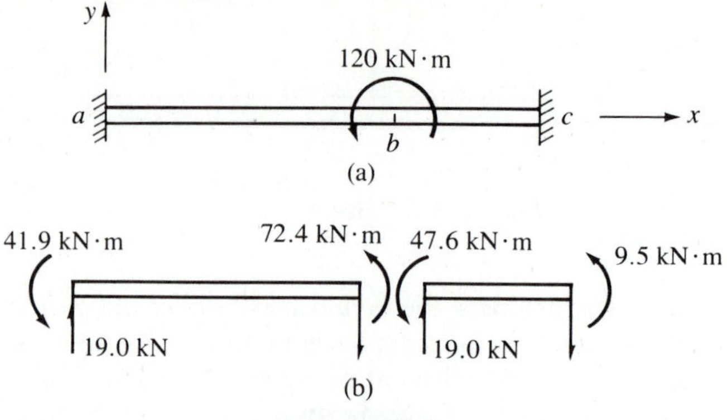

Figure E4.4

Solution

See Example 4.2 for dimensions, etc. From Example 4.2 (using appropriate entries from rows and columns 3 and 4 of **K**),

$$\begin{bmatrix} 0 \\ 120 \end{bmatrix} = 10^3 \begin{bmatrix} 16 & 12 \\ 12 & 72 \end{bmatrix} \begin{bmatrix} V_b \\ \Theta_b \end{bmatrix}$$

Thus

$$\begin{bmatrix} V_b \\ \Theta_b \end{bmatrix} = 10^{-3} \begin{bmatrix} -1.4286 \text{ m} \\ 1.9048 \text{ rad} \end{bmatrix}$$

Using Eq. (4.48) and the displacements, the element forces are

$$\begin{bmatrix} V \\ M_a \\ M_b \end{bmatrix}^{ab} = 8000 \begin{bmatrix} 0.3333 & 1.0000 & -0.3333 & 1.0000 \\ 1.0000 & 4.0000 & -1.0000 & 2.0000 \\ 1.0000 & 2.0000 & -1.0000 & 4.0000 \end{bmatrix} \begin{bmatrix} 0 \\ 0 \\ V_b \\ \Theta_b \end{bmatrix}$$

$$= \{19.0 \text{ kN} \quad 41.9 \text{ kN}\cdot\text{m} \quad 72.4 \text{ kN}\cdot\text{m}\}$$

$$\begin{bmatrix} V \\ M_b \\ M_c \end{bmatrix}^{bc} = 10{,}000 \begin{bmatrix} 1.3333 & 2.0000 & -1.3333 & 2.0000 \\ 2.0000 & 4.0000 & -2.0000 & 2.0000 \\ 2.0000 & 2.0000 & -2.0000 & 4.0000 \end{bmatrix} \begin{bmatrix} V_b \\ \Theta_b \\ 0 \\ 0 \end{bmatrix}$$

$$= \{19.0 \text{ kN} \quad 47.6 \text{ kN}\cdot\text{m} \quad 9.5 \text{ kN}\cdot\text{m}\}$$

See Fig. E4.4b.

EXAMPLE 4.5

Calculate the nodal displacements and element forces for the beam of Example 4.2 if it is supported as shown in Fig. E4.5.

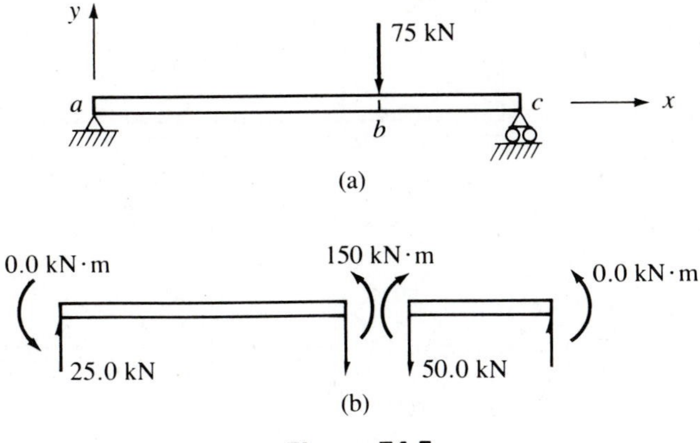

Figure E4.5

Solution

See Example 4.2 for dimensions, etc. From Example 4.2 (using appropriate entries from rows and columns 2, 3, 4, and 6 of **K**),

$$\begin{bmatrix} 0 \\ -75 \\ 0 \\ 0 \end{bmatrix} = 10^3 \begin{bmatrix} 32.0000 & -8.0000 & 16.0000 & 0.0000 \\ -8.0000 & 16.0000 & 12.0000 & 20.0000 \\ 16.0000 & 12.0000 & 72.0000 & 20.0000 \\ 0.0000 & 20.0000 & 20.0000 & 40.0000 \end{bmatrix} \begin{bmatrix} \Theta_a \\ V_b \\ \Theta_b \\ \Theta_c \end{bmatrix}$$

Thus

$$\begin{bmatrix} \Theta_a \\ V_b \\ \Theta_b \\ \Theta_c \end{bmatrix} = 10^{-3} \begin{bmatrix} -6.875 \text{ rad} \\ -22.50 \text{ mm} \\ 2.500 \text{ rad} \\ 10.00 \text{ rad} \end{bmatrix}$$

Using Eq. (4.48) and the displacements, the element forces are

$$\{V \quad M_a \quad M_b\}^{ab} = \{25.0 \text{ kN} \quad 0.0 \text{ kN}\cdot\text{m} \quad 150.0 \text{ kN}\cdot\text{m}\}$$
$$\{V \quad M_b \quad M_c\}^{bc} = \{-50.0 \text{ kN} \quad -150.0 \text{ kN}\cdot\text{m} \quad 0.0 \text{ kN}\cdot\text{m}\}$$

See Fig. E4.5b.

EXAMPLE 4.6

Calculate the nodal displacements and element forces for the beam of Example 4.3 if it is supported as shown in Fig. E4.6.

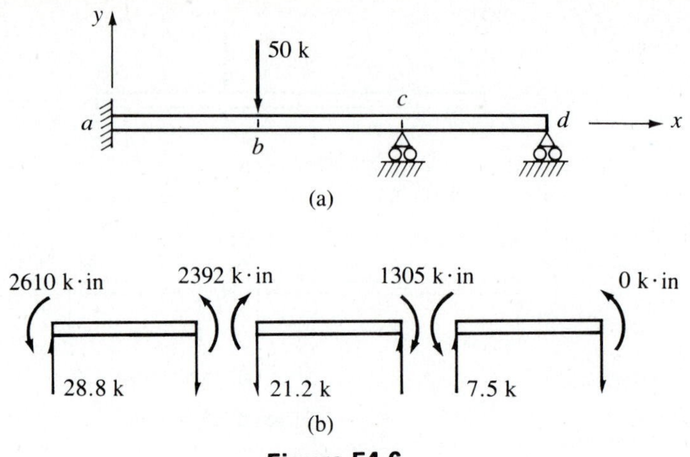

Figure E4.6

Solution

See Example 4.3 for dimensions, etc. From Example 4.3 (using appropriate entries from rows and columns 3, 4, 6, and 8 of **K**),

$$\begin{bmatrix} -50 \\ 0 \\ 0 \\ 0 \end{bmatrix} = 10^2 \begin{bmatrix} 1.58542 & 0000.0 & 68.966 & 0000.0 \\ 0000.0 & 16000.0 & 4000.0 & 0000.0 \\ 68.966 & 4000.0 & 16000.0 & 4000.0 \\ 0000.0 & 0000.0 & 4000.0 & 8000.0 \end{bmatrix} \begin{bmatrix} V_b \\ \Theta_b \\ \Theta_c \\ \Theta_d \end{bmatrix}$$

Thus

$$\begin{bmatrix} V_b \\ \Theta_b \\ \Theta_c \\ \Theta_d \end{bmatrix} = \begin{bmatrix} -0.40999 \text{ in.} \\ -0.00054 \text{ rad} \\ 0.00218 \text{ rad} \\ -0.00109 \text{ rad} \end{bmatrix}$$

Using Eq. (4.48) and the displacements, the element forces are

$$\begin{bmatrix} V \\ M_a \\ M_b \end{bmatrix}^{ab} = 10^2 \begin{bmatrix} 0.79271 & 68.966 & -0.79271 & 68.966 \\ 68.966 & 4000.0 & -68.966 & 2000.0 \\ 68.966 & 2000.0 & -68.966 & 4000.0 \end{bmatrix} \begin{bmatrix} V_a \\ \Theta_a \\ V_b \\ \Theta_b \end{bmatrix}$$

$$= \{28.8 \text{ k} \quad 2610 \text{ k} \cdot \text{in.} \quad 2392 \text{ k} \cdot \text{in.}\}$$

$$\{V \quad M_b \quad M_c\}^{bc} = \{-21.2 \text{ k} \quad -2392 \text{ k} \cdot \text{in.} \quad -1305 \text{ k} \cdot \text{in.}\}$$

$$\{V \quad M_c \quad M_d\}^{cd} = \{7.5 \text{ k} \quad 1305 \text{ k} \cdot \text{in.} \quad 0 \text{ k} \cdot \text{in.}\}$$

See Fig. E4.6b.

EXAMPLE 4.7

Calculate the nodal displacements and element forces for the beam of Example 4.3 if it is supported as shown in Fig. E4.7.

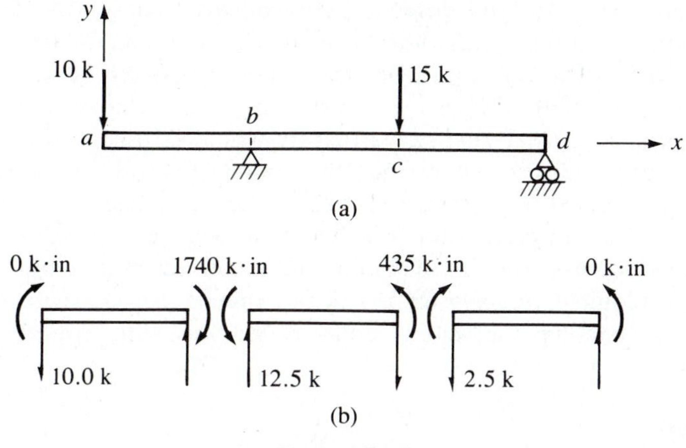

Figure E4.7

Solution

See Example 4.3 for dimensions, etc. From Example 4.3 (using appropriate entries from rows and columns 1, 2, 4, 5, 6, and 8 of **K**),

$$\begin{bmatrix} -10 \\ 0 \\ 0 \\ -15 \\ 0 \\ 0 \end{bmatrix} = 10^2 \begin{bmatrix} 0.79271 & 68.966 & 68.966 & 0000.0 & 0000.0 & 0000.0 \\ 68.966 & 8000.0 & 4000.0 & 0000.0 & 0000.0 & 0000.0 \\ 68.966 & 4000.0 & 16000.0 & -68.966 & 4000.0 & 0000.0 \\ 0000.0 & 0000.0 & -68.966 & 1.58542 & 0000.0 & 68.966 \\ 0000.0 & 0000.0 & 4000.0 & 0000.0 & 16000.0 & 4000.0 \\ 0000.0 & 0000.0 & 0000.0 & 68.966 & 4000.0 & 8000.0 \end{bmatrix} \begin{bmatrix} V_a \\ \Theta_a \\ \Theta_b \\ V_c \\ \Theta_c \\ \Theta_d \end{bmatrix}$$

Thus

$$\begin{bmatrix} V_a \\ \Theta_a \\ \Theta_b \\ V_c \\ \Theta_c \\ \Theta_d \end{bmatrix} = \begin{bmatrix} -0.94612 \text{ in.} \\ 0.00689 \text{ rad} \\ 0.00254 \text{ rad} \\ 0.00000 \text{ in.} \\ -0.00073 \text{ rad} \\ 0.00036 \text{ rad} \end{bmatrix}$$

Using Eq. (4.48) and the displacements, the element forces are

$$\begin{aligned} \{V \quad M_a \quad M_b\}^{ab} &= \{-10.0 \text{ k} \quad 0 \text{ k}\cdot\text{in.} \quad -1740 \text{ k}\cdot\text{in.}\} \\ \{V \quad M_b \quad M_c\}^{bc} &= \{12.5 \text{ k} \quad 1740 \text{ k}\cdot\text{in.} \quad 435 \text{ k}\cdot\text{in.}\} \\ \{V \quad M_c \quad M_d\}^{cd} &= \{-2.5 \text{ k} \quad -435 \text{ k}\cdot\text{in.} \quad 0 \text{ k}\cdot\text{in.}\} \end{aligned}$$

See Fig. E4.7b.

4.3 EQUIVALENT NODAL FORCES

The beam element described in the previous sections can be used to analyze structures in which the various segments display constant shear and linearly varying bending moment. The beam in Fig. 4.6a can be analyzed using three beam elements (one element between each pair of the four labeled nodal points). It appears that the beam in Fig. 4.6b cannot be analyzed using the beam element. Using a single beam element between nodes b and c violates the condition that the element can only characterize constant shear. One approach to analyzing this structure would be to use several beam elements between nodes b and c; thus the actual linear shear distribution would be approximated by a series of constant shear segments. The more elements that are used, the closer the true behavior will be approximated. This method would increase the size of the structural stiffness matrix and the computational time required to solve for the displacements. An alternate approach involves approximating the actual uniform loading with concentrated forces and

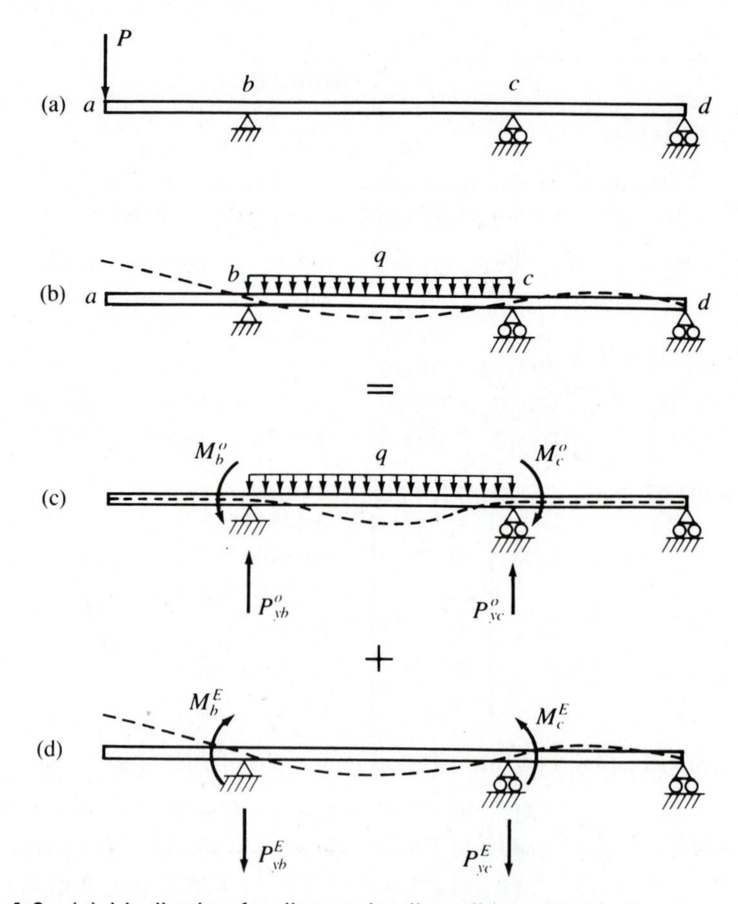

Figure 4.6 (a) Idealization for discrete loading; (b) to (d) idealization and solution strategy for distributed loading

moments at nodal points b and c; these equivalent forces must yield the same nodal displacements as the uniform loading. Thus a single element could be used between nodes b and c, and the analysis would be more efficient than the first solution strategy described.

Equivalent nodal forces can be envisaged in two ways. That is, they are either: (a) those forces required to initially constrain all nodal points of the loaded structure from moving (*fixed-end forces*) or (b) the forces that are equivalent in an energy sense to the loads between nodal points (*consistent forces*). Both approaches yield the same nodal displacements and element forces.

4.3.1 FIXED-END FORCES The solution of Fig. 4.6b with a uniformly distributed load can be visualized as a three-step process as follows:

1. Apply initial fixed-end forces to element bc to prevent displacements and rotations at all nodal points when the applied uniform load is on the beam. Note that these initial forces produce shear and bending moment in element bc.
2. Remove the actual distributed loading plus the initial forces used in step 1 and apply forces at nodal points b and c that are equal in magnitude but opposite in direction to the initial fixed-end forces applied in step 1. Calculate the nodal displacements. Note that these applied forces will introduce displacements throughout the structure.
3. Superpose the results from steps 1 and 2 to obtain the actual solution for the uniformly loaded structure.

This procedure for linearly elastic response is similar to that used in Ch. 2 for initial forces on trusses resulting from thermal effects, fabrication errors, etc. Since this approach is based on the principle of superposition, its validity need not be discussed in greater detail. Note that the nodal displacements for the substitute loading will closely approximate the actual loading, but the shears and moments along the element with uniform loading will be greatly in error. Calculated element forces for other parts of the structure will usually closely approximate actual internal forces.

This three-step procedure can be concisely written using the matrix notation of the stiffness method. The initial fixed-end forces for each individual beam element relate to the displacements through the element stiffness matrix

$$\bar{\mathbf{p}} = \bar{\mathbf{k}}\bar{\mathbf{u}} + \bar{\mathbf{p}}^o \tag{4.61}$$

where $\bar{\mathbf{p}}^o$ is the column matrix containing the initial fixed-end forces in the usual order, that is, $\bar{\mathbf{p}}^o = \{\bar{p}^o_{\bar{y}i} \quad \bar{m}_i \quad \bar{p}^o_{\bar{y}j} \quad \bar{m}_j\}$. For element bc of the beam in Fig. 4.6b these forces can be obtained using any of the methods of classical structural analysis, and for a beam of length L they are

$$\bar{m}^o_b = -\bar{m}^o_c = \frac{1}{12}qL^2 \qquad \text{and} \qquad \bar{p}^o_{yb} = \bar{p}^o_{yc} = \frac{1}{2}qL$$

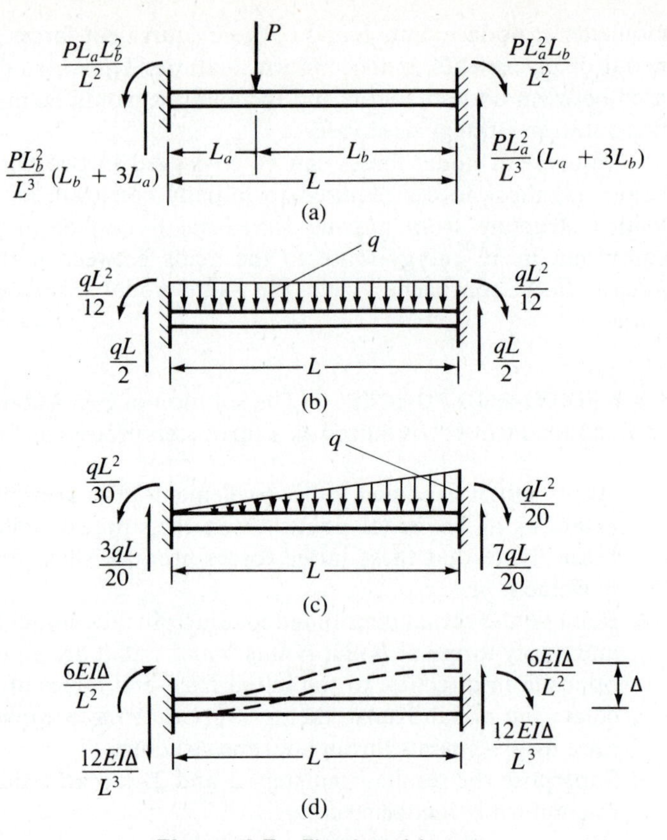

Figure 4.7 Fixed-end forces

Fixed-end forces for several common loadings are shown in Fig. 4.7; you are encouraged to derive these results using basic structural analysis concepts. For element *bc* of Fig. 4.6c,

$$\bar{\mathbf{p}}^{o(bc)} = \{qL/2 \quad qL^2/12 \quad qL/2 \quad -qL^2/12\}$$

For a structure requiring fixed-end forces for several elements, the initial fixed-end forces for the individual elements are combined at the nodal points when enforcing equilibrium for the assemblage. This process gives the matrix equation for the structure

$$\mathbf{P} = \mathbf{K}\mathbf{U} + \mathbf{P}^o \tag{4.62}$$

where $\mathbf{P}^o$ is the column matrix containing all the initial fixed-end forces referred to in step 1 of the solution procedure. This matrix equation can be written in partitioned form as

$$\begin{bmatrix} \mathbf{P}_f \\ \hline \mathbf{P}_s \end{bmatrix} = \left[\begin{array}{c|c} \mathbf{K}_{ff} & \mathbf{K}_{fs} \\ \hline \mathbf{K}_{sf} & \mathbf{K}_{ss} \end{array}\right] \begin{bmatrix} \mathbf{U}_f \\ \hline \mathbf{U}_s \end{bmatrix} + \begin{bmatrix} \mathbf{P}_f^o \\ \hline \mathbf{P}_s^o \end{bmatrix} \tag{4.63}$$

For the loaded structure in Fig. 4.6b this equation becomes

$$\begin{bmatrix} P_{ya}=0 \\ M_a=0 \\ M_b=0 \\ M_c=0 \\ M_d=0 \\ \hline P_{yb} \\ P_{yc} \\ P_{yd} \end{bmatrix} = \left[\begin{array}{c|c} \mathbf{K}_{ff} & \mathbf{K}_{fs} \\ \hline \mathbf{K}_{sf} & \mathbf{K}_{ss} \end{array}\right] \begin{bmatrix} V_a \\ \Theta_a \\ \Theta_b \\ \Theta_c \\ \Theta_d \\ \hline V_b=0 \\ V_c=0 \\ V_d=0 \end{bmatrix} + \begin{bmatrix} P^o_{ya}=0 \\ M^o_a=0 \\ M^o_b \\ M^o_c \\ M^o_d=0 \\ \hline P^o_{yb} \\ P^o_{yc} \\ P^o_{yd}=0 \end{bmatrix} \tag{4.63a}$$

Equation (4.62) can also be rewritten as

$$\mathbf{P} - \mathbf{P}^o = \mathbf{P} + \mathbf{P}^E = \mathbf{KU} \tag{4.64}$$

The physical interpretation of subtracting $\mathbf{P}^o$ from both sides of the equation is equivalent to applying the initial fixed-end forces (those required to maintain zero displacement at the nodes) to the structure in the opposite direction (step 2). Therefore, these forces are sometimes referred to as the *equivalent forces*; they are contained in $\mathbf{P}^E$. For the structure of Fig. 4.6b,

$$\mathbf{P} + \mathbf{P}^E = \{0 \quad 0 \quad -qL^2/12 \quad qL^2/12 \quad 0 \quad (P_{yb} - qL/2) \quad (P_{yc} - qL/2) \quad P_{yd}\}$$

After calculating the unknown displacements $\mathbf{U}_f$ using Eq. (4.63), the element forces can be found according to step 3 in the solution procedure, in which the initial fixed-end forces are combined with the forces produced by the nodal displacements. This superposition is attained by using the complete force-displacement relationship for the elements [see Eq. (4.61)]; that is, the forces defined by $\bar{\mathbf{k}}\bar{\mathbf{u}}$ are those from step 2 of the solution process and the forces $\bar{\mathbf{p}}^o$ are those associated with step 1 of the solution (see Fig. 4.6b). The element forces can also be calculated using the element force-transformation matrix $\bar{\mathbf{e}}$ and the corresponding initial forces to give

$$\mathbf{s}^{ij} = \bar{\mathbf{e}}\bar{\mathbf{u}}^{ij} + \bar{\mathbf{s}}^o \tag{4.65}$$

where the initial shear $\bar{p}^o_{yi}$ and the initial moments m^o_i and $\bar{m}^o_j$ are combined into $\bar{\mathbf{s}}^o$ so that

$$\bar{\mathbf{s}}^o = \{\bar{p}^o_{\bar{y}i} \quad \bar{m}^o_i \quad \bar{m}^o_j\} \tag{4.66}$$

A uniformly loaded beam is analyzed in Example 4.8 to illustrate the use of initial fixed-end forces.

EXAMPLE 4.8

Use fixed-end forces to calculate the displacements and element forces for the uniformly loaded beam in Fig. E4.8. EI = constant.

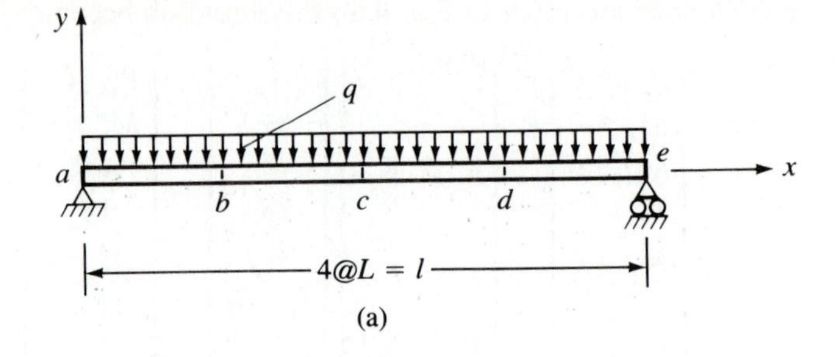

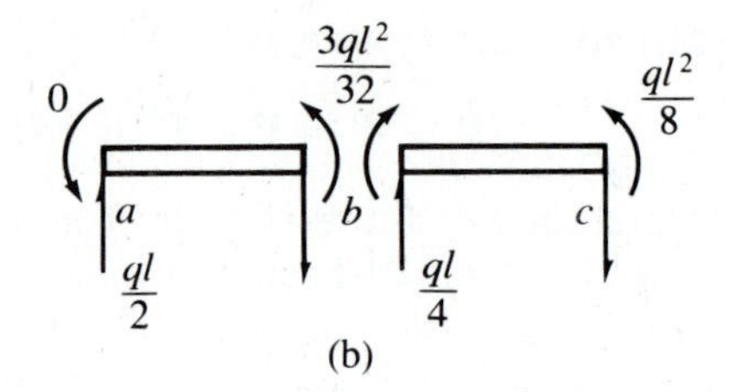

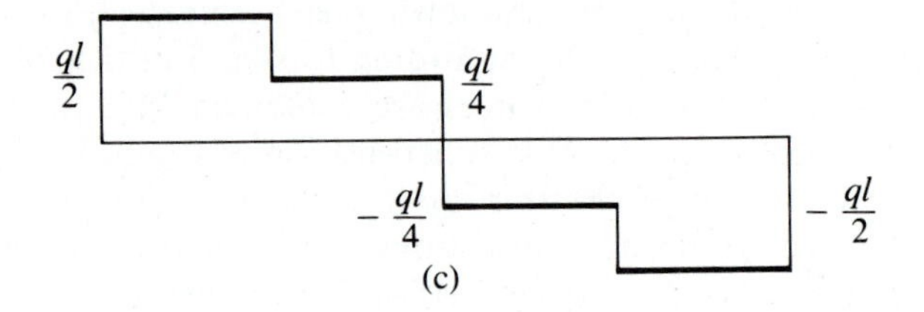

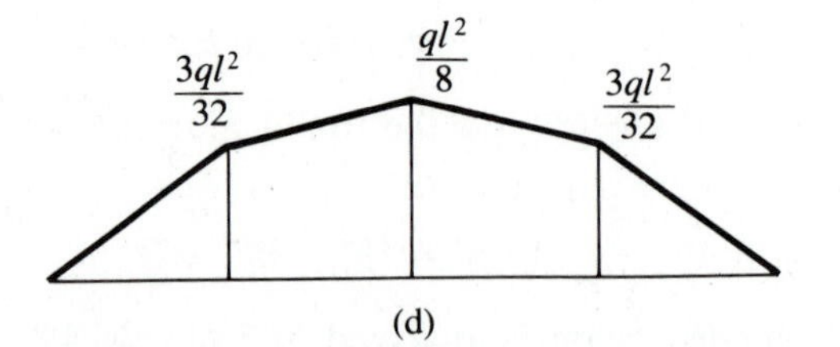

Figure E4.8 (a) The beam with uniform loading; (b) free-body diagrams of *ab* and *bc*; (c) shear diagram; (d) moment diagram

Solution

For all elements $\mathbf{k} = \bar{\mathbf{k}}$,

$$\mathbf{K}_{ff} = \frac{EI}{L}\begin{bmatrix} 4 & -\frac{6}{L} & 2 & 0 & 0 & 0 & 0 & 0 \\ -\frac{6}{L} & \frac{24}{L^2} & 0 & -\frac{12}{L^2} & \frac{6}{L} & 0 & 0 & 0 \\ 2 & 0 & 8 & -\frac{6}{L} & 2 & 0 & 0 & 0 \\ 0 & -\frac{12}{L^2} & -\frac{6}{L} & \frac{24}{L^2} & 0 & -\frac{12}{L^2} & \frac{6}{L} & 0 \\ 0 & \frac{6}{L} & 2 & 0 & 8 & -\frac{6}{L} & 2 & 0 \\ 0 & 0 & 0 & -\frac{12}{L^2} & -\frac{6}{L} & \frac{24}{L^2} & 0 & \frac{6}{L} \\ 0 & 0 & 0 & \frac{6}{L} & 2 & 0 & 8 & 2 \\ 0 & 0 & 0 & 0 & 0 & \frac{6}{L} & 2 & 4 \end{bmatrix} \quad P_f^o = qL\begin{bmatrix} \frac{L}{12} \\ 1 \\ 0 \\ 1 \\ 0 \\ 1 \\ 0 \\ -\frac{L}{12} \end{bmatrix}$$

Solving the force-displacement equations and noting that $L = l/4$ gives

$$\Theta_a = -\Theta_e = -\frac{8qL^3}{3EI} = -\frac{ql^3}{24EI}$$

$$\Theta_b = -\Theta_d = -\frac{11qL^3}{6EI} = -\frac{11ql^3}{384EI}$$

$$\Theta_c = 0$$

$$V_b = V_d = -\frac{19qL^4}{8EI} = -\frac{19ql^4}{2048EI}$$

$$V_c = -\frac{10qL^4}{3EI} = -\frac{5ql^4}{384EI}$$

Element forces are obtained using Eq. (4.65). Thus

$$\begin{bmatrix} \bar{p}_{ya} \\ \bar{m}_a \\ \bar{m}_b \end{bmatrix}^{ab} = \frac{EI}{L}\begin{bmatrix} \frac{12}{L^2} & \frac{6}{L} & -\frac{12}{L^2} & \frac{6}{L} \\ \frac{6}{L} & 4 & -\frac{6}{L} & 2 \\ \frac{6}{L} & 2 & -\frac{6}{L} & 4 \end{bmatrix} \frac{qL^3}{EI}\begin{bmatrix} 0 \\ -\frac{8}{3} \\ -\frac{19L}{8} \\ -\frac{11}{6} \end{bmatrix} + \begin{bmatrix} \frac{qL}{2} \\ \frac{qL^2}{12} \\ -\frac{qL^2}{12} \end{bmatrix}$$

$$= \begin{bmatrix} 2qL \\ 0 \\ \frac{3qL^2}{2} \end{bmatrix}$$

$$\{\bar{p}_{yb} \quad \bar{m}_b \quad \bar{m}_c\}^{bc} = \{qL \quad -3qL^2/2 \quad 2qL^2\}$$

The structure is symmetrical; therefore $s^{cd} = s^{bc}$ and $s^{de} = \bar{s}^{ab}$. The element forces are shown in Fig. E4.8; note that $L = l/4$. The shear and moment diagrams are plotted in Fig. E4.8c and d, respectively.

Discussion The matrix of initial forces $\mathbf{P}^o$ contains zero entries for the moments at b, c, and d since the left and right beam element at a typical node contribute equal fixed-end moments that are opposite in sign. Because the matrix $\mathbf{P}_f$ is null, the first matrix equation of Eq. (4.63) is as follows:

$$\mathbf{0} = \mathbf{K}_{ff}\mathbf{U}_f + \mathbf{P}_f^o$$

Since the displacements and rotations obtained at the nodes are exact, this idealization using four beam elements appears sufficient for these calculations.

The element forces are calculated by the superposition of initial fixed-end forces and the forces resulting from the displacements [see Eq. (4.65)]. The true straight-line shear diagram is approximated by the step function shown, but the value of the shear is exact at both ends of each element. The actual moment diagram is a parabola, which is approximated with straight-line segments. The values of the moments at the nodes are exact.

If equivalent forces are used, the nodal displacements obtained will be exact regardless of the number of beam elements. The element end moments will also be exact, but in general there will be a discontinuity in the shear from element to element (across the nodal points). In addition, for each element the moment is linear and the shear is constant; therefore, it is necessary to use enough beam elements to approximate the exact moment and shear diagrams closely.

4.3.2 CONSISTENT FORCES For elements that are loaded between nodal points such as bc of the structure in Fig. E4.6b, the equivalent nodal loads can be calculated using the theorem of virtual work. These loads are equivalent to those obtained as fixed-end forces, but the motivation for understanding the approach requires less visualization.

For the uniformly loaded beam segment in Fig. 4.8a, the virtual work of the loads is

$$\delta W_e = \int_0^L \boldsymbol{\delta}\mathbf{u}^T q \, dx = \int_0^L \boldsymbol{\delta}\bar{\mathbf{u}}^T \mathbf{N}^T q \, dx \qquad (4.67)$$

where the substitution $\boldsymbol{\delta}\mathbf{u} = \mathbf{N}\,\boldsymbol{\delta}\bar{\mathbf{u}}$ comes from Eq. (3.39). The virtual work of the nodal loads in Fig. 4.8b is

$$\delta W_e = \boldsymbol{\delta}\bar{\mathbf{u}}^T \bar{\mathbf{p}}^E \qquad (4.68)$$

where $\bar{\mathbf{p}}^E = \{\bar{p}_{\bar{y}i} \quad \bar{m}_i \quad \bar{p}_{\bar{y}j} \quad \bar{m}_j\}$. The shape functions for the beam element are shown in Eq. (4.23). Equating Eqs. (4.67) and (4.68) and noting that $\boldsymbol{\delta}\bar{\mathbf{u}}$ is

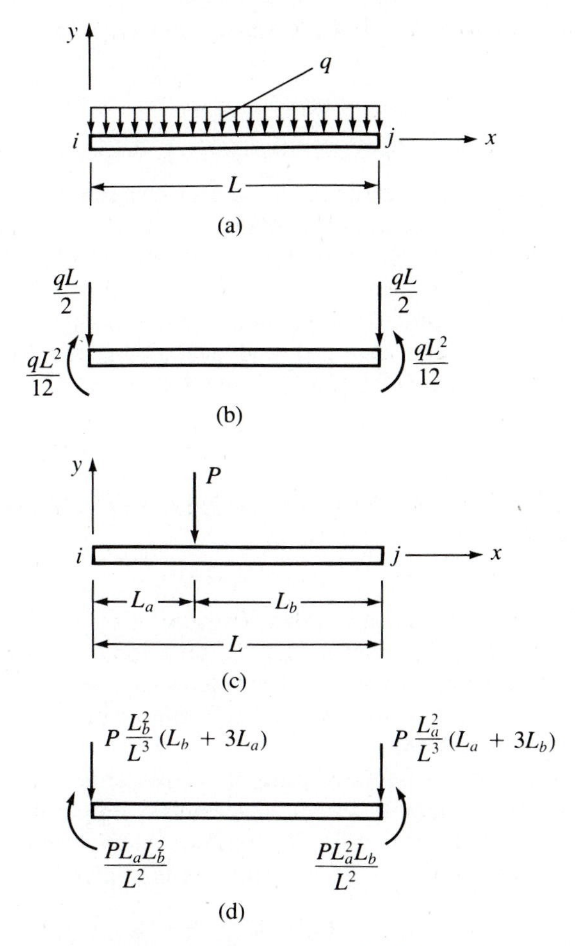

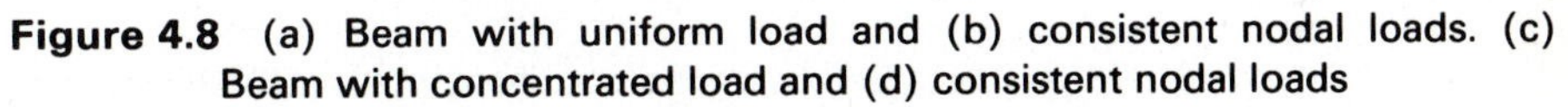
Figure 4.8 (a) Beam with uniform load and (b) consistent nodal loads. (c) Beam with concentrated load and (d) consistent nodal loads

nonzero and arbitrary yields

$$\bar{\mathbf{p}}^E = \int_0^L \mathbf{N}^T q \, dx \tag{4.69}$$

Substituting the shape functions for the beam element [Eq. (4.23)] in Eq. (4.69) we have

$$\bar{\mathbf{p}}^E = \{-qL/2 \quad -qL^2/12 \quad -qL/2 \quad qL^2/12\} \tag{4.70}$$

These equivalent forces are illustrated in Fig. 4.8b.

For the beam with a single concentrated force in Fig. 4.8c, the virtual work of the load is

$$\delta W_e = \delta u(x = L_a)\, P = \boldsymbol{\delta}\bar{\mathbf{u}}^T\, \mathbf{N}^T(x = L_a)\, P \tag{4.71}$$

Equating Eqs. (4.68) and (4.71) and noting that $\delta\bar{\mathbf{u}}$ is nonzero and arbitrary yields

$$\begin{aligned}\bar{\mathbf{p}}^E &= \{\bar{p}_{yi} \quad \bar{m}_i \quad \bar{p}_{yj} \quad \bar{m}_j\} = \mathbf{N}^T(x = L_a)P \\ &= P\{-L_b^2(L_b + 3L_a)/L^3 \quad -L_a L_b^2/L^2 \quad -L_a^2(L_a + 3L_b)/L^3 \quad L_a^2 L_b/L^2\} \end{aligned} \tag{4.72}$$

Verification of these results is left as an exercise for the reader. These equivalent forces are illustrated in Fig. 4.8d.

The equivalent forces in Eqs. (4.70) and (4.72), which are illustrated in Figs. 4.8b and 4.8d, are different from the fixed-end forces derived in the previous section. We note from Eq. (4.64) that equivalent forces are the fixed-end forces with the sign changed and placed on the load side of the force-displacement equation (cf. Figs. 4.7 and 4.8).

4.4 SETTLEMENT, INITIAL, AND THERMAL STRAINS

The calculations for beams with self-straining are similar in form to those used for trusses in Sec. 2.7; refer to the computational algorithm in Fig. 2.13. Alternatively, this general category of problems can be solved using the concepts of initial strains as described in Sec. 3.7. Both approaches will be used in the subsequent discussion.

A beam with differential support displacements can be analyzed by partitioning the structural force-displacement equations [see Eq. (2.31)] and noting that $\mathbf{U}_s$ is not null. The unknown displacements are computed by solving the first of the partitioned equations to give

$$\mathbf{U}_f = \mathbf{K}_{ff}^{-1}(\mathbf{P}_f - \mathbf{K}_{fs}\mathbf{U}_s) \tag{4.73}$$

The second of Eq. (2.31) yields the forces associated with the prescribed displacements, that is,

$$\mathbf{P}_s = \mathbf{K}_{sf}\mathbf{U}_f + \mathbf{K}_{ss}\mathbf{U}_s \tag{4.74}$$

The element forces are calculated using Eq. (4.48) in the usual fashion. Generally, the calculations in Eq. (4.74) are not carried out explicitly since the global forces associated with prescribed displacements can be obtained by combining the appropriate element forces.

Beam structures with settlement can also be analyzed using initial forces; the approach is similar to the three-step procedure outlined in Sec. 4.3.1. First, nodal forces are imposed on each element to prevent all but the known displacements from occurring. Second, these forces are applied in the opposite direction and nodal displacements are calculated. Third, element forces are computed by combining the forces resulting from the two previous loadings. This approach is summarized by Eqs. (4.62), (4.64), and (4.65); that is,

$$\mathbf{P} = \mathbf{KU} + \mathbf{P}^o$$

The partitioned equations are used to solve for the unknown displacements $\mathbf{U}_f$ in the usual fashion:

$$\mathbf{U}_f = \mathbf{K}_{ff}^{-1}(\mathbf{P}_f - \mathbf{P}_f^o)$$

Finally the element forces are computed with the equation

$$\mathbf{s}^{ij} = \bar{\mathbf{e}}\bar{\mathbf{u}}^{ij} + \bar{\mathbf{s}}_o$$

that is, Eq. (4.65). The approach is illustrated in Example 4.9.

EXAMPLE 4.9

The beam of Example 4.2 is supported as shown in Fig. E4.9. The support at b settles 20 mm. Compute the displacements and element forces.

Solution

If all degrees of freedom are constrained with $V_b = -20$ mm, the structure assumes the configuration in Fig. E4.9b. The fixed-end forces for each of the elements are computed using the expressions in Fig. 4.7d. Thus

$$M_a^{o(ab)} = M_b^{o(ab)} = \frac{6EI\Delta}{L^2} = \frac{6(200 \times 10^6 \text{ kPa})(240 \times 10^{-6} \text{ m}^4)(0.020 \text{ m})}{6^2}$$

$$= 160 \text{ kN}\cdot\text{m}$$

$$P_{ya}^{o(ab)} = -P_{yb}^{o(ab)} = 53.33 \text{ kN}$$

$$M_b^{o(bc)} = M_c^{o(bc)} = -\frac{6EI\Delta}{L^2} = -\frac{6(200 \times 10^6 \text{ kPa})(150 \times 10^{-6} \text{ m}^4)(0.020 \text{ m})}{3^2}$$

$$= -400 \text{ kN}\cdot\text{m}$$

$$P_{yb}^{o(bc)} = -P_{yc}^{o(bc)} = -266.67 \text{ kN}$$

The structural force-displacement equations with boundary conditions imposed are (see the equations in Example 4.2)

$$\begin{bmatrix} 0 \\ 0 \end{bmatrix} = 10^3 \begin{bmatrix} 72 & 20 \\ 20 & 40 \end{bmatrix} \begin{bmatrix} \Theta_b \\ \Theta_c \end{bmatrix} + \begin{bmatrix} -240 \\ -400 \end{bmatrix}$$

Solving yields

$$\begin{bmatrix} \Theta_b \\ \Theta_c \end{bmatrix} = 10^{-3} \begin{bmatrix} 0.64516 \\ 9.6774 \end{bmatrix} \text{rad}$$

Using Eqs. (4.48) and (4.65) the element forces are

$$\begin{bmatrix} V \\ M_a \\ M_b \end{bmatrix}^{ab} = 8000 \begin{bmatrix} 1 & 1 \\ 4 & 2 \\ 2 & 4 \end{bmatrix} 10^{-3} \begin{bmatrix} 0.00000 \\ 0.64516 \end{bmatrix} + \begin{bmatrix} 53.33 \\ 160 \\ 160 \end{bmatrix}$$

$$= \begin{bmatrix} 58.5 \text{ kN} \\ 170.3 \text{ kN}\cdot\text{m} \\ 180.6 \text{ kN}\cdot\text{m} \end{bmatrix}$$

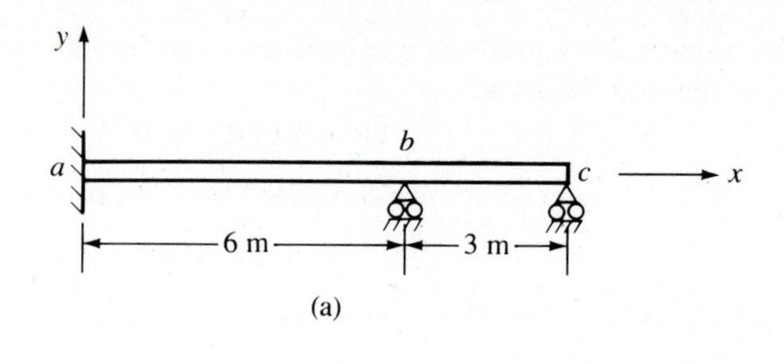

(a)

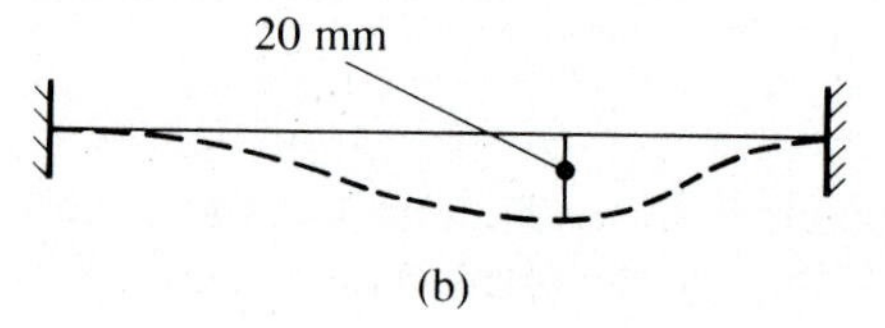

(b)

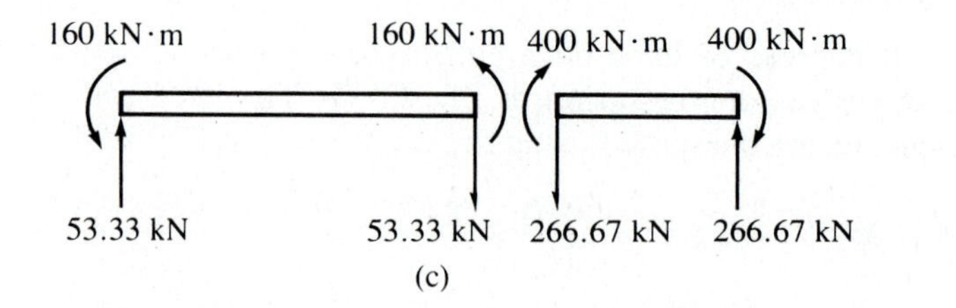

(c)

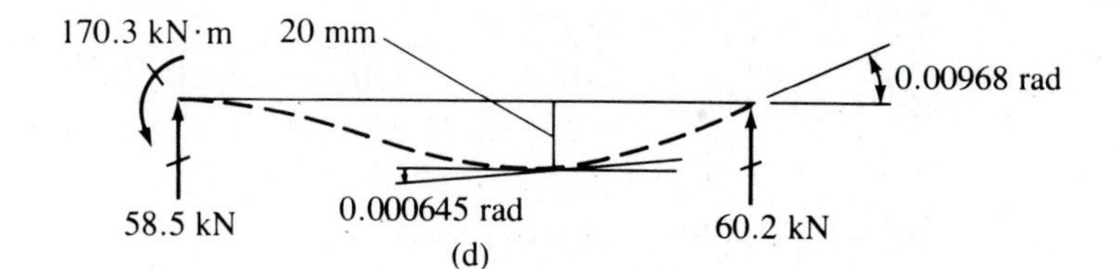

(d)

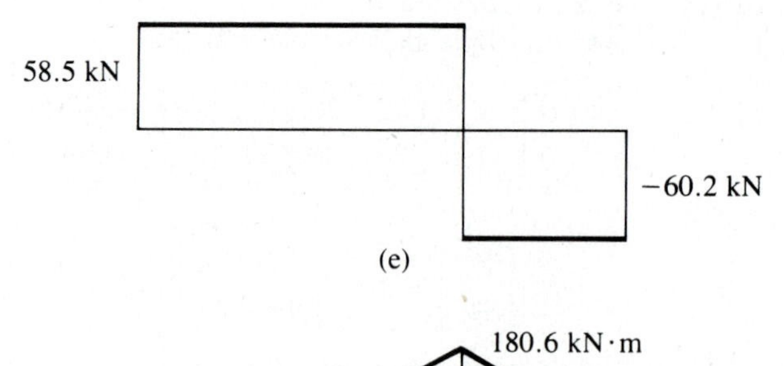

(e)

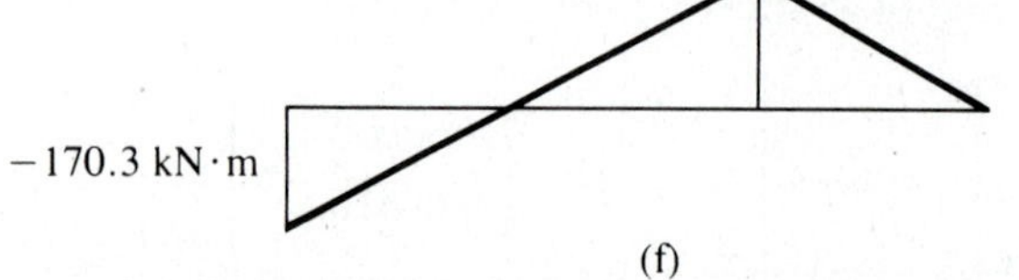

(f)

Figure E4.9 (a) Beam; (b) deflections with fixed-end moments; (c) free-body diagram with fixed-end moments; (d) final deflections; (e) shear diagram; (f) moment diagram

$$\begin{bmatrix} V \\ M_b \\ M_c \end{bmatrix}^{bc} = 10{,}000 \begin{bmatrix} 2 & 2 \\ 4 & 2 \\ 2 & 4 \end{bmatrix} 10^{-3} \begin{bmatrix} 0.64516 \\ 9.6774 \end{bmatrix} + \begin{bmatrix} -266.7 \\ -400 \\ -400 \end{bmatrix}$$

$$= \begin{bmatrix} -60.2 \text{ kN} \\ -180.6 \text{ kN} \cdot \text{m} \\ 0 \text{ kN} \cdot \text{m} \end{bmatrix}$$

The reactions can be calculated from the element forces. The reactions and deflected shape are illustrated in Fig. E4.9d, and the shear and moment diagrams appear in Fig. E4.9e and f, respectively.

The method of initial forces can also be used when the structure is subjected to thermal effects, fabrication errors, precambering, etc. [see Eqs. (4.62), (4.64), and (4.65)]. These self-straining phenomena can be envisaged as initial-force problems in the context of the equations; they do not all occur chronologically from the inception of a structure.

The fixed-end forces for a heated beam element are applied at the nodes to give zero rotations and displacements. These can be calculated using either basic considerations or the theorem of virtual work; both approaches will be demonstrated. Consider the beam element in Fig. 4.9 that is heated uniformly along its length with a thermal distribution that varies linearly from the lower to the upper surface of the cross section. The temperature at any position on the cross section is

$$T(y) = T_m + \frac{\Delta T y}{h} \tag{4.75}$$

where

$$T_m = \tfrac{1}{2}(T_l + T_u) \qquad \text{and} \qquad \Delta T = T_u - T_l$$

Thus the initial strain and stress are, respectively,

$$\varepsilon^o = \alpha T(y) \tag{4.76a}$$

and

$$\sigma^o = E\varepsilon^o = E\alpha T(y) \tag{4.76b}$$

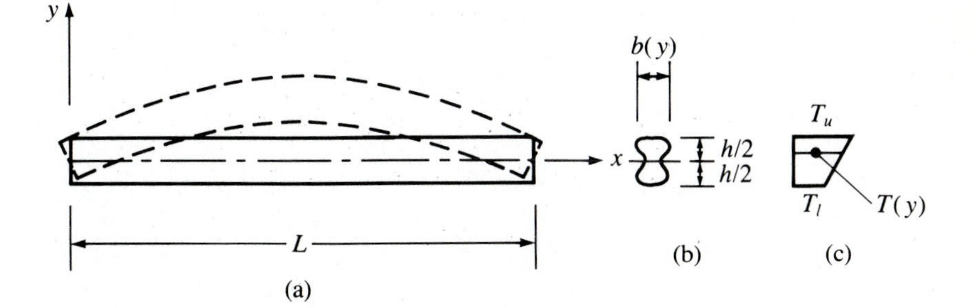

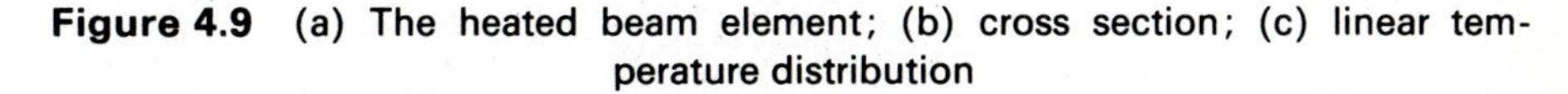
Figure 4.9 (a) The heated beam element; (b) cross section; (c) linear temperature distribution

The axial force required at each end of the element to maintain the element in an undeformed position is

$$P_x^o = \int_{-h/2}^{+h/2} \sigma^o b \, dy = \int_{-h/2}^{+h/2} E\alpha\left(T_m + \frac{\Delta T}{h} y\right) b \, dy = AE\alpha T_m \tag{4.77}$$

The moment that must be applied at the ends of the element to prevent rotation of the cross section is

$$M^o = \int_{-h/2}^{+h/2} E\alpha T(y) b y \, dy = E\alpha \int_{-h/2}^{+h/2} \left(T_m + \frac{\Delta T}{h} y\right) b y \, dy = \frac{EI\alpha\Delta T}{h} \tag{4.78}$$

The beam would deform into the shape in Fig. 4.9 with a constant curvature; therefore, there will be neither rotations nor displacements if M^o is applied to the left and right ends in a clockwise and counterclockwise direction, respectively. The axial force P_y^o will be neglected, since it is assumed that the neutral axis remains undeformed for conventional beam theory. Hence, the initial force matrix for the beam element is

$$\bar{\mathbf{p}}^o = \begin{bmatrix} \bar{p}_{yi}^o \\ \bar{m}_i^o \\ \bar{p}_{yj}^o \\ \bar{m}_j^o \end{bmatrix} = \frac{EI\alpha\Delta T}{h} \begin{bmatrix} 0 \\ -1 \\ 0 \\ 1 \end{bmatrix} \tag{4.79}$$

and the corresponding initial force matrix for calculating final element forces is

$$\bar{\mathbf{s}}^o = \begin{bmatrix} \bar{p}_{yi}^o \\ \bar{m}_i^o \\ \bar{m}_j^o \end{bmatrix} = \frac{EI\alpha\Delta T}{h} \begin{bmatrix} 0 \\ -1 \\ 1 \end{bmatrix} \tag{4.80}$$

EXAMPLE 4.10

The beam of Example 4.3 is fixed at nodes a and d. The structure in Fig. E4.10 is uniformly heated along its length with $T_u - T_l = 120°\text{F}$. Calculate the nodal displacements and the element forces. $h = 12$ in.; $\alpha = 6.5 \times 10^{-6}$ in./(in. · °F).

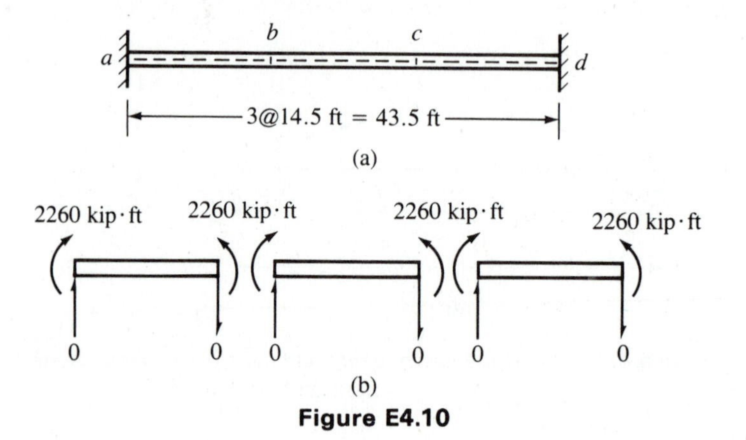

Figure E4.10

Solution

$$M^o = \frac{EI\alpha\Delta T}{h} = \frac{(29 \times 10^3 \text{ ksi})(1200 \text{ in.}^4)[6.5 \times 10^{-6} \text{ in./(in.} \cdot {}^\circ\text{F})](120^\circ\text{F})}{12 \text{ in.}}$$

$$= 2260 \text{ kip} \cdot \text{in.}$$

Hence for each element

$$\begin{bmatrix} \bar{p}^o_{yi} \\ \bar{m}^o_i \\ \bar{p}^o_{yj} \\ \bar{m}^o_j \end{bmatrix} = 2260 \begin{bmatrix} 0 \\ -1 \\ 0 \\ 1 \end{bmatrix}$$

These initial fixed-end forces are merged to give $\mathbf{P}^o_f$, which in this case is a null matrix. Since $\mathbf{P}_f$ is also null, solving the structural force-displacement equations yields $\mathbf{U}_f = \mathbf{0}$. That is, all the nodal points of this structure remain in their undisplaced positions. For all elements

$$\begin{bmatrix} \bar{p} \\ \bar{m}_i \\ \bar{m}_j \end{bmatrix}^{ij} = \bar{\mathbf{e}}\mathbf{0} + \bar{\mathbf{s}}^o = 2260 \text{ k} \cdot \text{in.} \begin{bmatrix} 0 \\ -1 \\ 1 \end{bmatrix}$$

See Fig. E4.10b.

Discussion Initial forces and the procedure embodied in Eqs. (4.62), (4.63), and (4.65) are used. This thermal distribution results in zero displacements, but since the structure is statically indeterminate, bending moments are imposed along its length.

The initial forces for self-straining effects can also be obtained using the theorem of virtual work. In Sec. 3.7 [see Eq. (3.92)] we calculated that

$$\bar{\mathbf{p}}^o = -\int_{\text{vol}} \mathbf{B}^T E\varepsilon^o \, dV + \int_{\text{vol}} \mathbf{B}^T \sigma^o \, dV$$

In the case of a prismatic beam element $\mathbf{B}$ is given in Eq. (4.27). For a thermal distribution that is uniform along the beam axis and varies linearly from the lower to the upper surface (see Fig. 4.9a), ε^o is given by Eqs. (4.75) and (4.76a). Substituting $\mathbf{B}$ and $T(y)$ into Eq. (3.92) yields

$$\bar{\mathbf{p}}^o = -\int_{\text{vol}} y \begin{bmatrix} \bar{B}_{11} \\ \bar{B}_{12} \\ \bar{B}_{13} \\ \bar{B}_{14} \end{bmatrix} E\alpha\left(T_m + \frac{\Delta T y}{h} \right) dV$$

where $B_{ij} = y\bar{B}_{ij}$. All the matrix elements $\bar{B}_{ij}$ are functions of $\bar{x}$ only; therefore this reduces to

$$\bar{\mathbf{p}}^o = -\frac{EI\alpha\Delta T}{h}\int_0^L \begin{bmatrix} \bar{B}_{11} \\ \bar{B}_{12} \\ \bar{B}_{31} \\ \bar{B}_{14} \end{bmatrix} d\bar{x}$$

Integration of the matrix term by term yields a result identical to the fixed-end forces in Eq. (4.79). Example 4.11 is an additional illustration of the analysis of a beam structure subjected to thermal effects.

EXAMPLE 4.11

The beam of Example 4.3 is simply supported at nodes a and d. The structure in Fig. E4.11 is uniformly heated along its length with $T_u - T_l = 120°\text{F}$. Calculate the nodal displacements and the element forces. $h = 12$ in.; $\alpha = 6.5 \times 10^{-6}$ in./(in. · °F).

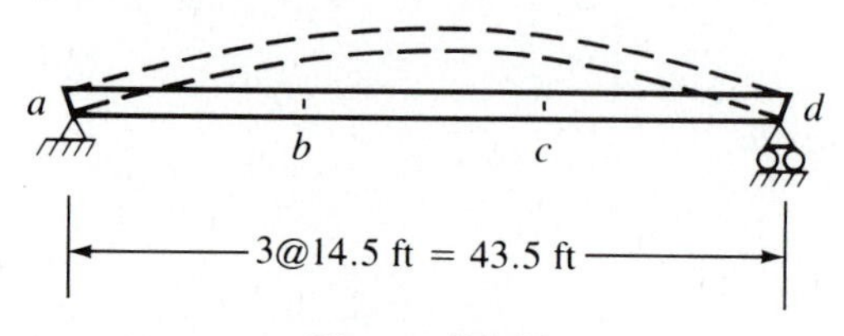

Figure E4.11

Solution

From Example 4.10, $M^o = 2260$ kip · in. and $\bar{\mathbf{p}}^o$ for each of the three elements is the same as shown there. For this simply supported structure, $\mathbf{P}_f^o = \{-2260\ 0\ 0\ 0\ 0\ 2260\}$:

$$\begin{bmatrix} 2260 \\ 0 \\ 0 \\ 0 \\ 0 \\ -2260 \end{bmatrix} = 10^2 \begin{bmatrix} 8000.0 & -68.966 & 4000.0 & 0000.0 & 0000.0 & 0000.0 \\ -68.966 & 1.58542 & 0000.0 & -0.79271 & 68.966 & 0000.0 \\ 4000.0 & 0000.0 & 16000.0 & -68.966 & 4000.0 & 0000.0 \\ 0000.0 & -0.79271 & -68.966 & 1.58542 & 0000.0 & 68.966 \\ 0000.0 & 68.966 & 4000.0 & 0000.0 & 16000.0 & 4000.0 \\ 0000.0 & 0000.0 & 0000.0 & 68.966 & 4000.0 & 8000.0 \end{bmatrix} \begin{bmatrix} \Theta_a \\ V_b \\ \Theta_b \\ V_c \\ \Theta_c \\ \Theta_d \end{bmatrix}$$

Solution gives

$$\begin{bmatrix} \Theta_a \\ V_b \\ \Theta_b \\ V_c \\ \Theta_c \\ \Theta_d \end{bmatrix} = \begin{bmatrix} 0.01695 \text{ rad} \\ 1.966 \text{ in.} \\ 0.00565 \text{ rad} \\ 1.966 \text{ in.} \\ -0.00565 \text{ rad} \\ -0.01695 \text{ rad} \end{bmatrix}$$

The element forces are calculated using Eq. (4.65); note that $\bar{\mathbf{e}}$ is given in Example 4.6. This computation gives

$$\mathbf{s}^{ab} = \mathbf{s}^{bc} = \mathbf{s}^{cd} = \{0 \quad 0 \quad 0\}$$

Discussion This structure is heated like that in the previous example. The same initial-force solution procedure is used. Since this structure is statically determinate, the nonzero displacements are accompanied by zero element forces.

4.5 THE PRISMATIC PLANE FRAME ELEMENT

The beam element can also be used to analyze plane rigid frames, but for these structures the local and global coordinate systems do not generally coincide. This is an additional aspect that we did not encounter in the analysis of beam structures. That is, it is necessary to transform the individual element stiffness matrices and element initial forces into a common global coordinate system prior to invoking nodal equilibrium for the total structure. Also, the nodal displacements must be transformed back to the local coordinate system of each element before the element end forces are calculated.

An additional consideration enters into plane frame analysis: the axial deformations of the elements. The axial stiffness of the beam element in the previous sections has not been considered; we have only included flexural behavior in our calculations of the relationships between the generalized forces and displacements. Classical equilibrium approaches, such as the slope-deflection method and moment distribution, implicitly incorporate the assumption that the members are inextensible along their longitudinal axes. That is, the deformed and undeformed member lengths are considered to be nominally equal, and the governing equations apply to both the initial and displaced geometry. In some situations axial deformations can significantly influence the structural behavior (e.g., the lower stories of high-rise buildings). Both elastic and rigid axial behavior of elements are described in the subsequent sections.

4.5.1 FLEXURAL AND AXIAL DEFORMATIONS The plane frame element in Fig. 4.10 has three degrees of freedom per node and differs from the beam element since axial displacements are included. The element is uniform along its length with a cross-sectional area A, a moment of inertia I, and it is made from a homogeneous, linearly elastic material with a modulus of elasticity E. We assume that small deformation theory is applicable, so that flexural and axial behavior are uncoupled. Thus, second-order phenomena cannot be investigated with this element. Such nonlinear behavior is typified by beam-column action in which large axial loads induce bending moments. The stiffness matrix for this element in local coordinates can be obtained by combining the stiffness matrix of the axial force element [Eq. (2.22)] with that for

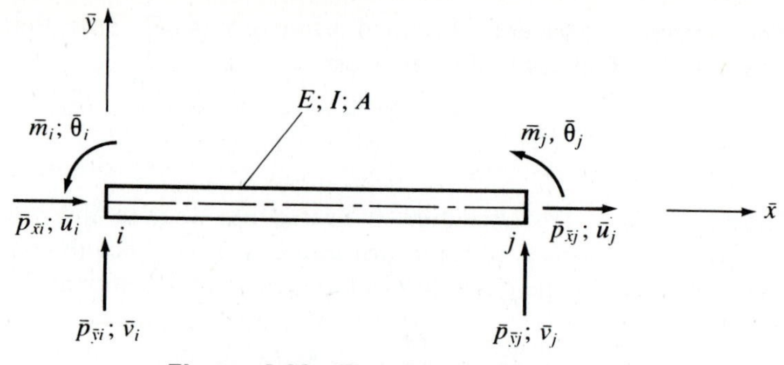

Figure 4.10 The plane frame element

the beam element [Eq. (4.19)]. This yields

$$\begin{bmatrix} \bar{p}_{\bar{x}i} \\ \bar{p}_{\bar{y}i} \\ \bar{m}_i \\ \bar{p}_{\bar{x}j} \\ \bar{p}_{\bar{y}j} \\ \bar{m}_j \end{bmatrix} = \frac{E}{L} \begin{bmatrix} A & 0 & 0 & -A & 0 & 0 \\ 0 & \dfrac{12I}{L^2} & \dfrac{6I}{L} & 0 & -\dfrac{12I}{L^2} & \dfrac{6I}{L} \\ 0 & \dfrac{6I}{L} & 4I & 0 & -\dfrac{6I}{L} & 2I \\ -A & 0 & 0 & A & 0 & 0 \\ 0 & -\dfrac{12I}{L^2} & -\dfrac{6I}{L} & 0 & \dfrac{12I}{L^2} & -\dfrac{6I}{L} \\ 0 & \dfrac{6I}{L} & 2I & 0 & -\dfrac{6I}{L} & 4I \end{bmatrix} \begin{bmatrix} \bar{u}_i \\ \bar{v}_i \\ \bar{\theta}_i \\ \bar{u}_j \\ \bar{v}_j \\ \bar{\theta}_j \end{bmatrix} \tag{4.81}$$

$$\bar{\mathbf{p}} = \bar{\mathbf{k}} \quad \bar{\mathbf{u}}$$

Usually, each node in a plane rigid frame has three independent generalized displacements that must be considered (two translations and one rotation), and usually it is necessary to transform the generalized element forces and displacements into the common global coordinate system. For the beam element in Fig. 4.11 the generalized forces at node i are transformed as follows:

$$\begin{aligned} \bar{p}_{\bar{x}i} &= p_{xi} \cos \phi + p_{yi} \sin \phi \\ \bar{p}_{\bar{y}i} &= -p_{xi} \sin \phi + p_{\bar{y}i} \cos \phi \\ \text{and} \qquad \bar{m}_i &= m_i \end{aligned} \tag{4.82}$$

Similar transformations can be expressed for the generalized forces at node j; thus for both nodes of the element

$$\begin{bmatrix} \bar{p}_{\bar{x}i} \\ \bar{p}_{\bar{y}i} \\ \bar{m}_i \\ \bar{p}_{\bar{x}j} \\ \bar{p}_{\bar{y}j} \\ \bar{m}_j \end{bmatrix} = \begin{bmatrix} \cos\phi & \sin\phi & 0 & 0 & 0 & 0 \\ -\sin\phi & \cos\phi & 0 & 0 & 0 & 0 \\ 0 & 0 & 1 & 0 & 0 & 0 \\ 0 & 0 & 0 & \cos\phi & \sin\phi & 0 \\ 0 & 0 & 0 & -\sin\phi & \cos\phi & 0 \\ 0 & 0 & 0 & 0 & 0 & 1 \end{bmatrix} \begin{bmatrix} p_{xi} \\ p_{yi} \\ m_i \\ p_{xj} \\ p_{yj} \\ m_j \end{bmatrix} \tag{4.83}$$

$$\bar{\mathbf{p}} = \mathbf{T} \quad \mathbf{p}$$

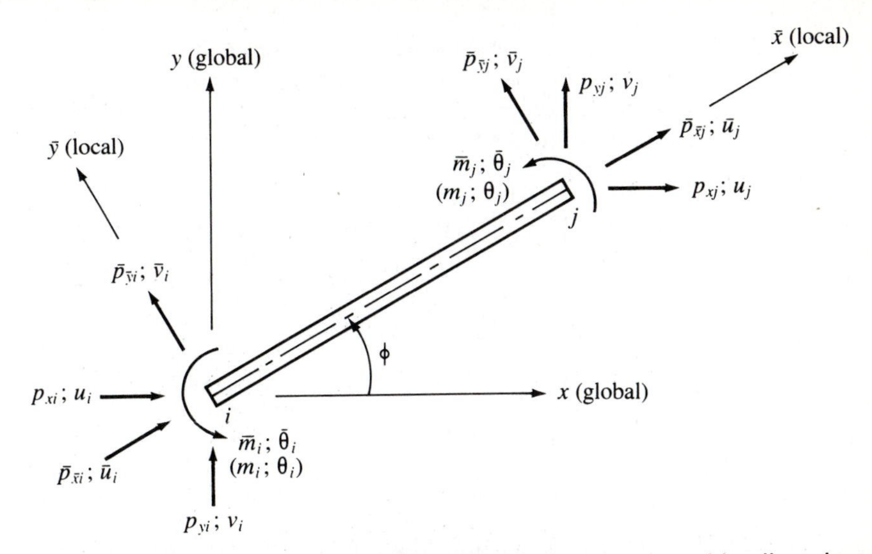

Figure 4.11 Generalized forces and displacements for an arbitrarily oriented beam element

where $\bar{\mathbf{p}}$ = column matrix containing nodal force components in local coordinates

$\mathbf{p}$ = column matrix containing nodal force components in global coordinates

$\mathbf{T}$ = coordinate transformation matrix for the beam element

In Ch. 2 we noted that the transformation matrix used in conjunction with the axial force element [Eq. (2.46)] is orthogonal; this is also true for $\mathbf{T}$ in Eq. (4.83). Thus $\mathbf{T}^T = \mathbf{T}^{-1}$ which implies that

$$\mathbf{p} = \mathbf{T}^T\bar{\mathbf{p}} \tag{4.84}$$

The same transformation can be applied to the corresponding column matrices of the generalized displacements $\bar{\mathbf{u}}$ and $\mathbf{u}$. Thus

$$\bar{\mathbf{u}} = \mathbf{T}\mathbf{u} \tag{4.85}$$

and the transformation from local to global coordinates is

$$\mathbf{u} = \mathbf{T}^T\bar{\mathbf{u}} \tag{4.86}$$

where the generalized nodal displacements in the local and global coordinate systems are represented, respectively, by the column matrices

$$\bar{\mathbf{u}} = \{\bar{u}_i \quad \bar{v}_i \quad \bar{\theta}_i \quad \bar{u}_j \quad \bar{v}_j \quad \bar{\theta}_j\}$$

and

$$\mathbf{u} = \{u_i \quad v_i \quad \theta_i \quad u_j \quad v_j \quad \theta_j\} \tag{4.87}$$

The element stiffness matrix in local coordinates can be transformed into the global coordinate system using Eq. (2.55) [also shown in Ch. 3 as Eq. (3.67)]; that is,

$$\mathbf{k} = \mathbf{T}^T\bar{\mathbf{k}}\mathbf{T} \tag{4.88}$$

The forces in individual elements are computed after nodal displacements are obtained from the structural force-displacement equations. This is done using an approach similar to that outlined in the general flowchart of the stiffness method in Fig. 2.13; that is,

$$\underset{\mathbf{s}^{(ij)}}{\begin{bmatrix} N \\ V \\ M_i \\ M_j \end{bmatrix}^{ij}} = \underset{\bar{\mathbf{e}}}{\frac{E}{L}\begin{bmatrix} -A & 0 & 0 & A & 0 & 0 \\ 0 & \frac{12I}{L^2} & \frac{6I}{L} & 0 & -\frac{12I}{L^2} & \frac{6I}{L} \\ 0 & \frac{6I}{L} & 4I & 0 & -\frac{6I}{L} & 2I \\ 0 & \frac{6I}{L} & 2I & 0 & -\frac{6I}{L} & 4I \end{bmatrix}} \underset{\bar{\mathbf{u}}^{(ij)}}{\begin{bmatrix} \bar{u}_i \\ \bar{v}_i \\ \bar{\theta}_i \\ \bar{u}_j \\ \bar{v}_j \\ \bar{\theta}_j \end{bmatrix}^{ij}} \tag{4.89}$$

or

$$\mathbf{s}^{(ij)} = \bar{\mathbf{e}}\mathbf{T}\mathbf{u}^{(ij)} \tag{4.90}$$

since $\bar{\mathbf{u}}^{(ij)} = \mathbf{T}\mathbf{u}^{(ij)}$. Thus $\mathbf{e} = \bar{\mathbf{e}}\mathbf{T}$, where $\mathbf{T}$ is defined in Eq. (4.83).

EXAMPLE 4.12

Calculate the nodal displacements and element forces for the rigid frame in Fig. E4.12. Include the effects of axial deformations of the elements. $I_{ab} = I_{cd} = 600$ in.^{4}; $I_{bc} = 1200$ in.^{4}; $A_{ab} = A_{cd} = 20$ in.^{2}; $A_{bc} = 30$ in.^{2}; and $E = 29 \times 10^3$ kips/in.^{2}.

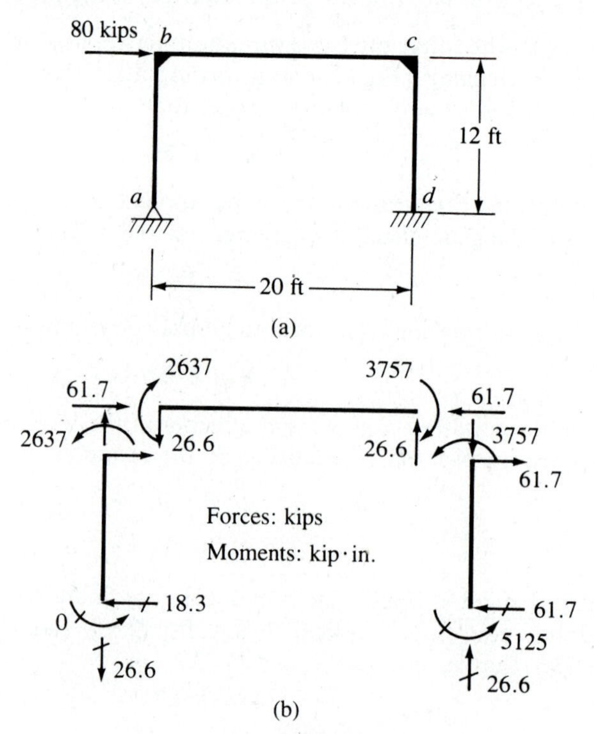

Figure E4.12

Solution

$$\mathbf{k}^{ab} = 10^2 \begin{bmatrix} 0.69927 & 0000.0 & -50.347 & -0.69927 & 0000.0 & -50.347 \\ 0000.0 & 40.278 & 0000.0 & 0.0000 & -40.278 & 0000.0 \\ -50.347 & 0000.0 & 4833.3 & 50.347 & 0000.0 & 2416.7 \\ -0.69927 & 0000.0 & 50.347 & 0.69927 & 0000.0 & 50.347 \\ 0000.0 & -40.278 & 0000.0 & 0000.0 & 40.278 & 0000.0 \\ -50.347 & 0000.0 & 2416.7 & 50.347 & 0000.0 & 4833.3 \end{bmatrix}$$

$$\mathbf{k}^{bc} = 10^2 \begin{bmatrix} 36.250 & 0000.0 & 0000.0 & -36.250 & 0000.0 & 0000.0 \\ 0000.0 & 0.30208 & 36.250 & 0000.0 & -0.30208 & 36.250 \\ 0000.0 & 36.250 & 5800.0 & 0000.0 & -36.250 & 2900.0 \\ -36.250 & 0000.0 & 0000.0 & 36.250 & 0000.0 & 0000.0 \\ 0000.0 & -0.30208 & -36.250 & 0000.0 & 0.30208 & -36.250 \\ 0000.0 & 36.250 & 2900.0 & 0000.0 & -36.250 & 5800.0 \end{bmatrix}$$

$$\mathbf{k}^{cd} = 10^2 \begin{bmatrix} 0.69927 & 0000.0 & 50.347 & -0.69927 & 0000.0 & 50.347 \\ 0000.0 & 40.278 & 0000.0 & 0000.0 & -40.278 & 0000.0 \\ 50.347 & 0000.0 & 4833.3 & -50.347 & 0000.0 & 2416.7 \\ -0.69927 & 0000.0 & -50.347 & 0.69927 & 0000.0 & -50.347 \\ 0000.0 & -40.278 & 0000.0 & 0000.0 & 40.278 & 0000.0 \\ 50.347 & 0000.0 & 2416.7 & -50.347 & 0000.0 & 4833.3 \end{bmatrix}$$

with $\mathbf{U}_f = \{\Theta_a \quad U_b \quad V_b \quad \Theta_b \quad U_c \quad V_c \quad \Theta_c\}$ the $\mathbf{K}_{ff}$ partition of the assembled stiffness matrix is

$$10^2 \begin{bmatrix} 4833.3 & 50.347 & 0000.0 & 2416.7 & 0000.0 & 0000.0 & 0000.0 \\ 50.347 & 36.949 & 0000.0 & 50.347 & -36.250 & 0000.0 & 0000.0 \\ 0000.0 & 0000.0 & 40.580 & 36.250 & 0000.0 & -0.30208 & 36.250 \\ 2416.7 & 50.347 & 36.250 & 10633. & 0000.0 & -36.250 & 2900.0 \\ 0000.0 & -36.250 & 0000.0 & 0000.0 & 36.949 & 0000.0 & 50.347 \\ 0000.0 & 0000.0 & -0.30208 & -36.250 & 0000.0 & 40.580 & -36.250 \\ 0000.0 & 0000.0 & 36.250 & 2900.0 & 50.347 & -36.250 & 10633. \end{bmatrix}$$

Solving the nodal equilibrium equations yields

$$\{\Theta_a \quad U_b \quad V_b \quad \Theta_b \quad U_c \quad V_c \quad \Theta_c\} = \{-0.01271 \text{ rad} \quad 1.3068 \text{ in.} \quad 0.00662 \text{ in.} \quad -0.00180 \text{ rad} \quad 1.2897 \text{ in.} \quad -0.00662 \text{ in.} \quad -0.00566 \text{ rad}\}$$

Using Eq. (4.89) the element forces are

$$\mathbf{s}^{ab} = \{26.64 \quad 18.32 \quad 0000 \quad 2637\}$$
$$\mathbf{s}^{bc} = \{-61.68 \quad -26.64 \quad -2637 \quad -3757\}$$
$$\mathbf{s}^{cd} = \{-26.64 \quad 61.68 \quad 3757 \quad 5125\}$$

where $\mathbf{s}^{ij} = \{N \quad V \quad M_i \quad M_j\}$ in kips and kip · in.

Discussion By designating the element with the labels at the two ends we imply the orientation of the local coordinate axis. Thus element ab has its local x axis pointing in the positive y global direction, while the local x axis for element cd is directed in the negative y global direction. A perusal of the stiffness matrices for these two elements reveals sign differences resulting from the different local coordinate axes orientations. The element forces are displayed in Fig. E4.12b. Note that all joints are in equilibrium as they must be from the nodal equilibrium equations (e.g., the shear in element ab is equal in magnitude and opposite in direction to the axial force in element bc at joint b).

4.5.2 FLEXURAL AND AXIAL RIGIDITY In various structures, the axial deformations of the elements may have no consequence; therefore, it is desirable to have the stiffness equations unencumbered by these spurious degrees of freedom. In the analysis of beams at the beginning of this chapter it was assumed that the element is axially inextensible, that is, the axial deformation is identically zero. For a girder in a rigid frame the floor system can be imagined as inhibiting axial deformations of the member so that the girder will displace with no deformation parallel to its longitudinal axis. Several different solution procedures can accommodate this type of behavior.

The analysis of the simple rigid plane frame in Fig. 4.12 offers one approach for applying the stiffness method when axially inextensible members are present. This rigid frame consists of four beam elements joined at node a by a moment connection. If the members are considered to be axially inextensible there is only one generalized unknown displacement (Θ_a) to consider;

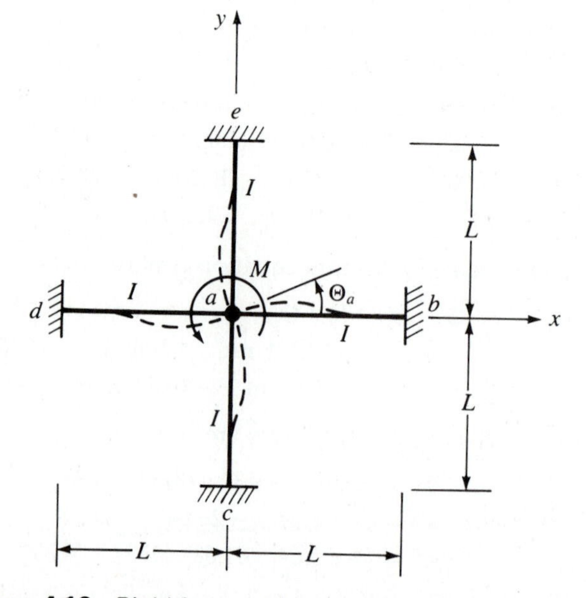

Figure 4.12 Rigid frame with one generalized displacement

therefore, $\mathbf{K}_{ff}$ will be a scalar. The stiffness terms contributed by each of the four elements is k_{33} from the stiffness matrix in Eq. (4.81); that is,

$$k^{ab} = k^{ac} = k^{ad} = k^{ad} = \frac{4EI}{L}$$

In this case the stiffness matrix of the total structure with boundary conditions enforced is the scalar

$$K_{ff} = \frac{16EI}{L}$$

The single force-displacement equation is

$$M = \frac{16EI}{L}\,\Theta_a$$

which gives

$$\Theta_a = \frac{ML}{16EI}$$

The element forces for a typical element such as ab are obtained using Eq. (4.48), yielding

$$\begin{bmatrix} V \\ M_a \\ M_b \end{bmatrix}^{ab} = \frac{EI}{L}\begin{bmatrix} \frac{12}{L^2} & \frac{6}{L} & -\frac{12}{L^2} & \frac{6}{L} \\ \frac{6}{L} & 4 & -\frac{6}{L} & 2 \\ \frac{6}{L} & 2 & -\frac{6}{L} & 4 \end{bmatrix}\begin{bmatrix} 0 \\ \frac{ML}{16EI} \\ 0 \\ 0 \end{bmatrix} = \begin{bmatrix} \frac{3M}{8L} \\ \frac{M}{4} \\ \frac{M}{8} \end{bmatrix}$$

These results, of course, satisfy overall equilibrium of the rigid frame and could have been calculated using one of the classical methods such as moment distribution.

Alternatively for more complex structures, the longitudinal inextensibility of the elements can be enforced by using the element stiffness matrix in Eq. (4.81) and giving the cross-sectional area a sufficiently large value so that $\bar{u}_i = \bar{u}_j$. This approach can cause $\mathbf{K}_{ff}$ to be ill conditioned with respect to solution if the large values of A are significantly larger than all other entries in the constrained stiffness matrix.

EXAMPLE 4.13

Calculate the nodal displacements and element forces for the rigid frame of Example 4.12 if the elements are axially rigid.

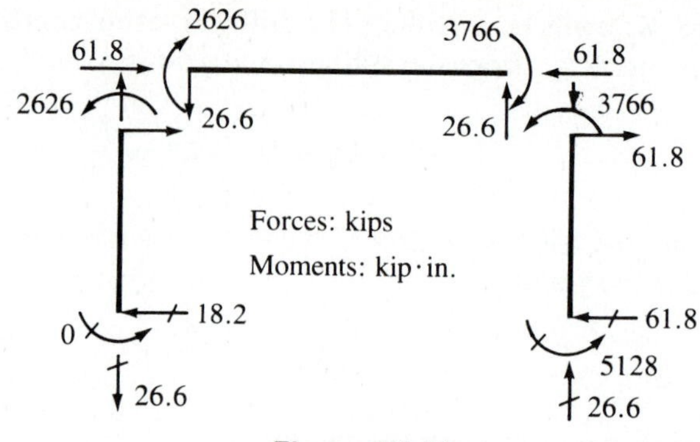

Figure E4.13

Solution

$$
\mathbf{k}^{ab} = 10^2 \begin{bmatrix}
0.69927 & 0000.0 & -50.347 & -0.69927 & 0000.0 & -50.347 \\
0000.0 & 201390000 & 0000.0 & 0.0000 & -201390000 & 0000.0 \\
-50.347 & 0000.0 & 4833.3 & 50.347 & 0000.0 & 2416.7 \\
-0.69927 & 0000.0 & 50.347 & 0.69927 & 0000.0 & 50.347 \\
0000.0 & -201390000 & 0000.0 & 0000.0 & 201390000 & 0000.0 \\
-50.347 & 0000.0 & 2416.7 & 50.347 & 0000.0 & 4833.3
\end{bmatrix}
$$

$$
\mathbf{k}^{bc} = 10^2 \begin{bmatrix}
120830000 & 0000.0 & 0000.0 & -120830000 & 0000.0 & 0000.0 \\
0000.0 & 0.30208 & 36.250 & 0000.0 & -0.30208 & 36.250 \\
0000.0 & 36.250 & 5800.0 & 0000.0 & -36.250 & 2900.0 \\
-120830000 & 0000.0 & 0000.0 & 120830000 & 0000.0 & 0000.0 \\
0000.0 & -0.30208 & -36.250 & 0000.0 & 0.30208 & -36.250 \\
0000.0 & 36.250 & 2900.0 & 0000.0 & -36.250 & 5800.0
\end{bmatrix}
$$

$$
\mathbf{k}^{cd} = 10^2 \begin{bmatrix}
0.69927 & 0000.0 & 50.347 & -0.69927 & 0000.0 & 50.347 \\
0000.0 & 201390000 & 0000.0 & 0000.0 & -201390000 & 0000.0 \\
50.347 & 0000.0 & 4833.3 & -50.347 & 0000.0 & 2416.7 \\
-0.69927 & 0000.0 & -50.347 & 0.69927 & 0000.0 & -50.347 \\
0000.0 & -201390000 & 0000.0 & 0000.0 & 201390000 & 0000.0 \\
50.347 & 0000.0 & 2416.7 & -50.347 & 0000.0 & 4833.3
\end{bmatrix}
$$

With $\mathbf{U}_f = \{\Theta_a \ \ U_b \ \ V_b \ \ \Theta_b \ \ U_c \ \ V_c \ \ \Theta_c\}$ the $\mathbf{K}_{ff}$ partition of the assembled stiffness matrix is

$$
10^2 \begin{bmatrix}
4833.3 & 50.347 & 0000.0 & 2416.7 & 0000.0 & 0000.0 & 0000.0 \\
50.347 & 120830000 & 0000.0 & 50.347 & -120830000 & 0000.0 & 0000.0 \\
0000.0 & 0000.0 & 201390000 & 36.250 & 0000.0 & -0.30208 & 36.250 \\
2416.7 & 50.347 & 36.250 & 10633. & 0000.0 & -36.250 & 2900.0 \\
0000.0 & -120830000 & 0000.0 & 0000.0 & 120830000 & 0000.0 & 50.347 \\
0000.0 & 0000.0 & -0.30208 & -36.250 & 0000.0 & 201390000 & -36.250 \\
0000.0 & 0000.0 & 36.250 & 2900.0 & 50.347 & -36.250 & 10633.
\end{bmatrix}
$$

Solving the nodal equilibrium equations yields

$$\{\Theta_a \quad U_b \quad V_b \quad \Theta_b \quad U_c \quad V_c \quad \Theta_c\} =$$
$$\{-0.01258 \text{ rad} \quad 1.2892 \text{ in.} \quad 0.0000 \text{ in.} \quad -0.00171 \text{ rad} \quad 1.2892 \text{ in.} \quad -0.0000 \text{ in.} \quad -0.00564 \text{ rad}\}$$

Using Eq. (4.89) the element forces are

$$\mathbf{s}^{ab} = \{26.63 \quad 18.24 \quad 0000 \quad 2626\}$$
$$\mathbf{s}^{bc} = \{-61.76 \quad -26.63 \quad -2626 \quad -3766\}$$
$$\mathbf{s}^{cd} = \{-26.63 \quad 61.76 \quad 3766 \quad 5128\}$$

where $\mathbf{s}^{ij} = \{N \quad V \quad M_i \quad M_j\}$ in kips and kip · in.

Discussion All element cross-sectional areas were prescribed as 10^8. This approach must be carefully used or $\mathbf{K}_{ff}$ may be ill conditioned with respect to solution. Note that the calculated displacements reveal that both V_b and V_c are zero and $U_b = U_c$; these are the results required for axially rigid elements. See Fig. E4.13 for free-body diagrams of the elements.

Constraint equations can also be used to enforce axial inextensibility of the elements. This method involves operating upon $\mathbf{K}_{ff}$. We will illustrate the approach heuristically here; the general equations are described in Sec. 6.2. For axial rigidity of the frame in Example 4.12 we require that $V_b = V_c = 0$ and $U_b = U_c$. The first condition can be enforced in a manner similar to that used for rigid supports; starting with $\mathbf{K}_{ff}$ of Example 4.12, this process yields the following nodal equilibrium equations:

$$\begin{bmatrix} 0 \\ 80 \\ 0 \\ 0 \\ 0 \end{bmatrix} = 10^2 \begin{bmatrix} 4833.3 & 50.347 & 2416.7 & 0000.0 & 0000.0 \\ 50.347 & 36.949 & 50.347 & -36.250 & 0000.0 \\ 2416.7 & 50.347 & 10633. & 0000.0 & 2900.0 \\ 0000.0 & -36.250 & 0000.0 & 36.949 & 50.347 \\ 0000.0 & 0000.0 & 2900.0 & 50.347 & 10633. \end{bmatrix} \begin{bmatrix} \Theta_a \\ U_b \\ \Theta_b \\ U_c \\ \Theta_c \end{bmatrix}$$

The inextensibility of girder *bc* (i.e., $U_b = U_c$) can be enforced in the above nodal equilibrium equations by: (a) adding columns two and four of the stiffness matrix; (b) replacing the second column with this sum; (c) deleting the fourth column; and (d) noting that $\mathbf{U}_f = \{\Theta_a \quad U_b \quad \Theta_b \quad \Theta_c\}$. This process yields a 5 by 4 stiffness matrix since there are now five nodal loads and four unknown generalized displacements. To obtain a square stiffness matrix we can either: (a) delete equation four or (b) add the second and fourth equations, replace the second equation with this summation, and delete the fourth equation. The first alternative gives a stiffness matrix that is not symmetric, while the second approach yields a symmetric stiffness matrix. We will use the second approach since most matrix structural analysis solution modules are designed to take advantage of the symmetry of the stiffness matrix. Note that it is imperative to add the forces in performing the addition of the second and fourth equations, but in our example there is no applied force at c in the x direction. Performing these steps gives

$$\begin{bmatrix} 0 \\ 80 \\ 0 \\ 0 \end{bmatrix} = 10^2 \begin{bmatrix} 4833.3 & 50.347 & 2416.7 & 0000.0 \\ 50.347 & 1.398 & 50.347 & 50.347 \\ 2416.7 & 50.347 & 10633. & 2900.0 \\ 0000.0 & 50.347 & 2900.0 & 10633. \end{bmatrix} \begin{bmatrix} \Theta_a \\ U_b \\ \Theta_b \\ \Theta_c \end{bmatrix}$$

Solving these equations gives the same nodal displacements obtained in Example 4.13.

4.5.3 SETTLEMENT, INITIAL, AND THERMAL STRAINS We have observed that the analyses of beams and planar frames are similar, but, unlike beams, the axial deformations of frames can play a significant role in their structural response. Therefore, we can envisage that the approach outlined for self-straining of beams is applicable to planar frames provided axial deformations are considered. In Sec. 4.4 we noted that structures with differential support displacement can be analyzed by: (a) partitioning the nodal equilibrium equations [see Eqs. (4.73) and (4.74)] or (b) using initial forces and the three-step approach summarized in Eqs. (4.62), (4.64), and (4.65). Structures with thermal effects, fabrication errors, precambering, etc., are also investigated with initial forces. The initial force matrix for the prismatic plane frame element must include axial forces, shear, and flexure. We will make the assumption that axial and bending effects are uncoupled. For example, consider a plane frame element as shown in Fig. 4.9 with a linear distribution of temperature between the bottom and top of the beam given by Eq. (4.75); that is,

$$T(y) = T_m + \frac{\Delta T y}{h}$$

where

$$T_m = \tfrac{1}{2}(T_l + T_u) \qquad \text{and} \qquad \Delta T = T_u - T_l$$

Combining the initial force matrix for the axial force element [Eq. (2.63)] with that for the beam element [Eq. (4.79)] yields

$$\bar{\mathbf{p}}^o = \begin{bmatrix} \bar{p}^o_{\bar{x}i} \\ p^o_{\bar{y}i} \\ \bar{m}^o_i \\ \bar{p}^o_{\bar{x}j} \\ \bar{p}^o_{\bar{y}j} \\ \bar{m}^o_j \end{bmatrix} = E\alpha \begin{bmatrix} AT_m \\ 0 \\ -\dfrac{I\Delta T}{h} \\ -AT_m \\ 0 \\ \dfrac{I\Delta T}{h} \end{bmatrix} \tag{4.91}$$

The corresponding initial force matrix for calculating final element forces is obtained by combining Eqs. (2.69a) and (4.80) to give

$$\bar{\mathbf{s}}^o = \begin{bmatrix} N^o \\ V^o \\ M_i^o \\ M_j^o \end{bmatrix} = \begin{bmatrix} \bar{p}_{\bar{x}j}^o \\ \bar{p}_{\bar{y}i}^o \\ \bar{m}_i^o \\ \bar{m}_j^o \end{bmatrix} = E\alpha \begin{bmatrix} -AT_m \\ 0 \\ -\dfrac{I\Delta T}{h} \\ \dfrac{I\Delta T}{h} \end{bmatrix} \tag{4.92}$$

EXAMPLE 4.14

The plane rigid frame of Example 4.12 has the applied load removed and is to be investigated for thermal effects. The structure was constructed when the ambient temperature was 70°F. The columns are kept at 70°F, while the top and bottom of beam *bc* are 150°F and 10°F, respectively, and the temperature varies linearly between the two surfaces. Beam *bc* has a depth of 12 in. and the frame is steel with $\alpha = 6.5 \times 10^{-6}$ in./(in. · °F). Calculate the nodal displacements and element forces caused by these temperatures.

Solution

The mean temperature of the neutral axis of *bc* is 80°F, $T_u = +80$°F, and $T_l = -60$°F. Therefore, $\Delta T = 140$°F, but the axial force results from a rise of 10°F over the original temperature of construction. Thus in Eqs. (4.91) and (4.92) $T_m = +10$°F. This gives for element *bc*,

$$\begin{aligned} \bar{p}_{\bar{x}b}^{o(bc)} &= -\bar{p}_{\bar{x}c}^{o(bc)} = AE\alpha T_m \\ &= (30 \text{ in.}^2)(29 \times 10^3 \text{ kips/in.}^2)[6.5 \times 10^{-6} \text{ in./(in.} \cdot {}^\circ\text{F)}](10^\circ\text{F}) \\ &= 56.55 \text{ kips} \end{aligned}$$

$$\begin{aligned} \bar{m}_c^{o(bc)} &= -\bar{m}_b^{o(bc)} = \frac{EI\alpha\Delta T}{h} \\ &= \frac{(29 \times 10^3 \text{ kips/in.}^2)(1200 \text{ in.}^4)[6.5 \times 10^{-6} \text{ in./(in.} \cdot {}^\circ\text{F)}](140^\circ\text{F})}{12 \text{ in.}} \\ &= 2639 \text{ kip} \cdot \text{in.} \end{aligned}$$

Using the cross-sectional areas of the elements as in Example 4.12 and including all associated degrees of freedom, the nodal equilibrium equations [see Eq. (4.63)] are

$$\begin{bmatrix} 0 \\ 0 \\ 0 \\ 0 \\ 0 \\ 0 \\ 0 \end{bmatrix} = \mathbf{K}_{ff} \begin{bmatrix} \Theta_a \\ U_b \\ V_b \\ \Theta_b \\ U_c \\ V_c \\ \Theta_c \end{bmatrix} + \begin{bmatrix} 0 \\ 56.55 \\ 0 \\ -2639 \\ -56.55 \\ 0 \\ 2639 \end{bmatrix}$$

See Example 4.12 for $\mathbf{K}_{ff}$. Solving for the nodal displacements gives

$$\mathbf{U}_f = \{-0.00308 \text{ rad} \quad 0.11536 \text{ in.} \quad 0.00034 \text{ in.} \quad 0.00376 \text{ rad} \quad 0.13413 \text{ in.} \quad -0.00034 \text{ in.} \quad -0.00415 \text{ rad}\}$$

Using Eq. (4.92), together with Eq. (4.65), the element forces are

$$\begin{aligned}
\mathbf{s}^{ab} &= \{1.36 \quad 11.50 \quad 0000 \quad 1655\} \\
\mathbf{s}^{bc} &= \{68.04 \quad -1.36 \quad 984 \quad -1310\} + \{-56.55 \quad 00 \quad -2639 \quad 2639\} \\
&= \{11.49 \quad -1.36 \quad -1655 \quad 1329\} \\
\mathbf{s}^{cd} &= \{-1.36 \quad -11.50 \quad -1329 \quad -327\}
\end{aligned}$$

where $\mathbf{s}^{ij} = \{N \quad V \quad M_i \quad M_j\}$ in kips and kip · in.

Discussion The temperatures experienced by beam *bc* are indicative of those encountered in the summer on the roof of a cold storage building. This statically indeterminate structure experiences significant displacements and internal forces from such a modest thermal loading. We urge the reader to sketch the free-body diagrams of all elements and check equilibrium of the joints, elements, and entire structure.

4.6 DISCUSSION

The approach for solving beams and rigid frames has been described in the preceding sections. We have followed the procedure outlined in the basic flowchart of Fig. 2.13, and appropriate matrices for the uniform beam and uniform plane frame element have been described. With the internal axial forces, shear, and bending in beams and frames it is necessary to account for nodal rotations as well as linear displacements. The reader who has studied the slope-deflection method will note the similarity between the equations of that approach and the moment-rotation equations of matrix structural analysis. Both methods stem from the principle of virtual work and are categorized as equilibrium methods; therefore, one should expect similarities. We have developed the capability of analyzing elements for axial deformations as well as shear and flexural behavior; this forms the basis for investigating rigid-frame response. Typically, member axial deformations can be ignored for frames, but in some cases this effect is paramount in influencing structural behavior. Settlement, temperature changes, fabrication errors, and other initial force effects are treated in a manner similar to that of trusses, but the basic matrices differ between axial force, beam, and rigid-frame elements.

We noted that accurate structural analysis usually requires an understanding of the behavior *a priori*. The location of nodes, definition of elements, and manner of assigning applied loads to nodal points for beams and rigid frames must be established by the structural engineer based upon his/her perceived behavior of the structure. This process of idealization consists of developing a model that accurately characterizes the actual structural behavior; it is an extremely important aspect of matrix structural analysis.

4.7 Problems

4.1 Calculate the displacements at point a in Fig. P4.1 using the principle of virtual displacements. Use $v = v_a \sin(\pi x/L)$ and a similar form for δv; EI is constant.

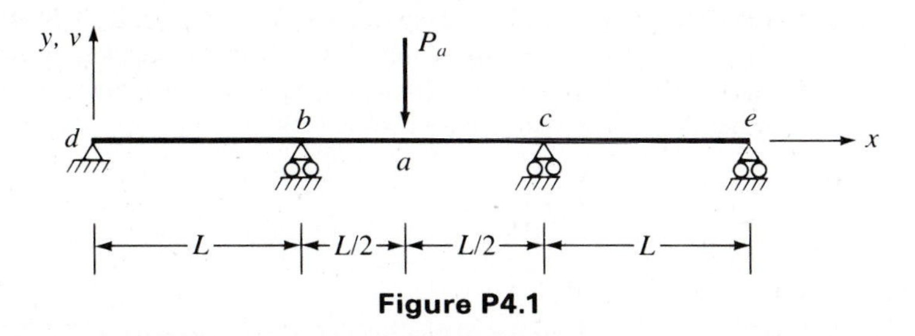

Figure P4.1

4.2 Calculate the displacement at point b and the moment and shear at point a in Fig. P4.2 using the principle of virtual displacements. Use an appropriate cosine function for the real and virtual displacements; EI is constant.

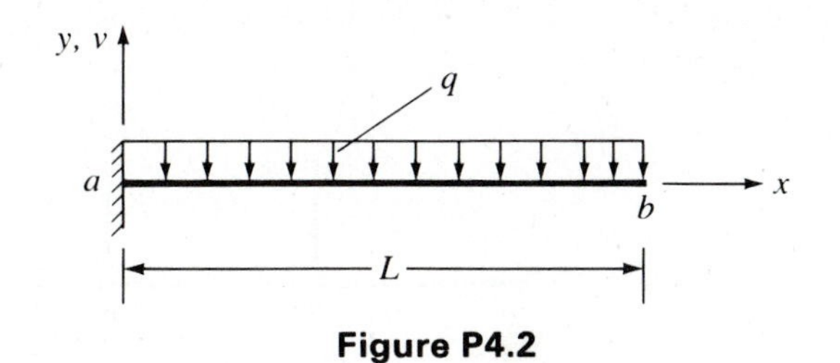

Figure P4.2

4.3 Using the degrees of freedom $\{V_b \quad \Theta_b \quad \Theta_c\}$, we have for the beam in Fig. P4.3,

$$\mathbf{K}_{ff} = \begin{bmatrix} 24 & 0 & 6 \\ 0 & 8 & 2 \\ 6 & 2 & 4 \end{bmatrix}$$

Describe the physical meaning of a single column of $\mathbf{K}_{ff}$. Sketch (and identify the magnitude) of the forces at nodes b and c necessary to deform the beam so that each of the degrees of freedom is alternatively given a unit value, while all other degrees of freedom are zero (i.e., this requires three sketches).

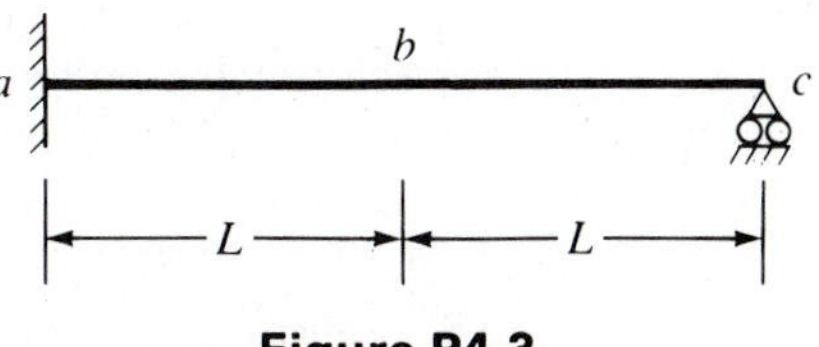

Figure P4.3

4.4 Using the degrees of freedom $\{V_b \quad \Theta_b \quad V_c \quad \Theta_c \quad \Theta_d\}$, we have for the beam in Fig. P4.4,

$$\mathbf{K}_{ff} = \begin{bmatrix} 24 & 0 & -12 & 6 & 0 \\ 0 & 8 & -6 & 2 & 0 \\ -12 & -6 & 24 & 0 & 6 \\ 6 & 2 & 0 & 8 & 2 \\ 0 & 0 & 6 & 2 & 4 \end{bmatrix}$$

Describe the physical meaning of a single column of $\mathbf{K}_{ff}$. Sketch (and identify the magnitude) of the forces at nodes b, c, and d necessary to deform the beam so that each of the degrees of freedom is alternately given a unit value, while all other degrees of freedom are zero (i.e., this requires five sketches).

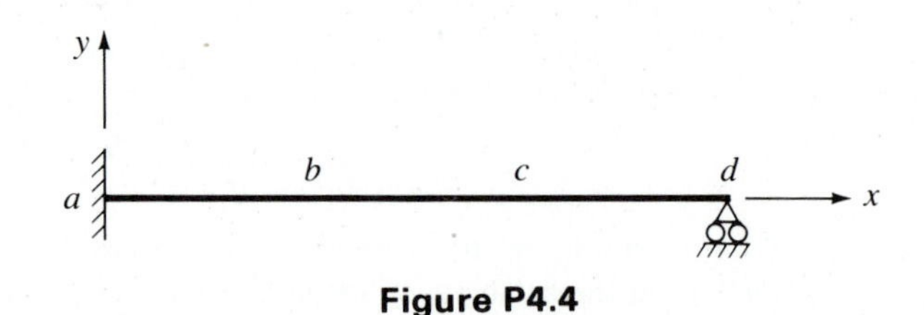

Figure P4.4

4.5 Calculate $\mathbf{K}_{ff}$ and $\mathbf{P}_f$ for the beam in Fig. P4.5; $I_{ab} = I$, $I_{bc} = 3I$, $I_{cd} = 2I$, and E = constant.

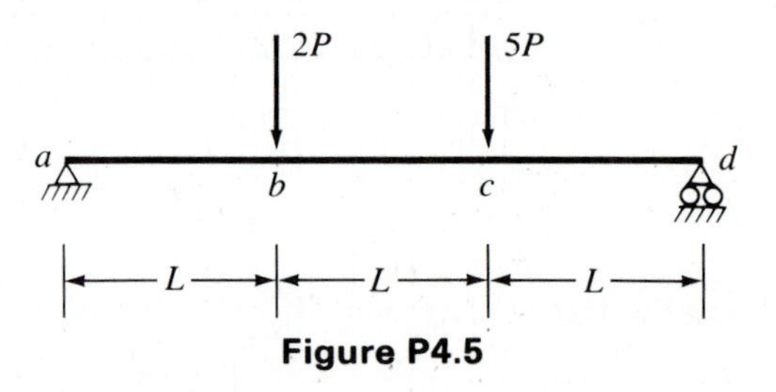

Figure P4.5

4.6 **(a)** Obtain $\mathbf{K}_{ff}$ for the uniform beam in Fig. P4.6; $I = 96 \times 10^{-6}$ m^4, $E = 200$ GPa, and $L = 5$ m.

(b) Calculate element forces if $\{V_b \quad \Theta_b \quad \Theta_c\} = (L^2/EI)\{L \quad 1 \quad -2\}$.

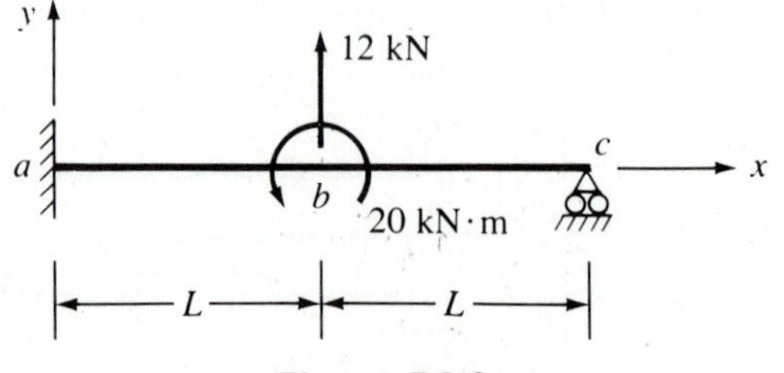

Figure P4.6

4.7 **(a)** Obtain $\mathbf{K}_{ff}$ for the beam shown in Fig. P4.7; $(EI)_{ab} = 2.0$, $(EI)_{bc} = 1.0$, and $L = 1.0$.

(b) Calculate element forces if $\{\Theta_b \quad V_c \quad \Theta_c\} = (-P/24)\{3 \quad 11 \quad 15\}$.

(c) Obtain $\mathbf{K}_{ff}$, nodal displacements, and element forces if node c is supported by a truss element with $(AE/L) = 3.0$ and a downward force of P is applied at c.

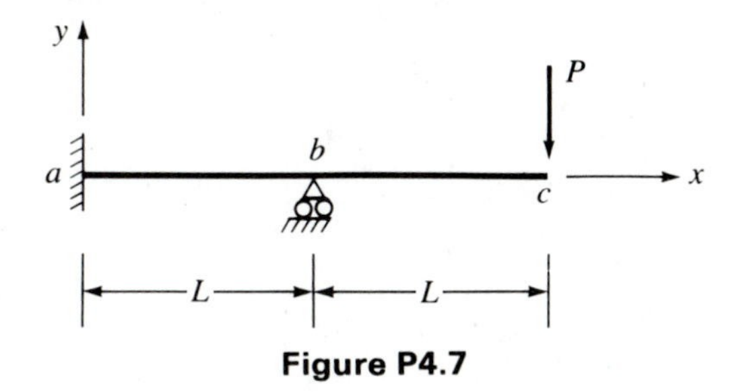

Figure P4.7

4.8 Calculate the nodal displacements and element forces for the beam in Fig. P4.8 if node b displaces an amount δ; $I_{bc} = 2I_{ab} = 2I$; E = constant.

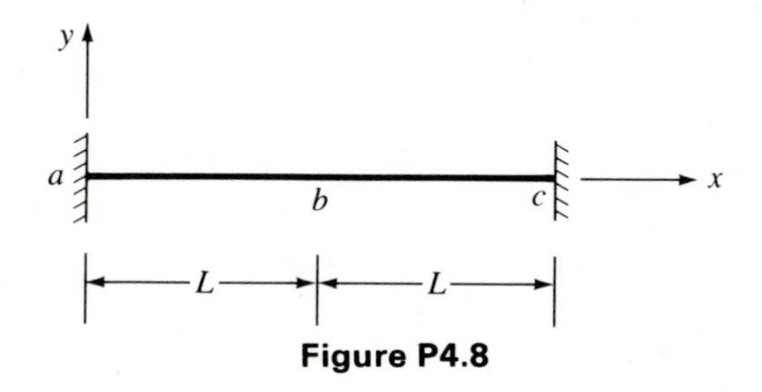

Figure P4.8

4.9 Calculate the nodal displacements and element forces for the beam in Fig. P4.9; EI = constant.

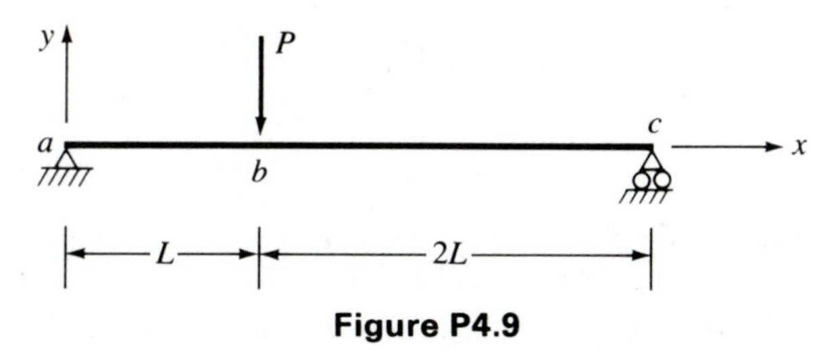

Figure P4.9

4.10 Calculate the nodal displacements and element forces for the beam in Fig. P4.10; $I = 862$ in.^{4} and $E = 29 \times 10^3$ ksi.

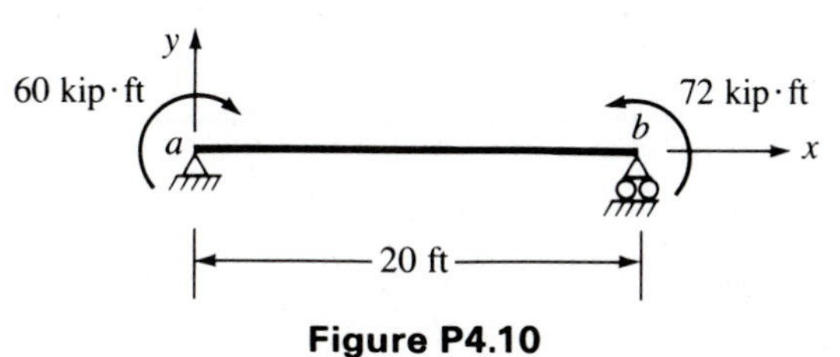

Figure P4.10

4.11 Calculate the nodal displacements and element forces for the beam in Fig. P4.11; $I = 776$ in.^{4} and $E = 29 \times 10^3$ ksi.

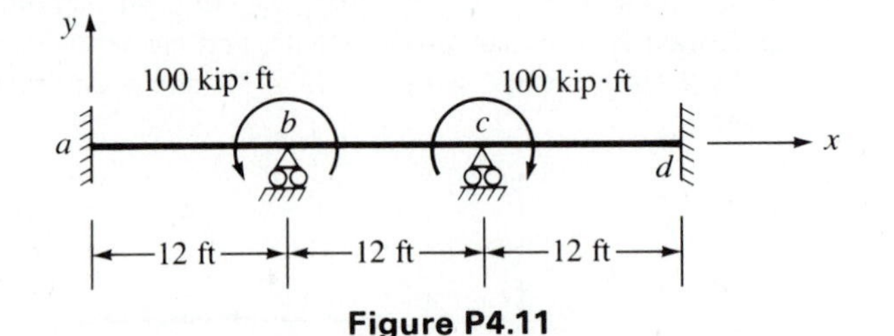

Figure P4.11

4.12 Calculate nodal displacements and element forces for the beam in Fig. P4.12; $I_{ab} = 200 \times 10^{-6}$ m^4; $I_{bc} = 300 \times 10^{-6}$ m^4, and $E = 200$ GPa.

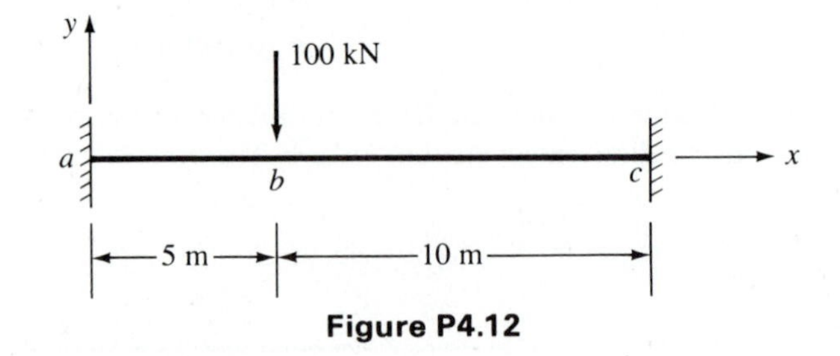

Figure P4.12

4.13 Calculate nodal displacements and element forces for the beam in Fig. P4.13; $I_{ab} = 2I$, $I_{bc} = I$, and $E =$ constant.

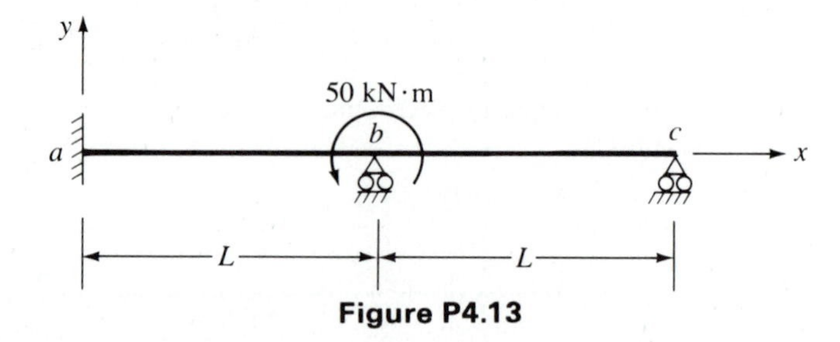

Figure P4.13

4.14 (a) Calculate the nodal displacements and element forces in Fig. P4.14 if the spring stiffness is $12EI/L^3$, $I_{ab} = I_{bc} = I$, and $E =$ constant.

(b) Repeat part (a) if the structure is modified so that $I_{bc} = 2I$.

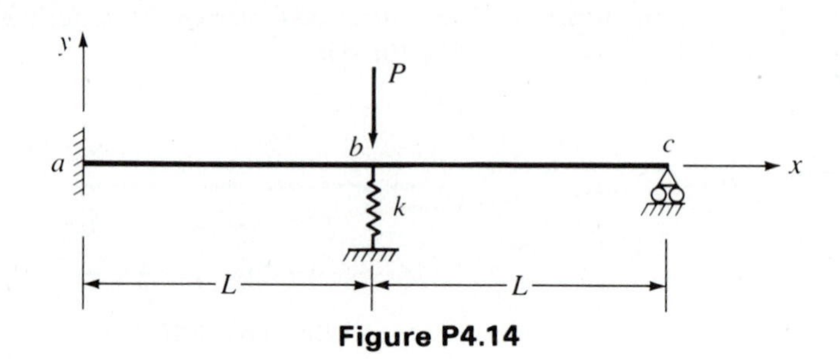

Figure P4.14

4.15 Use the principle of virtual work to obtain the consistent nodal loads corresponding to the fixed-end forces illustrated in Fig. 4.7a, b, and c.

4.16 Calculate the nodal displacements and element forces in Fig. P4.16. $I_{ab} = 200 \times 10^6$ mm^4 and $I_{bc} = 50 \times 10^6$ mm^4. Replace the distributed loading with consistent nodal forces.

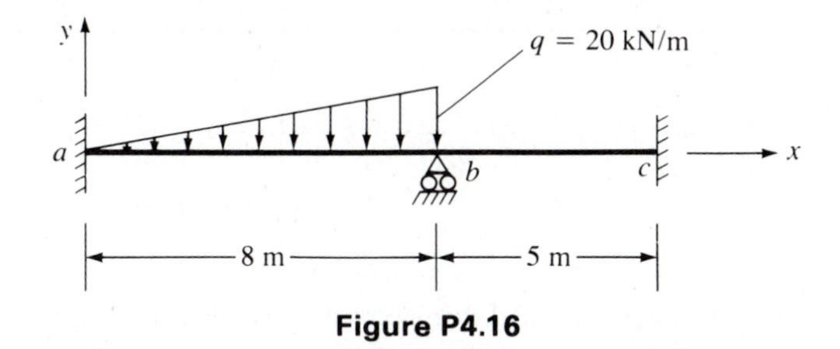

Figure P4.16

4.17 We do the following to establish the flexural properties of the structure in Fig. P4.17. First, a downward load of 500 lb is applied at point *b*, and we observe that $V_b = -0.375$ in. and $V_c = -0.50$ in. After removing the 500 lb load, we impose 600 lb at *c* in the downward direction, which gives $V_b = -0.60$ in. and $V_c = -1.50$ in.

(a) Does the generated data allow one to write the stiffness or flexibility matrix with minimal calculations?

(b) Is there any evidence that the data are consistent?

(c) Calculate both the stiffness and flexibility matrices from the data.

(d) Is the stiffness matrix obtained $\mathbf{K}$, $\mathbf{K}_{ff}$, or something else (identify if the latter is your answer)?

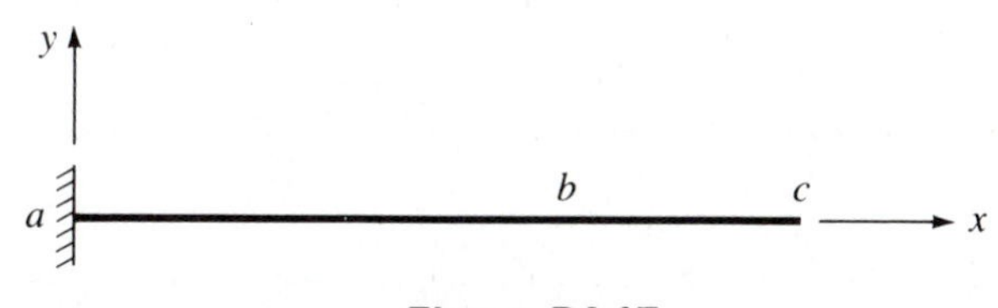

Figure P4.17

4.18 Calculate the reactions if supports *b* and *c* in Fig. P4.18 both settle an amount δ. Compare your results with the classical solution.

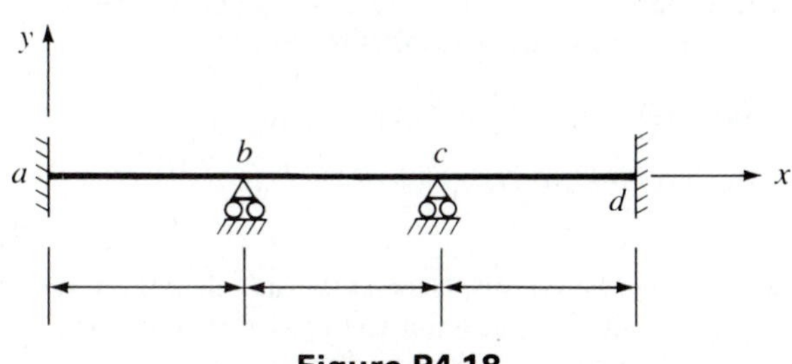

Figure P4.18

4.19 Calculate the nodal displacements and element forces if the right support in Fig. P4.19 settles 0.50 in.; $L = 162$ in., $I_{ef} = 998$ in.^{4}, $I = 960$ in.^{4} (all other elements); $E = 29 \times 10^3$ ksi.

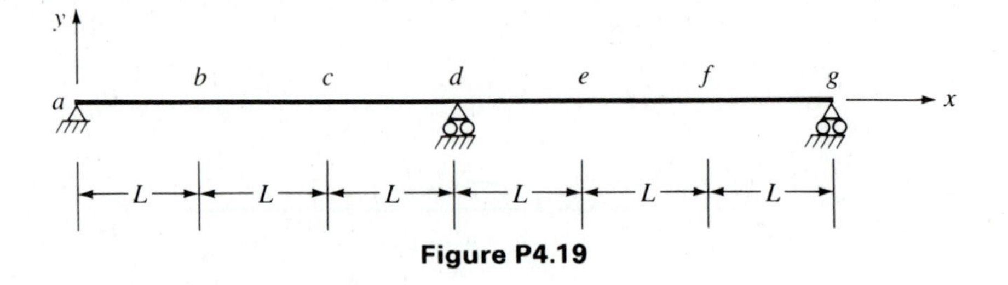

Figure P4.19

4.20 Calculate the nodal displacements and element forces for the uniform beam of Prob. 4.11 if the applied moments are removed and:

(a) Support d settles 0.33 in.

(b) The beam has a linear thermal gradient through its 12 in. depth, with $T_u = 120°F$ and $T_l = 75°F$ [$\alpha = 6.5 \times 10^{-6}$ in./(in. · °F)].

4.21 The beam in Fig. P4.21 has $I = 1000 \times 10^6$ mm^4 and $E = 200$ GPa. Calculate nodal displacements and element forces: **(a)** for the illustrated loading; **(b)** if the loads are removed and the segment between c and e has a linear temperature distribution through its 380-mm depth with $\Delta T = 40°C$ [the top is warmer than the lower surface; $\alpha = 1.2 \times 10^{-5}$ mm/(mm · °C)].

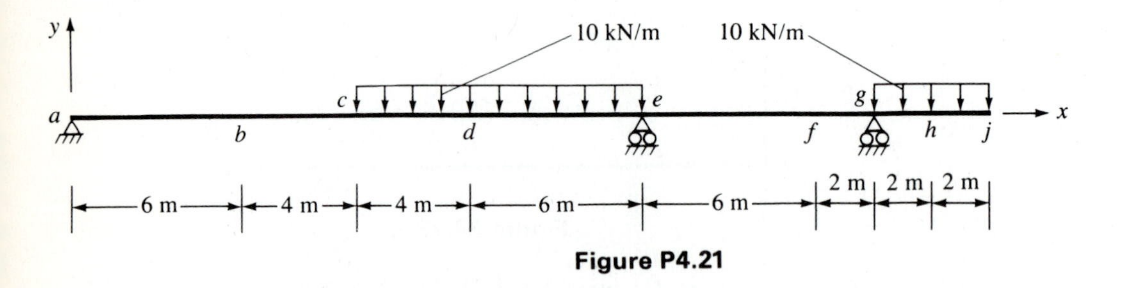

Figure P4.21

4.22 Calculate the fixed-end forces for the heated beam element of Fig. 4.9 if the illustrated thermal distribution is replaced by:

(a) $T(y) = T_m + (T_u - T_m)(y/h)^2$ and

(b) $T(y) = T_m + (T_u - T_m)(y/h)^3$.

4.23 Calculate nodal displacements and element forces for the king post truss in Fig. 4.23 if beam acd consists of a W16 × 45, post bc consists of 2-C7 × 9.8, and ab and bd are 2 in.-diameter rods. $E = 29 \times 10^3$ ksi.

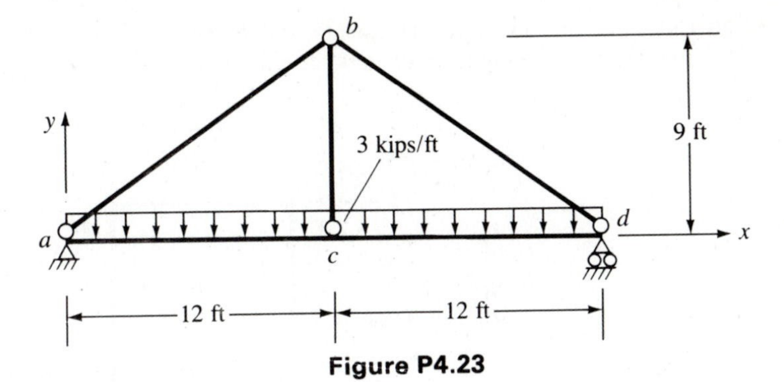

Figure P4.23

4.24 Calculate the bandwidth for the rigid frames in Fig. P4.24 assuming three degrees of freedom per node.

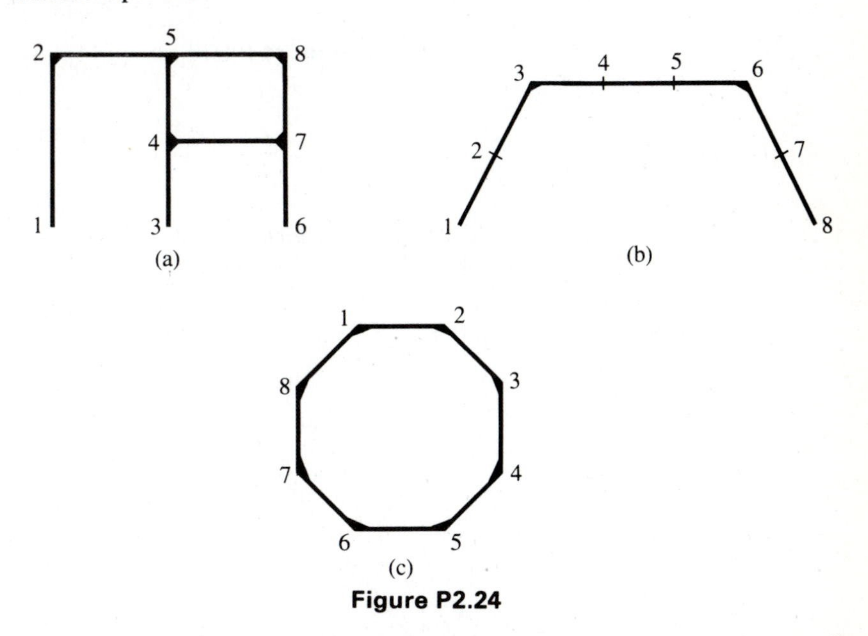

Figure P2.24

4.25 Calculate Θ_b and the element forces for the rigid frame in Fig. P4.25. Assume all members are axially rigid; $I_{ab} = I$, $I_{bc} = 2I$, $I_{bd} = 3I$, and E is constant.

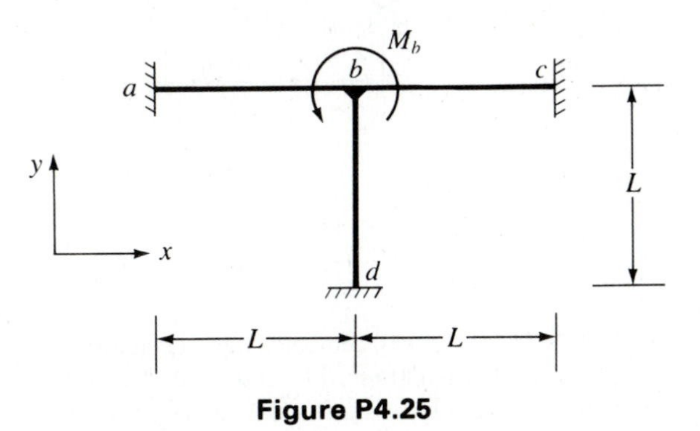

Figure P4.25

4.26 Calculate the nodal displacements and element forces for the rigid frame in Fig. P4.26 assuming the members are axially rigid; EI is constant.

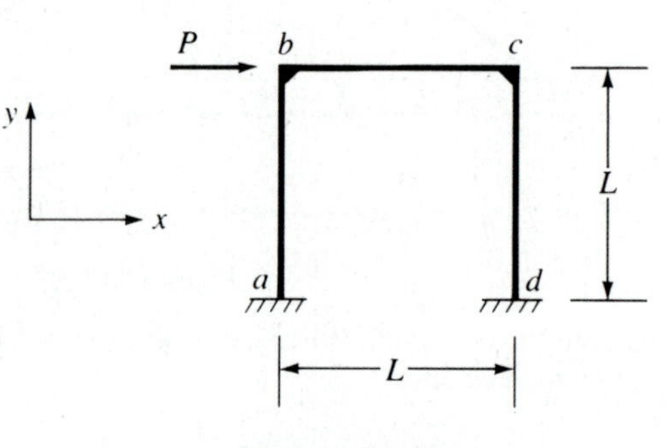

Figure P4.26

4.27 Calculate nodal displacements and element forces for the two-element, three-node rigid frame in Fig. P4.27; $I = 200 \times 10^{-6}$ m^4 and $E = 200$ GPa (assume axial deformations are zero).

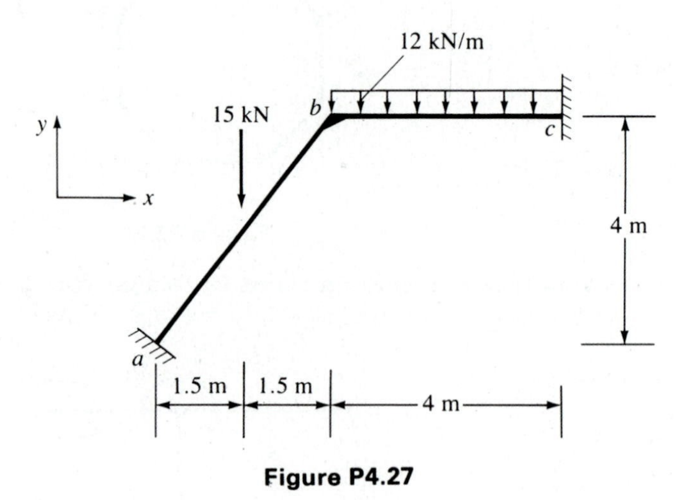

Figure P4.27

4.28 Calculate nodal displacements and element forces for the rigid frame in Fig. P4.28; $A = 0.030$ m^2, $I = 260 \times 10^{-6}$ m^4, $E = 200$ GPa. **(a)** Include axial deformations and **(b)** assume axial deformations are zero.

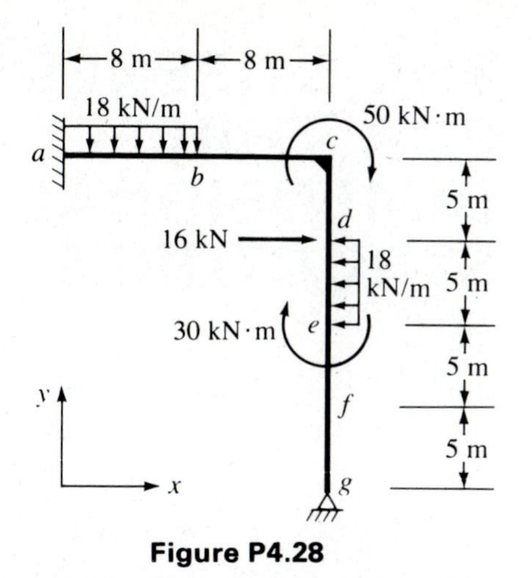

Figure P4.28

4.29 The properties of the plane rigid frame in Fig. P4.29 are: $E = 29 \times 10^3$ ksi, $\alpha = 6.5 \times 10^{-6}$ in./(in. · °F); $I_{ab} = I_{cd} = 1000$ in.^{4}, $A_{ab} = A_{cd} = 35$ in.^{2}, $h_{ab} = h_{cd} = 12$ in.; $I_{bc} = 1600$ in.^{4}, $A_{bc} = 25$ in.^{2}, $h_{bc} = 14$ in. Compute nodal displacements if: **(a)** $P_{yc} = -60$ kips (ignore axial deformations); **(b)** $P_{yc} = -60$ kips (include axial deformations); **(c)** support d settles 0.25 in. (include axial deformations); and **(d)** the top of beam bc has a linear temperature distribution through its depth with the top 150°F greater than the bottom and the mean temperature equal to 0°F (include axial deformations).

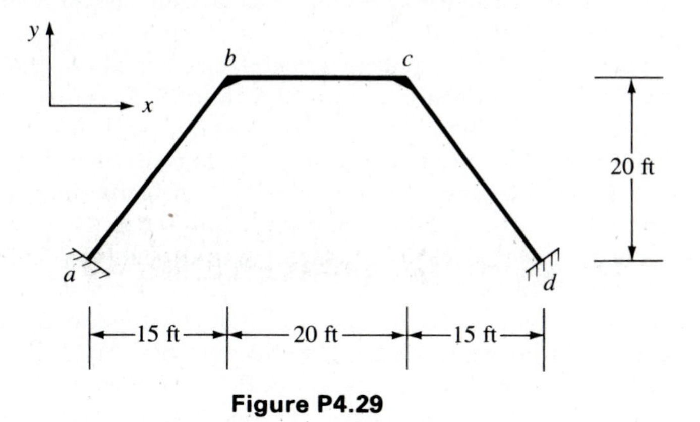

Figure P4.29

5 THE STIFFNESS METHOD FOR THREE-DIMENSIONAL STRUCTURES

The planar truss, beam, and frame components discussed in previous chapters can be analyzed separately as a part of the total structural system. In many structures the individual constituents act independently of each other; the forces are transmitted between parts, but no additional interaction occurs. For some structures it is impossible to resolve the load-transfer mechanism into a system of individual planar structural components. Many radar telescopes, power transmission towers, and steel offshore oil-drilling platforms consist of composites of trusses arranged in multiple planes; their structural integrity depends upon the three-dimensional interaction of the members. Some three-dimensional structures are constructed using flexural members. Interdependent flexural elements arranged in a plane form a grid structure, and space frames have their members oriented in general three-dimensional space. A relatively common use of three-dimensional structures consists of the space frames used in buildings as the horizontal structure to support the gravity loads and the vertical structural system to resist lateral loads.

If possible, one should decompose a structure into its planar components and analyze the resulting individual subsystems. If a structure displays true three-dimensional behavior there is no alternative other than performing an analysis of the entire assemblage. A three-dimensional analysis will show the actual behavior of the system, but these investigations are fraught with inherent difficulties. The structural stiffness matrix is inevitably large with a sizable bandwidth (the measure of bandwidth and how a significant size adversely influences solution efficiency are described in App. A). Three-dimensional coordinate transformations are more time consuming than those in a plane. Also, preparing the input data and interpreting the analysis results require extreme care to avoid errors; computer graphics are almost mandatory to efficiently and conveniently perform these pre- and postanalysis tasks.

This chapter begins with the analysis of space trusses and ends by discussing complex space frames. Grid structures, a special case of space frames, are investigated after the general three-dimensional beam element and its transformations are derived. With this progression of topics, the reader is able to visualize the complex physical behavior exhibited by three-dimensional structures.

5.1 SPACE TRUSSES

A space truss is composed of individual axial force elements, transmitting forces in either tension or compression. The members meeting at a joint are carefully aligned so that their centroidal axes intersect at a common point, and they are welded, bolted, or riveted so that the member forces at a joint form a concurrent force system. For the analysis we assume that members are connected with ideal spherical hinges, and that all loads are applied at the points of connection. This latter assumption implies that members are treated as weightless. Secondary stresses due to shear forces, bending moments, and/or torsion are ignored, but the analysis for these effects can be performed by treating the structure as a space frame (see Sec. 5.5). The three translational degrees of freedom at each node are required to describe structural behavior.

The stiffness matrix for the axial force element of Fig. 5.1 was obtained in Ch. 2 using the basic equations, and it was also derived using the theorem of virtual work in Ch. 3. It is repeated below for convenience:

$$\begin{bmatrix} \bar{p}_{\bar{x}i} \\ \bar{p}_{\bar{x}j} \end{bmatrix} = \frac{AE}{L} \begin{bmatrix} 1 & -1 \\ -1 & 1 \end{bmatrix} \begin{bmatrix} \bar{u}_i \\ \bar{u}_j \end{bmatrix} \tag{5.1}$$

The axial force element, with its local coordinate system, is shown oriented in the global coordinate system in Fig. 5.2. The relationships between local and global displacements are

$$\begin{aligned} \bar{u}_i &= u_i \cos \alpha_{\bar{x}} + v_i \cos \beta_{\bar{x}} + w_i \cos \gamma_{\bar{x}} \\ \bar{u}_j &= u_j \cos \alpha_{\bar{x}} + v_j \cos \beta_{\bar{x}} + w_j \cos \gamma_{\bar{x}} \end{aligned} \tag{5.2}$$

where u_i, v_i, and w_i are the displacements in the global x, y, and z directions, respectively, at end i; these same symbols with a j subscript denote similar degrees of freedom at end j of the element. The angles $\alpha_{\bar{x}}$, $\beta_{\bar{x}}$, and $\gamma_{\bar{x}}$ are the angles between the local $\bar{x}$ axis and the x, y, and z global axes, respectively. Denoting the direction cosines as

$$\lambda_{\bar{x}} = \cos \alpha_{\bar{x}}, \qquad \mu_{\bar{x}} = \cos \beta_{\bar{x}}, \qquad \nu_{\bar{x}} = \cos \gamma_{\bar{x}} \tag{5.3}$$

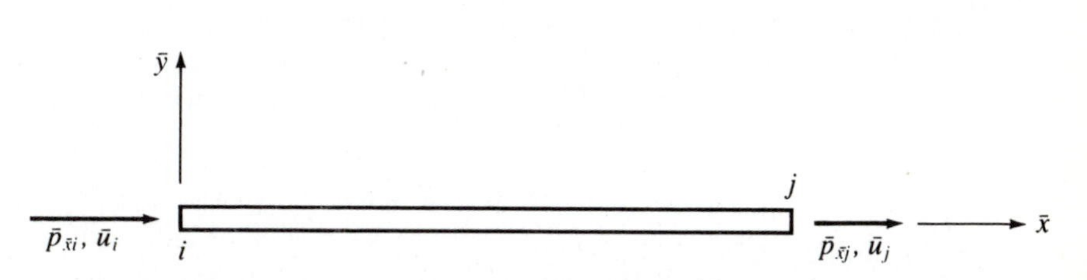

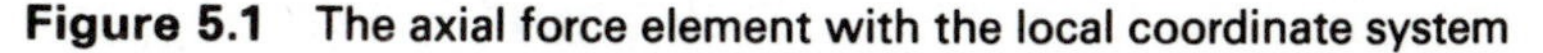
Figure 5.1 The axial force element with the local coordinate system

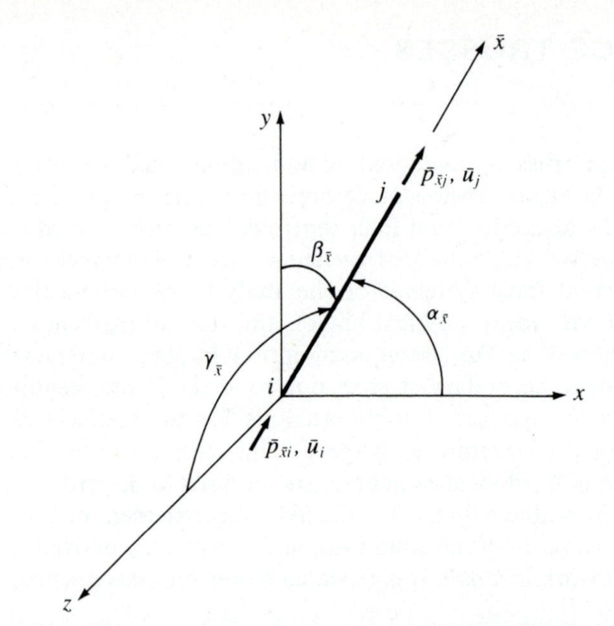

Figure 5.2 The axial force element oriented in the global coordinates

and using matrix notation we have

$$
\begin{bmatrix} \bar{u}_i \\ \bar{u}_j \end{bmatrix} = \begin{bmatrix} \lambda_{\bar{x}} & \mu_{\bar{x}} & \nu_{\bar{x}} & 0 & 0 & 0 \\ 0 & 0 & 0 & \lambda_{\bar{x}} & \mu_{\bar{x}} & \nu_{\bar{x}} \end{bmatrix} \begin{bmatrix} u_i \\ v_i \\ w_i \\ u_j \\ v_j \\ w_j \end{bmatrix} \tag{5.4}
$$

$$
\bar{\mathbf{u}} = \mathbf{T} \quad \mathbf{u}
$$

In Ch. 3 we used the theorem of virtual work to show that for a transformation of the type in Eq. (5.4) there is a corresponding contragredient transformation of forces such that

$$
\mathbf{p} = \mathbf{T}^T \bar{\mathbf{p}} \tag{5.5}
$$

where $\mathbf{p} = \{p_{xi} \quad p_{yi} \quad p_{zi} \quad p_{xj} \quad p_{yj} \quad p_{zj}\}$ and $\bar{\mathbf{p}} = \{\bar{p}_{\bar{x}i} \quad \bar{p}_{\bar{x}j}\}$. All symbols denote forces, with the first subscript indicating the coordinate direction and the second subscript specifying the end of the element upon which the force acts; that is, Eqs. (5.4) and (5.5) indicate contragredient transformations.

Using the theorem of virtual work in Ch. 3 we showed that [see Eq. (3.67)]

$$
\mathbf{k} = \mathbf{T}^T \bar{\mathbf{k}} \mathbf{T} \tag{5.6}
$$

and from Eq. (3.74),

$$
\bar{\mathbf{k}} = \mathbf{T} \mathbf{k} \mathbf{T}^T \tag{5.7}
$$

The element forces can be calculated by transforming the displacements for the element into local coordinates and premultiplying by $\bar{\mathbf{e}}$ according to Eq. (2.57); that is,

$$s^{ij} = \bar{\mathbf{e}}\bar{\mathbf{u}}^{ij} \tag{5.8}$$

where

$$\bar{\mathbf{e}} = \frac{AE}{L}[-1 \quad 1] \qquad \text{and} \qquad \bar{\mathbf{u}}^{ij} = \{\bar{u}_i \quad \bar{u}_j\}$$

Substituting Eq. (5.4) into Eq. (5.8) yields

$$s^{ij} = \bar{\mathbf{e}}\bar{\mathbf{u}}^{ij} = \bar{\mathbf{e}}\mathbf{T}\mathbf{u}^{ij} = \mathbf{e}\mathbf{u}^{ij} \tag{5.9}$$

where

$$\mathbf{e} = \bar{\mathbf{e}}\mathbf{T} = \frac{AE}{L}[-\lambda_{\bar{x}} \quad -\mu_{\bar{x}} \quad -\nu_{\bar{x}} \quad \lambda_{\bar{x}} \quad \mu_{\bar{x}} \quad \nu_{\bar{x}}] \tag{5.10}$$

The matrix $\mathbf{e}$ is the global element force-transformation matrix, and corresponds to the $\mathbf{e}$ matrix in Eq. (2.59) for planar trusses. Examples 5.1 and 5.2 illustrate the analysis of two space trusses using the stiffness method.

EXAMPLE 5.1

Analyze the space truss in Fig. E5.1. $A_{ad} = A_{bd} = A_{de} = 10$ in.^{2}; $A_{cd} = A_{ce} =$ 16 in.^{2}; $E = 29 \times 10^3$ kips/in.^{2}.

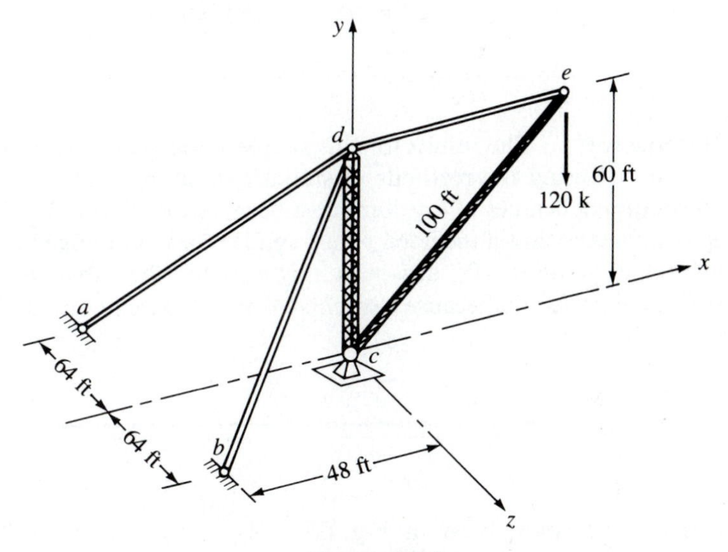

Figure E5.1

Solution

The element stiffness matrices are obtained using Eq. (5.1). Invoking nodal equilibrium for the structure and enforcing the support conditions yields

$$\begin{bmatrix} 0 \\ 0 \\ 0 \\ 0 \\ -120 \end{bmatrix} = \begin{bmatrix} 413.443 & 139.200 & 0.000 & -302.083 & 0.000 \\ 139.200 & 818.444 & 0.000 & 0.000 & 0.000 \\ 0.000 & 0.000 & 197.973 & 0.000 & 0.000 \\ -302.083 & 0.000 & 0.000 & 549.550 & 185.600 \\ 0.000 & 0.000 & 0.000 & 185.600 & 139.200 \end{bmatrix} \begin{bmatrix} U_d \\ V_d \\ W_d \\ U_e \\ V_e \end{bmatrix}$$

Hence

$$\begin{bmatrix} U_d \\ V_d \\ W_d \\ U_e \\ V_e \end{bmatrix} = \begin{bmatrix} 1.825 \\ -0.310 \\ 0.000 \\ 2.354 \\ -4.001 \end{bmatrix} \text{ in.}$$

Using Eq. (5.9) the element axial forces are

$$\begin{aligned} \mathbf{s}^{ad} &= \mathbf{eu}^{ad} = +167 \text{ kips (t)} \\ \mathbf{s}^{bd} &= \mathbf{eu}^{bd} = +167 \text{ kips (t)} \\ \mathbf{s}^{de} &= \mathbf{eu}^{de} = +160 \text{ kips (t)} \\ \mathbf{s}^{cd} &= \mathbf{eu}^{cd} = -200 \text{ kips (c)} \\ \mathbf{s}^{ce} &= \mathbf{eu}^{ce} = -200 \text{ kips (c)} \end{aligned}$$

Discussion The results for this simple statically determinate structure can be checked using the methods of statics and strength of materials. Note that all rotational degrees of freedom have been excluded from $\mathbf{K}_{ff}$ since they have zero stiffnesses and if included would render the equations ill conditioned with respect to solution; W_e has been excluded for the same reason. There is a stiffness term for W_d because elements *ad* and *bd* extend out of the *x-y* plane.

EXAMPLE 5.2

Analyze the space truss in Fig. E5.2. $A_{de} = A_{ad} = 10 \times 10^{-4}$ m^2; all other element areas are 30×10^{-4} m^2; $E = 200$ GPa.

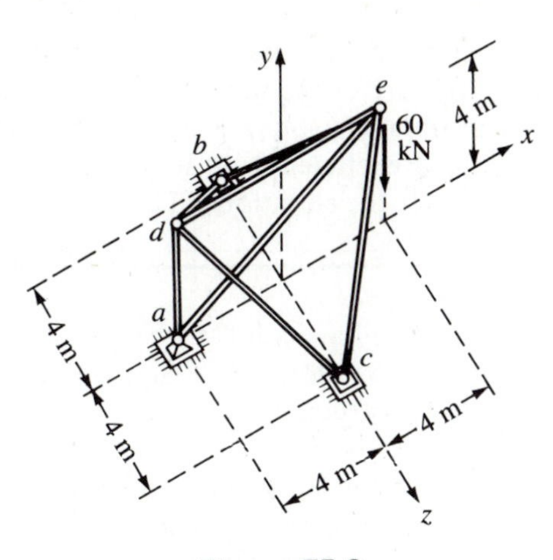

Figure E5.2

Solution

The element stiffness matrices are obtained using Eq. (5.1). Invoking nodal equilibrium for the structure and enforcing the support conditions yields

$$\begin{bmatrix} 0 \\ 0 \\ 0 \\ 0 \\ -60 \\ 0 \end{bmatrix} = \begin{bmatrix} 82735 & -57735 & 00000 & -25000 & 00000 & 00000 \\ -57735 & 107740 & 00000 & 00000 & 00000 & 00000 \\ 00000 & 00000 & 57735 & 00000 & 00000 & 00000 \\ -25000 & 00000 & 00000 & 118510 & 75624 & 00000 \\ 00000 & 00000 & 00000 & 75624 & 66679 & 00000 \\ 00000 & 00000 & 00000 & 00000 & 00000 & 57735 \end{bmatrix} \begin{bmatrix} U_d \\ V_d \\ W_d \\ U_e \\ V_e \\ W_e \end{bmatrix}$$

Hence

$$\begin{bmatrix} U_d \\ V_d \\ W_d \\ U_e \\ V_e \\ W_e \end{bmatrix} = 10^{-3} \begin{bmatrix} 1.588 \\ -0.851 \\ 0.000 \\ 3.291 \\ -4.632 \\ 0.000 \end{bmatrix} \text{ m}$$

Using Eq. (5.9) the element axial forces are

$$\mathbf{s}^{ad} = \mathbf{eu}^{ad} = +42.6 \text{ kN (t)}$$
$$\mathbf{s}^{de} = \mathbf{eu}^{de} = +42.6 \text{ kN (t)}$$
$$\mathbf{s}^{ae} = \mathbf{eu}^{ae} = +39.0 \text{ kN (t)}$$
$$\mathbf{s}^{bd} = \mathbf{eu}^{bd} = -36.9 \text{ kN (c)}$$
$$\mathbf{s}^{be} = \mathbf{eu}^{be} = -67.1 \text{ kN (c)}$$
$$\mathbf{s}^{cd} = \mathbf{eu}^{cd} = -36.9 \text{ kN (c)}$$
$$\mathbf{s}^{ce} = \mathbf{eu}^{ce} = -67.1 \text{ kN (c)}$$

Discussion This space truss is statically indeterminate to the first degree, and unlike the structure of Example 5.1 it would require more than the equations of statics to obtain a solution for member forces. All rotational degrees of freedom have been excluded from $\mathbf{K}_{ff}$ since they have zero stiffnesses and if included would render the equations ill conditioned with respect to solution.

Many space trusses have a form wherein the load path is not easily envisaged. As a result there are some notable cases where unstable structures have been designed and constructed. These critical forms can be detected for relatively simple structures by inspection or by using the *zero-load test.* For an unstable configuration, the coefficient matrix for the equilibrium equations (see Sec. 7.9) has a zero determinant, which implies there is no unique nontrivial solution for the member forces. The zero-load test is performed by attempting to find some nonzero member forces for zero applied loads.

An unstable configuration also implies that the global stiffness matrix equation $\mathbf{P}_f = \mathbf{K}_{ff}\mathbf{U}_f$ has no unique nontrivial solution. That is, there are unbounded displacements for nonzero loads. This situation can be studied by recalling some basic information from linear algebra. In the system of equations

$$\mathbf{AX} + \mathbf{B} = \mathbf{0}$$

where $\mathbf{A}$ is the coefficient matrix, $\mathbf{X}$ is the matrix of unknowns, and $\mathbf{B}$ is the matrix of known quantities. Then $\mathbf{A} \vdots \mathbf{B}$ is the augmented matrix. If the ranks of the coefficient matrix $\mathbf{A}$ and the augmented matrix $\mathbf{A} \vdots \mathbf{B}$ are the same, the system of equations is *consistent.* Let R denote the rank of a consistent set of equations and let N denote the number of unknowns.

The general solution of a set of consistent equations is any particular solution of $\mathbf{AX} + \mathbf{B} = \mathbf{0}$, plus the general solution of the corresponding homogeneous equations $\mathbf{AX} = \mathbf{0}$. If $R = N$, there is no nontrivial solution of $\mathbf{AX} = \mathbf{0}$; the solution of $\mathbf{AX} + \mathbf{B} = \mathbf{0}$ is the general solution. If $R < N$, there are $N - R$ independent solutions of $\mathbf{AX} = \mathbf{0}$ and an infinite number of possible solutions of $\mathbf{AX} + \mathbf{B} = \mathbf{0}$.

Consider a geometrically unstable structure. For certain loadings, the equations will be consistent; since $R < N$, there are nontrivial solutions of $\mathbf{AX} = \mathbf{0}$. For other loadings, the rank of the augmented matrix will be greater than that of the coefficient matrix; the equations will be inconsistent and there will be no solution.

EXAMPLE 5.3

Use the test involving the consistency of the set of equations to find the configuration(s) of the loaded structure in Fig. E5.3 (i.e., values of α) that is (are) unstable and those that are stable. Members ab and bc are axial force elements with length L, and AE/L = constant for both elements.

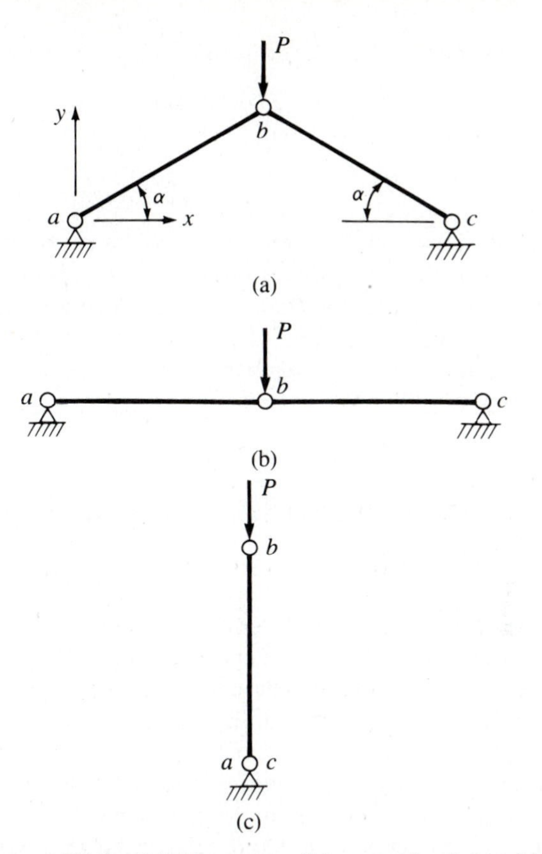

Figure E5.3 (a) Plan truss; (b) unstable; (c) stable

Solution

The element stiffness matrices are given by Eq. (2.22), with $\phi = \alpha$ for ab and $\phi = \pi - \alpha$ for cb; those partitions associated with $\mathbf{K}_{ff}$ are

$$\mathbf{k}_{ff}^{ab} = \frac{AE}{L}\begin{bmatrix} \cos^2\alpha & \sin\alpha\cos\alpha \\ \sin\alpha\cos\alpha & \sin^2\alpha \end{bmatrix}$$

$$\mathbf{k}_{ff}^{cb} = \frac{AE}{L}\begin{bmatrix} \cos^2\alpha & -\sin\alpha\cos\alpha \\ -\sin\alpha\cos\alpha & \sin^2\alpha \end{bmatrix}$$

Therefore

$$\begin{bmatrix} P_{xb} \\ P_{yb} \end{bmatrix} = \begin{bmatrix} 0 \\ -P \end{bmatrix} = \frac{2AE}{L}\begin{bmatrix} \cos^2\alpha & 0 \\ 0 & \sin^2\alpha \end{bmatrix}\begin{bmatrix} U_b \\ V_b \end{bmatrix}$$

The augmented matrix is

$$\left[\begin{array}{cc|c} \cos^2\alpha & 0 & 0 \\ 0 & \sin^2\alpha & -P \end{array}\right]$$

The rank of the stiffness matrix is one if $|\mathbf{K}_{ff}| = 0$, that is, if

$$\cos^2 \alpha \sin^2 \alpha = 0$$

or if

$$(\tfrac{1}{2} \sin 2\alpha)^2 = 0$$

This is true if $\alpha = 0$, $\pi/2$, π, etc. If $\alpha = 0$, the rank of the augmented matrix is two; therefore the equations are inconsistent and we have a mechanism (see Fig. E5.3b). If $\alpha = \pi/2$, the rank of the augmented matrix is one; therefore the equations are consistent and the structure is stable since it consists of two coincidental axial force elements (see Fig. E5.3c).

5.2 THE THREE-DIMENSIONAL BISYMMETRIC BEAM ELEMENT

Three-dimensional structures constructed from flexural members exhibit a more complex behavior than a space truss. In the most general situation there are six degrees of freedom per node, and each element is subjected to axial forces, torsion, shear, and bending about two orthogonal axes. Consider the prismatic beam element in Fig. 5.3; its cross section is symmetric with respect to each of its two principal axes, $\bar{y}$ and $\bar{z}$. When arbitrarily oriented in three-dimensional space, at each node the element must sustain six generalized forces ($\bar{p}_{\bar{x}}, \bar{p}_{\bar{y}}, \bar{p}_{\bar{z}}, \bar{m}_{\bar{x}}, \bar{m}_{\bar{y}}, \bar{m}_{\bar{z}}$) and will exhibit the corresponding six degrees of freedom ($\bar{u}, \bar{v}, \bar{w}, \bar{\theta}_{\bar{x}}, \bar{\theta}_{\bar{y}}, \bar{\theta}_{\bar{z}}$). All these quantities are shown in Fig. 5.3 directed in their respective positive local coordinate directions, with the moments and their corresponding rotations indicated as vectors with double arrowheads. This linear elastic element has four distinct response modes: (1) tension-compression in the $\bar{x}$ direction; (2) flexure in the $\bar{x}$-$\bar{y}$ plane; (3) flexure in the

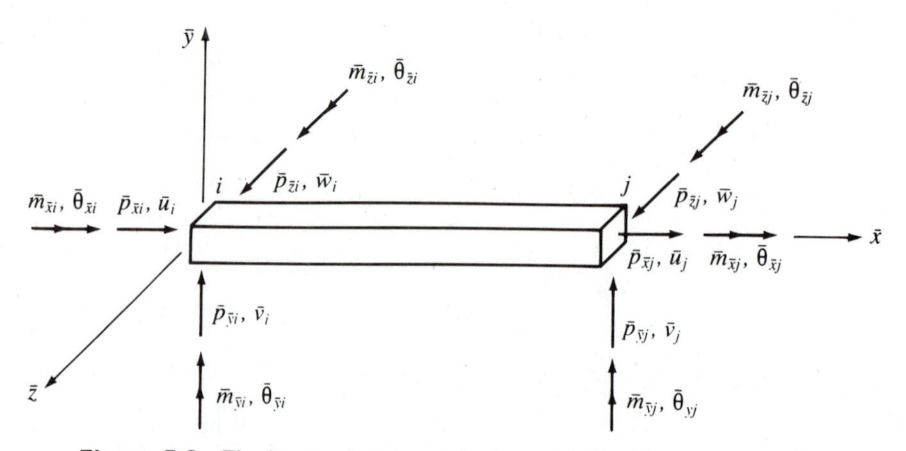

Figure 5.3 The beam element with six generalized forces per node

$\bar{x}$-$\bar{z}$ plane; and (4) torsion about the $\bar{x}$ axis. For small displacements these four individual responses are uncoupled; that is, the generalized forces from each individual response mode produce neither forces nor displacements associated with the other modes.

5.2.1 AXIAL AND BENDING EFFECTS The generalized force-displacement relationship for the axial-force response is shown in Eq. (5.1). The force-displacement equations for flexure in the $\bar{x}$-$\bar{y}$ plane are given in Eq. (4.19), but the moment of inertia must now be noted more explicitly as $I_{\bar{z}}$, where

$$I_{\bar{z}} = \int_{\text{area}} \bar{y}^2 \, dA \tag{5.11}$$

and the moments and rotations must also be subscripted with a $\bar{z}$. Combining the stiffness matrices for the axial force and prismatic beam elements yields the equivalent of the prismatic plane frame element, which is described in Ch. 4.

The generalized force-displacement equations for flexure in the $\bar{x}$-$\bar{z}$ plane (see Fig. 5.4) are similar to those expressed for the $\bar{x}$-$\bar{y}$ plane; thus

$$\begin{aligned}
\bar{p}_{\bar{z}i} &= \frac{EI_{\bar{y}}}{L}\left(\frac{12}{L^2}\bar{w}_i - \frac{6}{L}\bar{\theta}_{\bar{y}i} - \frac{12}{L^2}\bar{w}_j - \frac{6}{L}\bar{\theta}_{\bar{y}j}\right)\\
\bar{m}_{\bar{y}i} &= \frac{EI_{\bar{y}}}{L}\left(-\frac{6}{L}\bar{w}_i + 4\bar{\theta}_{\bar{y}i} + \frac{6}{L}\bar{w}_j + 2\bar{\theta}_{\bar{y}j}\right)\\
\bar{p}_{\bar{z}j} &= \frac{EI_{\bar{y}}}{L}\left(-\frac{12}{L^2}\bar{w}_i + \frac{6}{L}\bar{\theta}_{\bar{y}i} + \frac{12}{L^2}\bar{w}_j + \frac{6}{L}\bar{\theta}_{\bar{y}j}\right)\\
\bar{m}_{\bar{y}j} &= \frac{EI_{\bar{y}}}{L}\left(-\frac{6}{L}\bar{w}_i + 2\bar{\theta}_{\bar{y}i} + \frac{6}{L}\bar{w}_j + 4\bar{\theta}_{\bar{y}j}\right)
\end{aligned} \tag{5.12}$$

where

$$I_{\bar{y}} = \int_{\text{area}} \bar{z}^2 \, dA \tag{5.13}$$

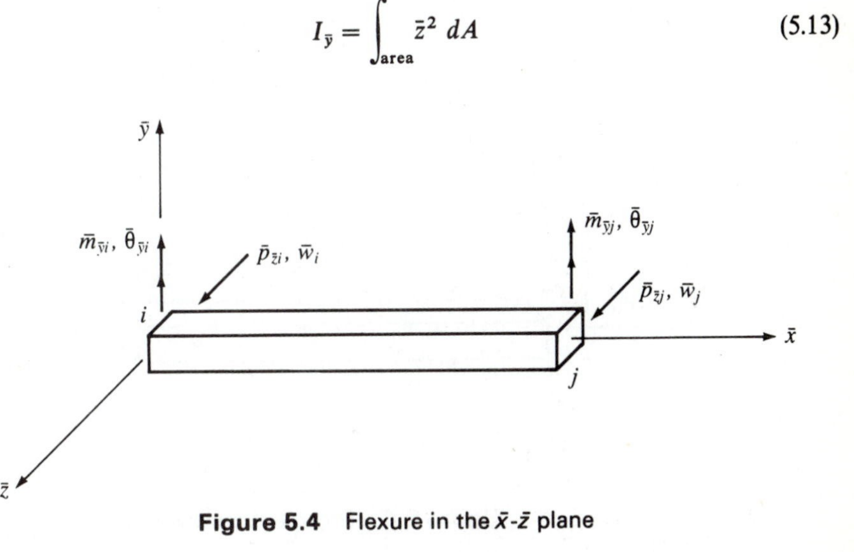

Figure 5.4 Flexure in the $\bar{x}$-$\bar{z}$ plane

Hence, we have for the beam bending about its $\bar{y}$ axis

$$\begin{bmatrix} \bar{p}_{\bar{z}i} \\ \bar{m}_{\bar{y}i} \\ \bar{p}_{\bar{z}j} \\ \bar{m}_{\bar{y}j} \end{bmatrix} = \frac{EI_{\bar{y}}}{L} \begin{bmatrix} \frac{12}{L^2} & -\frac{6}{L} & -\frac{12}{L^2} & -\frac{6}{L} \\ -\frac{6}{L} & 4 & \frac{6}{L} & 2 \\ -\frac{12}{L^2} & \frac{6}{L} & \frac{12}{L^2} & \frac{6}{L} \\ -\frac{6}{L} & 2 & \frac{6}{L} & 4 \end{bmatrix} \begin{bmatrix} \bar{w}_i \\ \bar{\theta}_{\bar{y}i} \\ \bar{w}_j \\ \bar{\theta}_{\bar{y}j} \end{bmatrix} \tag{5.14}$$

$$\bar{\mathbf{p}} = \bar{\mathbf{k}} \quad \bar{\mathbf{u}}$$

It is left as an exercise for the reader to verify the sign changes that occur between the sets of equations for flexural behavior in the two planes.

All the equations for bending have been obtained using the assumption that plane sections remain plane after distortion, but this can be altered by shear deformations in certain cases (e.g., those involving beams with perforated or laced webs or when the depth-to-span ratio is large). The additional displacements due to web shear strain are negligible for most civil engineering structures.

5.2.2 TORSION The generalized force-displacement relationships for the torsional effects of the element can be investigated using the basic governing equations. The element shown in Fig. 5.5 is subjected to end moments $\bar{m}_{\bar{x}i}$ and $\bar{m}_{\bar{x}j}$. A uniform element that is not restrained against longitudinal deformation (warping) and is subjected to pure torsion along its length experiences a twist per unit length of

$$\frac{d\bar{\theta}_{\bar{x}}}{d\bar{x}} = \frac{\bar{m}_{\bar{x}}}{GJ} \tag{5.15}$$

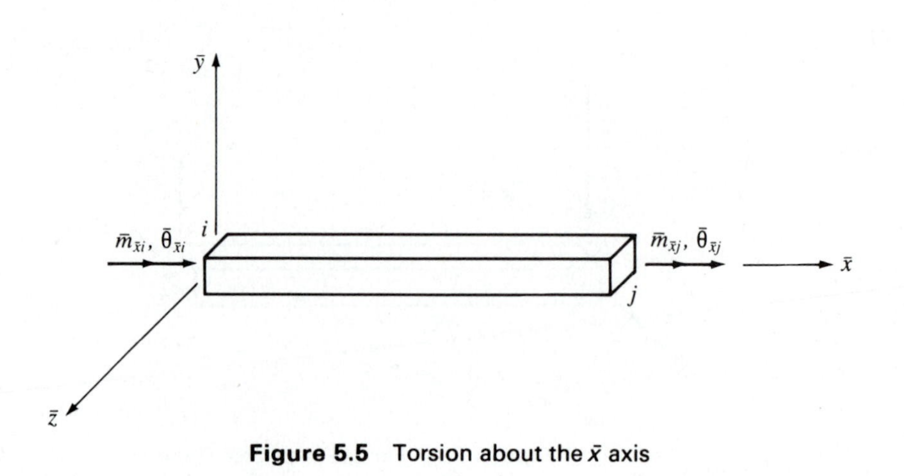

Figure 5.5 Torsion about the $\bar{x}$ axis

where $\bar{m}_{\bar{x}}$ = twisting moment at a generic section
G = modulus of rigidity = $E/2(1 + \nu)$
ν = Poisson's ratio
J = torsion constant

The torsion constant, also referred to as the *St. Venant torsion constant*, depends upon the geometry of the cross section. For a circular cross section J is the polar moment of inertia, but for other shapes, J must be computed using methods from the theory of elasticity. Although the derivation of J is beyond the scope of this book, approximate expressions for several common shapes are presented in Fig. 5.6. Integrating Eq. (5.15) over the length of the element gives

$$\bar{\theta}_{\bar{x}j} - \bar{\theta}_{\bar{x}i} = \frac{L}{GJ}\,\bar{m}_{\bar{x}j} \tag{5.16}$$

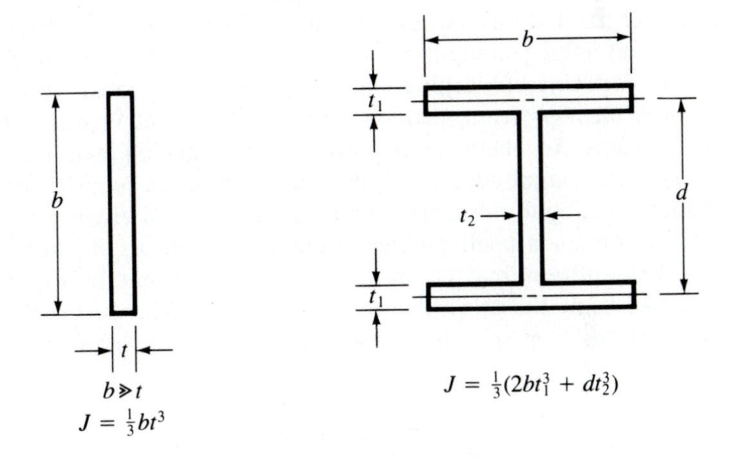

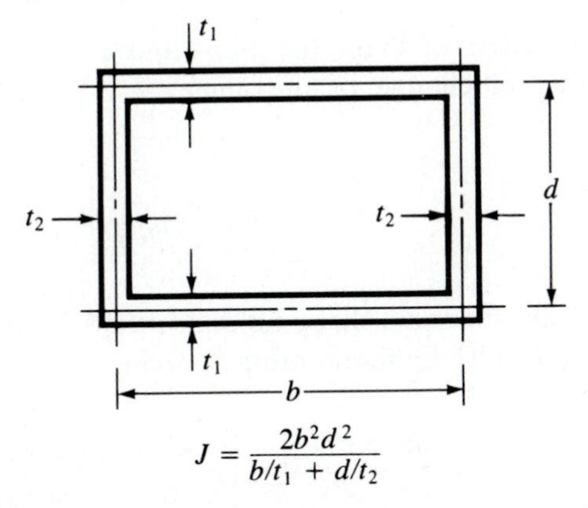

Figure 5.6 St. Venant's torsion constants (J) for some common shapes

Moment equilibrium about the $\bar{x}$ axis requires that

$$\bar{m}_{\bar{x}i} = -\bar{m}_{\bar{x}j} \tag{5.17}$$

Thus, the relations between the generalized forces and displacements for torsion are

$$\begin{aligned} \bar{m}_{\bar{x}i} &= \frac{GJ}{L}(\bar{\theta}_{\bar{x}i} - \bar{\theta}_{\bar{x}j}) \\ \bar{m}_{\bar{x}j} &= \frac{GJ}{L}(-\bar{\theta}_{\bar{x}i} + \bar{\theta}_{\bar{x}j}) \end{aligned} \tag{5.18}$$

Torsion of members with circular cross sections results in only the uniform St. Venant stresses, and the distortions occur in the planes normal to the member axis. For other cross sections, deformations due to the shearing strains will occur in the longitudinal direction; this longitudinal deformation of the cross sections due to torsion is termed *warping*. We can envisage that if a member with a propensity for warping is rigidly confined at each end, considerable axial stresses will be produced. For closed cylindrical or rectangular tubes and solid rectangular sections (e.g., concrete members) normal stresses due to restraint of warping can generally be neglected; this may not be the case for member cross sections such as rolled-steel W sections and light-gage steel beams. In these cases, restraint of warping can produce significant stresses, the magnitudes of which depend upon the support conditions and the induced torsional moments. For most typical civil engineering structures, the applied torsion is small, and we can ignore the effects of constrained warping.

The stiffness matrix for the torsional element in Fig. 5.5 can also be obtained using the principle of virtual work as described in Sec. 3.2. The rotations of the element about the local $\bar{x}$ axis are described by the matrix equation

$$\bar{\theta}_{\bar{x}} = \mathbf{N}\bar{\mathbf{u}} \tag{5.19}$$

where $\mathbf{N}$ is the matrix of shape functions and $\bar{\mathbf{u}} = \{\bar{\theta}_{\bar{x}i} \quad \bar{\theta}_{\bar{x}j}\}$. For a prismatic element the rate of change of the *angle of twist* ($\alpha = d\bar{\theta}_{\bar{x}}/d\bar{x}$) is constant; therefore

$$\mathbf{N} = \left[\left(1 - \frac{\bar{x}}{L}\right) \quad \frac{\bar{x}}{L}\right] \tag{5.20}$$

Note that these are the same shape functions used for the two-node axial force element [see Eq. (3.39)]. Differentiating $\bar{\theta}_{\bar{x}}$ yields

$$\begin{aligned} \alpha &= \frac{d\bar{\theta}}{d\bar{x}} = \frac{dN}{d\bar{x}}\bar{\mathbf{u}} = \begin{bmatrix} -\frac{1}{L} & \frac{1}{L} \end{bmatrix} \begin{bmatrix} \bar{\theta}_{\bar{x}i} \\ \bar{\theta}_{\bar{x}j} \end{bmatrix} \\ \alpha &= \qquad\qquad \mathbf{B} \qquad\qquad \bar{\mathbf{u}} \end{aligned} \tag{5.21}$$

The internal virtual work for this element is

$$\delta W_i = \int_0^L \delta\alpha^T \bar{m}_{\bar{x}} \, d\bar{x} \tag{5.22}$$

The torsional moment is given by Eq. (5.15) in conjunction with Eq. (5.21) as

$$\bar{m}_{\bar{x}} = GJ\alpha = GJ\mathbf{B}\bar{\mathbf{u}} \tag{5.23}$$

The generalized virtual displacement will be selected to be of the same form as the real displacements in Eq. (5.21); thus

$$\delta\alpha = \mathbf{B} \, \boldsymbol{\delta}\bar{\mathbf{u}} \tag{5.24}$$

Substituting Eqs. (5.23) and (5.24) into Eq. (5.22) gives

$$\delta W_i = \int_0^L \boldsymbol{\delta}\bar{\mathbf{u}}^T \mathbf{B}^T GJ\mathbf{B}\bar{\mathbf{u}} \, dx \tag{5.25}$$

The virtual work of the generalized nodal forces is

$$\delta W_e = \boldsymbol{\delta}\bar{\mathbf{u}}^T \bar{\mathbf{p}} \tag{5.26}$$

where $\bar{\mathbf{p}} = \{\bar{m}_{\bar{x}i} \quad \bar{m}_{\bar{x}j}\}$. Applying the theorem of virtual work, substituting $\mathbf{B}$ from Eq. (5.21), carrying out the integration in Eq. (5.25), and noting that $\boldsymbol{\delta}\bar{\mathbf{u}}$ is nonzero and arbitrary gives the following force-displacement relations for the torsional element:

$$\underset{\bar{\mathbf{p}}}{\begin{bmatrix} \bar{m}_{\bar{x}i} \\ \bar{m}_{\bar{x}j} \end{bmatrix}} = \underset{\bar{\mathbf{k}}}{\frac{GJ}{L}\begin{bmatrix} 1 & -1 \\ -1 & 1 \end{bmatrix}} \underset{\bar{\mathbf{u}}}{\begin{bmatrix} \bar{\theta}_{\bar{x}i} \\ \bar{\theta}_{\bar{x}j} \end{bmatrix}} \tag{5.27}$$

These are the same results obtained by solving the basic equations [see Eqs. (5.18)].

5.2.3 THE ELEMENT STIFFNESS MATRIX The general 12 by 12 stiffness matrix for the three-dimensional bisymmetric beam element in Fig. 5.3 can be obtained by combining the force-displacement relations for the four individual response modes. Thus, by assembling the matrices from Eqs. (5.1), (4.19), (5.14), and (5.27), we get $\bar{\mathbf{k}}$ in local coordinates as follows:

$$\bar{\mathbf{p}} = \bar{\mathbf{k}}\bar{\mathbf{u}} \tag{5.28}$$

where

$$\bar{\mathbf{k}} = \left[\begin{array}{cccccc|cccccc}
\frac{AE}{L} & 0 & 0 & 0 & 0 & 0 & -\frac{AE}{L} & 0 & 0 & 0 & 0 & 0 \\
0 & \frac{12EI_{\bar{z}}}{L^3} & 0 & 0 & 0 & \frac{6EI_{\bar{z}}}{L^2} & 0 & -\frac{12EI_{\bar{z}}}{L^3} & 0 & 0 & 0 & \frac{6EI_{\bar{z}}}{L^2} \\
0 & 0 & \frac{12EI_{\bar{y}}}{L^3} & 0 & -\frac{6EI_{\bar{y}}}{L^2} & 0 & 0 & 0 & -\frac{12EI_{\bar{y}}}{L^3} & 0 & -\frac{6EI_{\bar{y}}}{L^2} & 0 \\
0 & 0 & 0 & \frac{GJ}{L} & 0 & 0 & 0 & 0 & 0 & -\frac{GJ}{L} & 0 & 0 \\
0 & 0 & -\frac{6EI_{\bar{y}}}{L^2} & 0 & \frac{4EI_{\bar{y}}}{L} & 0 & 0 & 0 & \frac{6EI_{\bar{y}}}{L^2} & 0 & \frac{2EI_{\bar{y}}}{L} & 0 \\
0 & \frac{6EI_{\bar{z}}}{L^2} & 0 & 0 & 0 & \frac{4EI_{\bar{z}}}{L} & 0 & -\frac{6EI_{\bar{z}}}{L^2} & 0 & 0 & 0 & \frac{2EI_{\bar{z}}}{L} \\
\hline
-\frac{AE}{L} & 0 & 0 & 0 & 0 & 0 & \frac{AE}{L} & 0 & 0 & 0 & 0 & 0 \\
0 & -\frac{12EI_{\bar{z}}}{L^3} & 0 & 0 & 0 & -\frac{6EI_{\bar{z}}}{L^2} & 0 & \frac{12EI_{\bar{z}}}{L^3} & 0 & 0 & 0 & -\frac{6EI_{\bar{z}}}{L^2} \\
0 & 0 & -\frac{12EI_{\bar{y}}}{L^3} & 0 & \frac{6EI_{\bar{y}}}{L^2} & 0 & 0 & 0 & \frac{12EI_{\bar{y}}}{L^3} & 0 & \frac{6EI_{\bar{y}}}{L^2} & 0 \\
0 & 0 & 0 & -\frac{GJ}{L} & 0 & 0 & 0 & 0 & 0 & \frac{GJ}{L} & 0 & 0 \\
0 & 0 & -\frac{6EI_{\bar{y}}}{L^2} & 0 & \frac{2EI_{\bar{y}}}{L} & 0 & 0 & 0 & \frac{6EI_{\bar{y}}}{L^2} & 0 & \frac{4EI_{\bar{y}}}{L} & 0 \\
0 & \frac{6EI_{\bar{z}}}{L^2} & 0 & 0 & 0 & \frac{2EI_{\bar{z}}}{L} & 0 & -\frac{6EI_{\bar{z}}}{L^2} & 0 & 0 & 0 & \frac{4EI_{\bar{z}}}{L}
\end{array}\right] \quad (5.28a)$$

$$\bar{\mathbf{p}} = \{\bar{p}_{\bar{x}i} \quad \bar{p}_{\bar{y}i} \quad \bar{p}_{\bar{z}i} \quad \bar{m}_{\bar{x}i} \quad \bar{m}_{\bar{y}i} \quad \bar{m}_{\bar{z}i} \quad \bar{p}_{\bar{x}j} \quad \bar{p}_{\bar{y}j} \quad \bar{p}_{\bar{z}j} \quad \bar{m}_{\bar{x}j} \quad \bar{m}_{\bar{y}j} \quad \bar{m}_{\bar{z}j}\}$$

$$\bar{\mathbf{u}} = \{\bar{u}_i \quad \bar{v}_i \quad \bar{w}_i \quad \bar{\theta}_{\bar{x}i} \quad \bar{\theta}_{\bar{y}i} \quad \bar{\theta}_{\bar{z}i} \quad \bar{u}_j \quad \bar{v}_j \quad \bar{w}_j \quad \bar{\theta}_{\bar{x}j} \quad \bar{\theta}_{\bar{y}j} \quad \bar{\theta}_{\bar{z}j}\}$$

Interpretation of element forces for general three-dimensional structures is difficult; therefore, it is advisable to compute all six forces at both nodes of each element. Thus, we transform element nodal displacements into local coordinates using the equations of Sec. 5.4 and premultiply these by the element stiffness matrix.

The structural engineer must decide when the various responses such as warping, bending about the $\bar{y}$ axis, and torsion are important; this raises the more general question of how to approach the analysis of complex structures to obtain an economical and accurate solution. Problem-solution methods can be broadly divided into two approaches: (a) all approximations are made at the beginning of the investigation to give a simplified model or (b) no simplifying assumptions are made, and all effects, no matter how insignificant, are included in the analysis. The structural engineer with sufficient experience is capable of making assumptions to yield a solution that characterizes the behavior of the structure with the required degree of accuracy. Making rational initial assumptions should not be confused with the totally unacceptable approach of neglecting effects that the engineer neither understands nor is able to calculate.

5.3 COORDINATE TRANSFORMATIONS IN THREE-DIMENSIONAL SPACE

The three-dimensional beam element, along with the local coordinate system, is shown oriented in the global coordinate system in Fig. 5.7. There are three independent orthogonal components of force and a similar number of moment

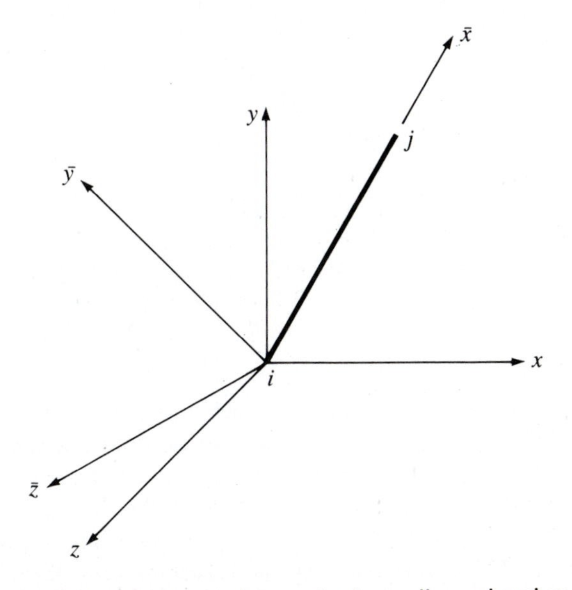

Figure 5.7 The beam element in three-dimensional space

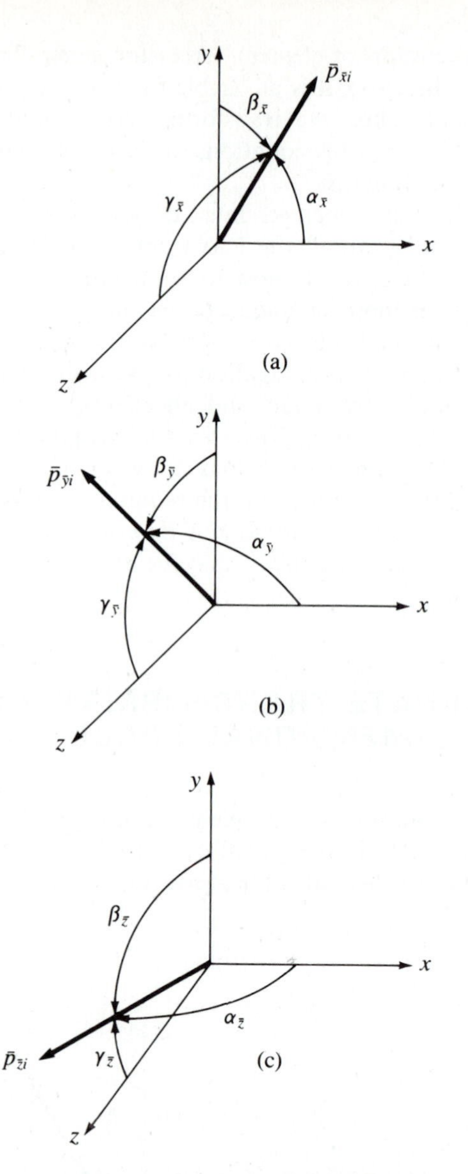

Figure 5.8 Transformation of forces from local to global coordinates: (a) $\bar{p}_{\bar{x}i}$; (b) $\bar{p}_{\bar{y}i}$; (c) $\bar{p}_{\bar{z}i}$

components per node, each of which must be transformed from the local to the global coordinate system. For clarity, only one component of force is displayed in Fig. 5.8a to c. Denoting the angles between the force $\bar{p}_{\bar{x}i}$ and the x, y, and z axes as $\alpha_{\bar{x}}$, $\beta_{\bar{x}}$, and $\gamma_{\bar{x}}$, respectively, the three components of this force in the respective global x, y, and z directions are $\bar{p}_{\bar{x}i} \cos \alpha_{\bar{x}}$, $\bar{p}_{\bar{x}i} \cos \beta_{\bar{x}}$, and $\bar{p}_{\bar{x}i} \cos \gamma_{\bar{x}}$. The components of the forces $\bar{p}_{\bar{y}i}$ and $\bar{p}_{\bar{z}i}$ in the global coordinates can be expressed in a similar fashion using the notation of Fig. 5.8b and c.

Combining the components of $\bar{p}_{\bar{x}i}$, $\bar{p}_{\bar{y}i}$, and $\bar{p}_{\bar{z}i}$ gives

$$\begin{aligned} p_{xi} &= \bar{p}_{\bar{x}i} \cos \alpha_{\bar{x}} + \bar{p}_{\bar{y}i} \cos \alpha_{\bar{y}} + \bar{p}_{\bar{z}i} \cos \alpha_{\bar{z}} \\ p_{yi} &= \bar{p}_{\bar{x}i} \cos \beta_{\bar{x}} + \bar{p}_{\bar{y}i} \cos \beta_{\bar{y}} + \bar{p}_{\bar{z}i} \cos \beta_{\bar{z}} \\ p_{zi} &= \bar{p}_{\bar{x}i} \cos \gamma_{\bar{x}} + \bar{p}_{\bar{y}i} \cos \gamma_{\bar{y}} + \bar{p}_{\bar{z}i} \cos \gamma_{\bar{z}} \end{aligned} \tag{5.29}$$

Let λ, μ, and ν with the appropriate subscripts denote the direction cosines; these transformation equations can be written as follows:

$$\begin{aligned} p_{xi} &= \bar{p}_{\bar{x}i} \lambda_{\bar{x}} + \bar{p}_{\bar{y}i} \lambda_{\bar{y}} + \bar{p}_{\bar{z}i} \lambda_{\bar{z}} \\ p_{yi} &= \bar{p}_{\bar{x}i} \mu_{\bar{x}} + \bar{p}_{\bar{y}i} \mu_{\bar{y}} + \bar{p}_{\bar{z}i} \mu_{\bar{z}} \\ p_{zi} &= \bar{p}_{\bar{x}i} \nu_{\bar{x}} + \bar{p}_{\bar{y}i} \nu_{\bar{y}} + \bar{p}_{\bar{z}i} \nu_{\bar{z}} \end{aligned} \tag{5.30}$$

or

$$\begin{bmatrix} p_{xi} \\ p_{yi} \\ p_{zi} \end{bmatrix} = \begin{bmatrix} \lambda_{\bar{x}} & \lambda_{\bar{y}} & \lambda_{\bar{z}} \\ \mu_{\bar{x}} & \mu_{\bar{y}} & \mu_{\bar{z}} \\ \nu_{\bar{x}} & \nu_{\bar{y}} & \nu_{\bar{z}} \end{bmatrix} \begin{bmatrix} \bar{p}_{\bar{x}i} \\ \bar{p}_{\bar{y}i} \\ \bar{p}_{\bar{z}i} \end{bmatrix} \tag{5.31}$$

$$\mathbf{p}_i = \mathbf{\Gamma}^T \bar{\mathbf{p}}_i$$

Since $\mathbf{\Gamma}^T$ is an orthogonal transformation,

$$(\mathbf{\Gamma}^T)^T = \mathbf{\Gamma} = (\mathbf{\Gamma}^T)^{-1}$$

or

$$\mathbf{\Gamma}^T = \mathbf{\Gamma}^{-1} \tag{5.32}$$

Therefore

$$\bar{\mathbf{p}}_i = \mathbf{\Gamma} \mathbf{p}_i \tag{5.33}$$

This same transformation can be carried out for the moments at node i and for all the generalized forces at node j, yielding

$$\left[\begin{array}{c} \bar{p}_{\bar{x}i} \\ \bar{p}_{\bar{y}i} \\ \bar{p}_{\bar{z}i} \\ \hline \bar{m}_{\bar{x}i} \\ \bar{m}_{\bar{y}i} \\ \bar{m}_{\bar{z}i} \\ \hline \bar{p}_{\bar{x}j} \\ \bar{p}_{\bar{y}j} \\ \bar{p}_{\bar{z}j} \\ \hline \bar{m}_{\bar{x}j} \\ \bar{m}_{\bar{y}j} \\ \bar{m}_{\bar{z}j} \end{array}\right] = \left[\begin{array}{c|c|c|c} \mathbf{\Gamma} & \mathbf{0} & \mathbf{0} & \mathbf{0} \\ \hline \mathbf{0} & \mathbf{\Gamma} & \mathbf{0} & \mathbf{0} \\ \hline \mathbf{0} & \mathbf{0} & \mathbf{\Gamma} & \mathbf{0} \\ \hline \mathbf{0} & \mathbf{0} & \mathbf{0} & \mathbf{\Gamma} \end{array}\right] \left[\begin{array}{c} p_{xi} \\ p_{yi} \\ p_{zi} \\ \hline m_{xi} \\ m_{yi} \\ m_{zi} \\ \hline p_{xj} \\ p_{yj} \\ p_{zj} \\ \hline m_{xj} \\ m_{yj} \\ m_{zj} \end{array}\right] \tag{5.34}$$

$$\bar{\mathbf{p}} = T \mathbf{p}$$

where $\mathbf{0}$ indicates a 3 by 3 null matrix. Just as $\mathbf{\Gamma}$ is orthogonal, so also is $\mathbf{T}$. Thus

$$\mathbf{T}^T = \mathbf{T}^{-1} \qquad \text{and} \qquad \mathbf{p} = \mathbf{T}^T\bar{\mathbf{p}} \tag{5.35}$$

In addition, the generalized displacements transform in the usual fashion with this 12 by 12 version of $\mathbf{T}$, that is,

$$\bar{\mathbf{u}} = \mathbf{T}\mathbf{u} \qquad \text{and} \qquad \mathbf{u} = \mathbf{T}^T\bar{\mathbf{u}} \tag{5.36}$$

Furthermore, the stiffness matrix in local coordinates as given in Eq. (5.28a) can be transformed into global coordinates using $\mathbf{k} = \mathbf{T}^T\bar{\mathbf{k}}\mathbf{T}$.

5.4 GRID STRUCTURES

Plane grid structures typically consist of a series of intersecting long beam elements with parallel upper and lower chords, an example of which is shown in Fig. 5.9. The joints at intersection points are constructed to be rigid. In plan view the structure is typically rectangular, but it can be of almost any geometry; members may be arranged to form either a rectangular or a skew grid. The loads are transmitted to the supports in the plane of the grid by bending and torsion. The elements are generally assumed to be axially rigid, thus precluding bending about the vertical (weak) axis. Therefore, there are three degrees of freedom per node. The general three-dimensional beam element can be used, but the axial stiffness terms, plus those for bending about

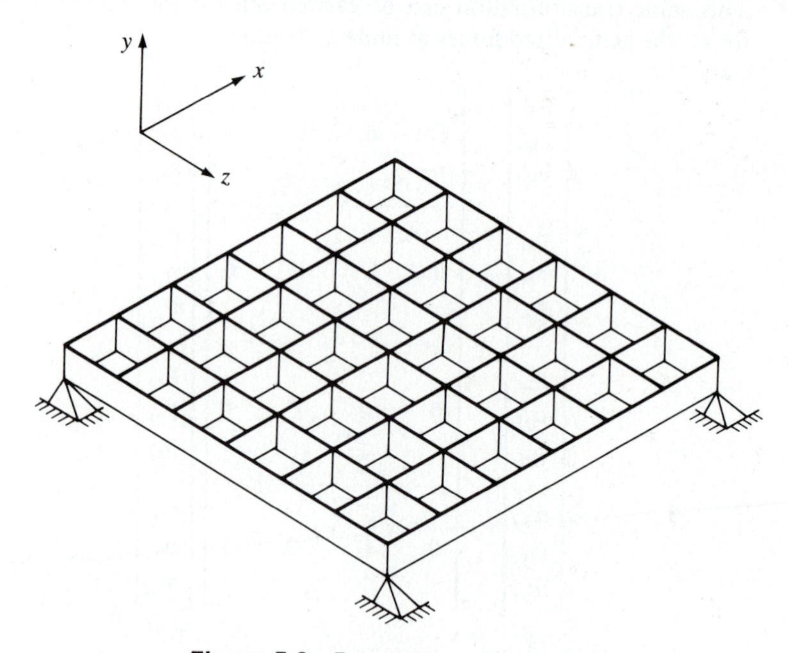

Figure 5.9 Rectangular grid structure

the $\bar{y}$ axis, must be given very large values. Alternatively, the torsional element and the flexural element for bending about the $\bar{z}$ axis can be combined to give an element specifically for the grid structures. The stiffness matrix for the grid element is 6 by 6, and the force-displacement equations are

$$\begin{bmatrix} \bar{p}_{\bar{y}i} \\ \bar{m}_{\bar{x}i} \\ \bar{m}_{\bar{z}i} \\ \bar{p}_{\bar{y}j} \\ \bar{m}_{\bar{x}j} \\ \bar{m}_{\bar{z}j} \end{bmatrix} = \begin{bmatrix} \frac{12EI_{\bar{z}}}{L^3} & 0 & \frac{6EI_{\bar{z}}}{L^2} & -\frac{12EI_{\bar{z}}}{L^3} & 0 & \frac{6EI_{\bar{z}}}{L^2} \\ 0 & \frac{GJ}{L} & 0 & 0 & -\frac{GJ}{L} & 0 \\ \frac{6EI_{\bar{z}}}{L^2} & 0 & \frac{4EI_{\bar{z}}}{L} & -\frac{6EI_{\bar{z}}}{L^2} & 0 & \frac{2EI_{\bar{z}}}{L} \\ -\frac{12EI_{\bar{z}}}{L^3} & 0 & -\frac{6EI_{\bar{z}}}{L^2} & \frac{12EI_{\bar{z}}}{L^2} & 0 & -\frac{6EI_{\bar{z}}}{L^2} \\ 0 & -\frac{GJ}{L} & 0 & 0 & \frac{GJ}{L} & 0 \\ \frac{6EI_{\bar{z}}}{L^2} & 0 & \frac{2EI_{\bar{z}}}{L} & -\frac{6EI_{\bar{z}}}{L^2} & 0 & \frac{4EI_{\bar{z}}}{L} \end{bmatrix} \begin{bmatrix} \bar{v}_i \\ \bar{\theta}_{\bar{x}i} \\ \bar{\theta}_{\bar{z}i} \\ \bar{v}_j \\ \bar{\theta}_{\bar{x}j} \\ \bar{\theta}_{\bar{z}j} \end{bmatrix} \tag{5.37}$$

Transforming the displacements from global to local coordinates for a grid structure is accomplished as follows:

$$\begin{bmatrix} \bar{v}_i \\ \bar{\theta}_{\bar{x}i} \\ \bar{\theta}_{\bar{z}i} \end{bmatrix} = \begin{bmatrix} \lambda_{\bar{y}} & \mu_{\bar{y}} & \nu_{\bar{y}} & 0 & 0 & 0 \\ 0 & 0 & 0 & \lambda_{\bar{x}} & \mu_{\bar{x}} & \nu_{\bar{x}} \\ 0 & 0 & 0 & \lambda_{\bar{z}} & \mu_{\bar{z}} & \nu_{\bar{z}} \end{bmatrix} \begin{bmatrix} u_i \\ v_i \\ w_i \\ \theta_{zi} \\ \theta_{yi} \\ \theta_{zi} \end{bmatrix} \tag{5.38}$$

$$\bar{\mathbf{u}}_i = \boldsymbol{\Gamma}_G \mathbf{u}_i$$

where $\boldsymbol{\Gamma}_G$ is a transformation matrix for grid structures. For the degrees of freedom at both ends of a grid element

$$\begin{bmatrix} \bar{v}_i \\ \bar{\theta}_{\bar{x}i} \\ \bar{\theta}_{\bar{z}i} \\ \bar{v}_j \\ \bar{\theta}_{\bar{x}j} \\ \bar{\theta}_{\bar{z}j} \end{bmatrix} = \begin{bmatrix} \boldsymbol{\Gamma}_G & \mathbf{0} \\ \mathbf{0} & \boldsymbol{\Gamma}_G \end{bmatrix} \begin{bmatrix} u_i \\ v_i \\ w_i \\ \theta_{xi} \\ \theta_{yi} \\ \theta_{zi} \\ u_j \\ v_j \\ w_j \\ \theta_{xj} \\ \theta_{yj} \\ \theta_{zj} \end{bmatrix} \tag{5.39}$$

$$\bar{\mathbf{u}} = \mathbf{T}_G \mathbf{u}$$

There is a corresponding contragredient transformation for the generalized forces, so that

$$\mathbf{p} = \mathbf{T}_G^T \bar{\mathbf{p}} \tag{5.40}$$

where $\mathbf{p} = \{\mathbf{p}_i \quad \mathbf{p}_j\}$, $\mathbf{p}_i = \{p_{xi} \quad p_{yi} \quad p_{zi} \quad m_{xi} \quad m_{yi} \quad m_{zi}\}$, $\mathbf{p}_j$ has the same generalized forces as $\mathbf{p}_i$ at the end j of the element; $\bar{\mathbf{p}} = \{\bar{\mathbf{p}}_i \quad \bar{\mathbf{p}}_j\}$, with $\bar{\mathbf{p}}_i = \{\bar{p}_{\bar{y}i} \quad \bar{m}_{\bar{x}i} \quad \bar{m}_{\bar{y}i}\}$, and $\bar{\mathbf{p}}_j$ has the same generalized forces as $\bar{\mathbf{p}}_i$ but at the end j of the element. The stiffness matrix can be transformed between coordinates in the usual fashion, that is,

$$\mathbf{k} = \mathbf{T}_G^T \bar{\mathbf{k}} \mathbf{T}_G \qquad \text{and} \qquad \bar{\mathbf{k}} = \mathbf{T}_G \mathbf{k} \mathbf{T}_G^T \tag{5.41}$$

Subsequent to calculating nodal displacements the element forces can be calculated by transforming element displacements to local coordinates using Eq. (5.39); these can then be premultiplied by the element stiffness matrix of Eq. (5.37) to give three generalized forces at each node. Alternatively, we know that $\bar{p}_{\bar{y}j} = -\bar{p}_{\bar{y}i}$ and $\bar{m}_{\bar{x}j} = -\bar{m}_{\bar{x}i}$; therefore, the element force matrix in local coordinates ($\bar{\mathbf{e}}$) can be obtained using the first, second, third, and sixth equations of Eq. (5.37). Thus using an argument similar to that described in Ch. 2 we have

$$\mathbf{s}^{ij} = \bar{\mathbf{e}}\bar{\mathbf{u}}^{ij} = \bar{\mathbf{e}}\mathbf{T}_G \mathbf{u}^{ij} = \mathbf{e}\mathbf{u}^{ij} \tag{5.42}$$

where

$$\mathbf{e} = \bar{\mathbf{e}}\mathbf{T}_G \tag{5.43}$$

and

$$\mathbf{s}^{ij} = \{\bar{p}_{\bar{y}i} \quad \bar{m}_{\bar{x}i} \quad \bar{m}_{\bar{z}i} \quad \bar{m}_{\bar{z}j}\} \tag{5.44}$$

A simple grid is analyzed in Example 5.4 to demonstrate the use of these equations. This structure can be envisaged as a part of an extensive orthogonal grid, but only two members are considered in the example so that numerical calculations do not mask the behavior.

EXAMPLE 5.4

Analyze the simple grid in Fig. E5.4. For all elements $I_{\bar{y}} = 350$ in.^{4}, $I_{\bar{z}} = 700$ in.^{4}; $J = 800$ in.^{4}; $E = 29 \times 10^3$ kips/in.^{2}; and $G = 11 \times 10^3$ kips/in.^{2}.

Solution

The element stiffness matrices are obtained using Eq. (5.37). Assuming axial inextensibility of all elements, we need consider only V_a, Θ_{xa}, and Θ_{za}. Note that this simplifies calculation of relevant element stiffness entries (we urge the reader to calculate the appropriate 3 by 3 partition for the four element stiffness matrices). Invoking nodal equilibrium for the structure with support conditions enforced yields (the reader is encourage to carry out this process)

$$\begin{bmatrix} -160 \\ 4000 \\ 1440 \end{bmatrix} = \begin{bmatrix} 264.35 & 6343.8 & -3263.2 \\ 6343.8 & 1116900 & 00000 \\ -3263.2 & 00000 & 1049800 \end{bmatrix} \begin{bmatrix} V_a \\ \Theta_{xa} \\ \Theta_{za} \end{bmatrix}$$

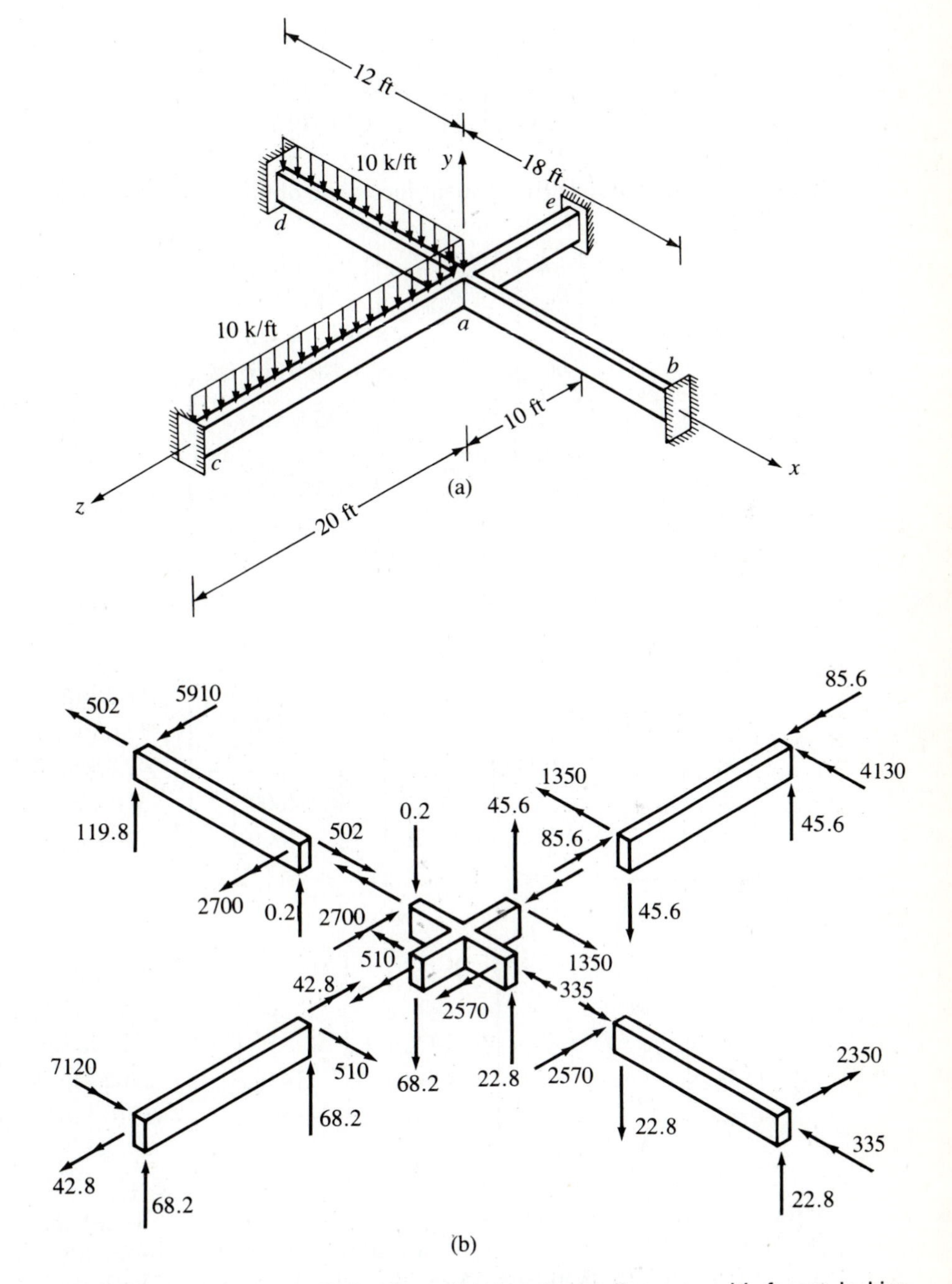

Figure E5.4 (a) Grid with loading; (b) free-body diagrams with forces in kips and moments in kip-inches (with forces in kips and moments in kip · in.). Hence

$$\begin{bmatrix} V_a \\ \Theta_{xa} \\ \Theta_{za} \end{bmatrix} = \begin{bmatrix} -0.8170 \text{ in.} \\ 8.222 \times 10^{-3} \text{ rad} \\ -1.168 \times 10^{-3} \text{ rad} \end{bmatrix}$$

For elements *ab* and *ae* the element forces are obtained using Eq. (5.42); thus

$$\begin{bmatrix} \bar{p}_{\bar{y}a} \\ \bar{m}_{\bar{x}a} \\ \bar{m}_{\bar{z}a} \\ \bar{m}_{\bar{z}b} \end{bmatrix}^{ab} = \mathbf{eu}^{ab} = \begin{bmatrix} -22.8 \text{ kips} \\ +335 \text{ kip}\cdot\text{in.} \\ -2570 \text{ kip}\cdot\text{in.} \\ -2350 \text{ kip}\cdot\text{in.} \end{bmatrix}$$

and

$$\begin{bmatrix} \bar{p}_{\bar{y}a} \\ \bar{m}_{\bar{x}a} \\ \bar{m}_{\bar{z}a} \\ \bar{m}_{\bar{z}e} \end{bmatrix}^{ae} = \mathbf{eu}^{ae} = \begin{bmatrix} -45.6 \text{ kips} \\ +85.6 \text{ kip}\cdot\text{in.} \\ -1350 \text{ kip}\cdot\text{in.} \\ -4130 \text{ kip}\cdot\text{in.} \end{bmatrix}$$

Elements *ac* and *ad* have initial forces applied; therefore using Eq. (4.65) we have

$$\begin{bmatrix} \bar{p}_{\bar{y}a} \\ \bar{p}_{\bar{y}c} \\ \bar{m}_{\bar{x}a} \\ \bar{m}_{\bar{z}a} \\ \bar{m}_{\bar{z}c} \end{bmatrix}^{ac} = \mathbf{eu}^{ae} + \bar{\mathbf{s}}^{o} = \begin{bmatrix} -31.8 \\ +31.8 \\ 42.8 \\ -4510 \\ -3120 \end{bmatrix} + \begin{bmatrix} 100 \\ 100 \\ 0 \\ 4000 \\ -4000 \end{bmatrix} = \begin{bmatrix} 68.2 \text{ kips} \\ 131.8 \text{ kips} \\ -42.8 \text{ kip}\cdot\text{in.} \\ -510 \text{ kip}\cdot\text{in.} \\ -7120 \text{ kip}\cdot\text{in.} \end{bmatrix}$$

and

$$\begin{bmatrix} \bar{p}_{\bar{y}a} \\ \bar{p}_{\bar{y}d} \\ \bar{m}_{\bar{x}a} \\ \bar{m}_{\bar{z}a} \\ \bar{m}_{\bar{z}d} \end{bmatrix}^{ad} = \mathbf{eu}^{ae} + \bar{\mathbf{s}}^{o} = \begin{bmatrix} -59.8 \\ +59.8 \\ -502 \\ -4140 \\ 4470 \end{bmatrix} + \begin{bmatrix} 60 \\ 60 \\ 0 \\ 1440 \\ -1440 \end{bmatrix} = \begin{bmatrix} +0.2 \text{ kips} \\ 119.8 \text{ kips} \\ -502 \text{ kip}\cdot\text{in.} \\ -2700 \text{ kip}\cdot\text{in.} \\ -5910 \text{ kip}\cdot\text{in.} \end{bmatrix}$$

Discussion Note that only the independent element forces were calculated using Eq. (5.42). The free-body diagrams for the four elements, plus that for node *a*, are shown in Fig. E5.4b. We urge the reader to check each of these for equilibrium. The stiffness method ensures element and nodal equilibrium, but it is a good idea to perform spot checks to guard against misinterpreting the results.

5.5 SPACE FRAMES

Space frames typically consist of a series of intersecting long rigid-beam elements that may be either straight or curved. They are capable of covering large areas with no intermediate supports. The joints at intersection points are constructed to be rigid. In plan view the structure is typically rectangular, but it can be of almost any geometry; members may intersect at either right angles or on a skew pattern (in the horizontal plane). An example of a lamella roof is shown in Fig. 5.10. The curved arches make an angle of approximately 20° in horizontal projection with the transverse axis of the structure. When curved in two directions, the space frames become domes or saddle roofs. In general, the elements are subjected to axial forces, torsion, and bending about the two principal axes. There are six degrees of freedom per node; thus the general three-dimensional beam element and the corresponding transformation matrices must be used.

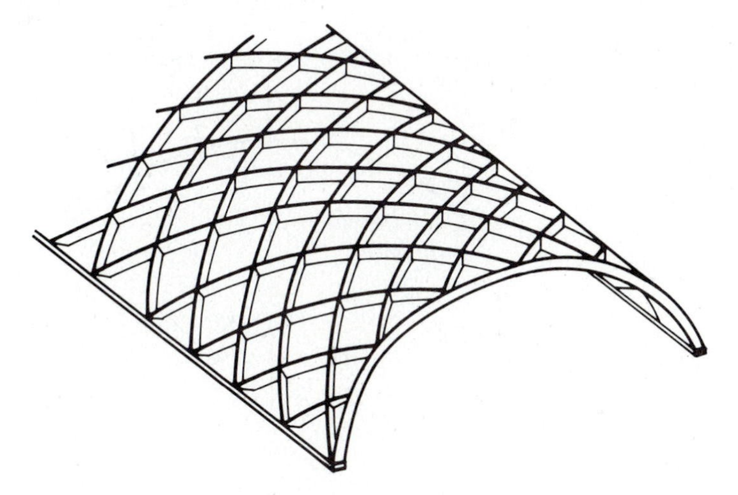

Figure 5.10 A lamella roof

Example 5.5 shows the response of two intersecting arches, which can be visualized as two isolated arches from a lamella roof. The results are not indicative of the forces in a lamella roof since loads are more widely distributed to adjacent arches.

EXAMPLE 5.5

Analyze the space frame in Fig. E5.5. For all elements $A = 15$ in.^{2}; $I_{\bar{y}} = 100$ in.^{4}; $I_{\bar{z}} = 550$ in.^{4}; $J = 270$ in^4; $E = 29 \times 10^3$ kips/in.^{2}; and $G = 11 \times 10^3$ kips/in.^{2}.

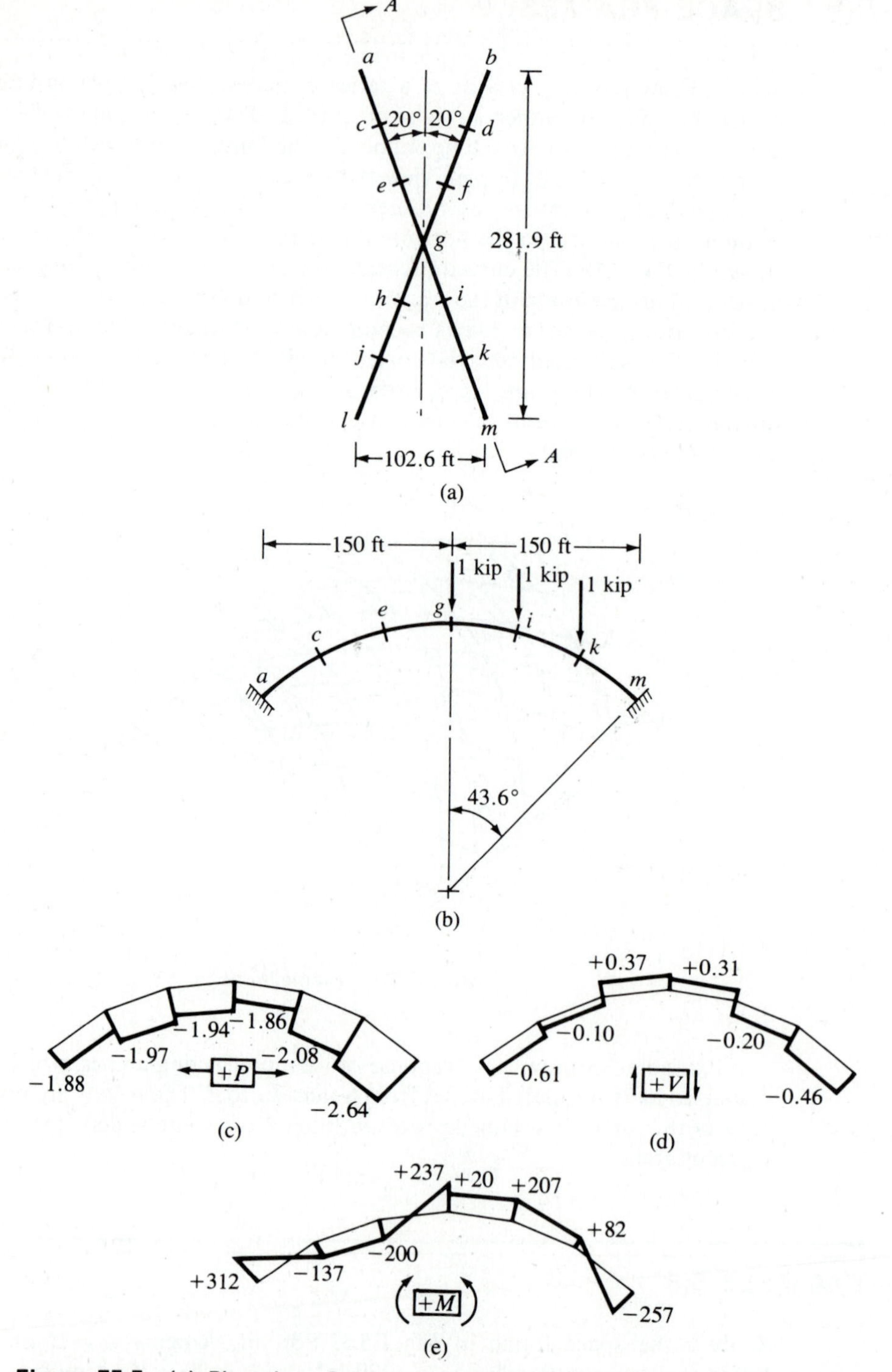

Figure E5.5 (a) Plan view. Section A-A showing (b) elevation; (c) axial force diagram (kips); (d) shear diagram (kips); (e) bending moment diagram (kip · in.)

Solution

Nodes *a*, *b*, *l*, and *m* have all degrees of freedom fixed while the other nodes each have six unconstrained degrees of freedom; therefore $\mathbf{K}_{ff}$ is 54 by 54. The structure has a load of 1 kip in the negative *y* direction at nodes *g*, *h*, *i*, *j*, and *k*. Solving the equation $\mathbf{P}_f = \mathbf{K}_{ff}\mathbf{U}_f$ yields

Node	U	V	W	Θ_x	Θ_y	Θ_z
c	−0.1569E + 01	0.2252E + 01	0.3947E + 00	0.1857E − 02	0.7168E − 03	0.3637E − 02
d	−0.1569E + 01	0.2252E + 01	−0.3947E + 00	−0.1857E − 02	−0.7168E − 03	0.3637E − 02
e	−0.1843E + 01	0.2781E + 01	0.2243E + 00	0.2948E − 03	0.6691E − 03	−0.2988E−02
f	−0.1843E + 01	0.2781E + 01	−0.2243E + 00	−0.2948E − 03	−0.6691E − 03	−0.2988E − 02
g	−0.1550E + 01	−0.2679E + 00	0.1210E − 12	−0.1284E − 15	−0.1064E − 14	−0.6432E − 02
h	−0.1808E + 01	−0.2954E + 01	−0.2123E + 00	−0.6282E − 03	0.6705E − 03	−0.2072E − 02
i	−0.1808E + 01	−0.2954E + 01	0.2123E + 00	0.6282E − 03	−0.6705E − 03	−0.2072E − 02
j	−0.1393E + 01	−0.2003E + 01	−0.3315E + 00	−0.1861E − 02	0.7161E − 03	0.3643E − 02
k	−0.1393E + 01	−0.2003E + 01	0.3315E + 00	0.1861E − 02	−0.7161E − 03	0.3643E − 02

(Translations are in inches; rotations are expressed in rads.)

Discussion Element forces are obtained using the calculated generalized displacements and the stiffness matrix in Eq. (5.28). The axial force, shear, and moment diagrams are shown in Fig. E5.5. The discontinuity in the moment diagram at the top (crown) occurs because of the existence of the other arch. We encourage the reader to draw a free-body diagram of node *g* to observe that it is in equilibrium.

5.6 DISCUSSION

The load path for some structures does not follow a system of distinct planar structural components. There are categories of structures that derive their structural integrity from three-dimensional interaction of the members. Space trusses, grid works, and space frames are examples of structures composed of discrete elements that must be analyzed by considering their true three-dimensional response to loading. Each of these systems has more degrees of freedom at each node than their planar counterpart; generally, their analysis yields systems of nodal equilibrium equations that are large. There are six degrees of freedom per node for the space frame and three degrees of freedom per node for the space truss and grid work. The element stiffness matrices for each of these different types of elements are described in the foregoing sections, along with the associated methods for obtaining element forces. Notwithstanding these differences, three-dimensional analysis of discrete-element structures are analyzed in the fashion used for planar trusses, beams, and rigid frames (see Fig. 2.13). Equivalent nodal forces (i.e., consistent forces) have not been explicitly discussed in this chapter because the approach outlined in Ch. 4 is applicable. Also, the analysis of three-dimensional structures for settlement, initial, and thermal strains is guided by the same precepts discussed in Chs. 2, 3, and 4. We urge the reader to review the appropriate methods for two-dimensional structures and infer their application for space trusses, grid works, and general space frames.

5.7 Problems

5.1 Three different matrix structural analysis computer programs exist. One solves plane trusses, another plane frames, and the third space trusses. What modules would be the same in the three programs? What ones would be different?

5.2 Calculate nodal displacements and element forces for the space truss in Fig. P5.2; $E = 29 \times 10^3$ ksi and $A = 4$ in.^{2} for all elements. Nodal coordinates (in feet) are: $a(0, 10, 0)$; $b(-7.5, 0, 0)$; $c(10, 0, 0)$; and $d(0, 10, -10)$. Applied forces (in kips) are: $P_{xa} = 50$; $P_{ya} = 80$; and $P_{za} = 100$.

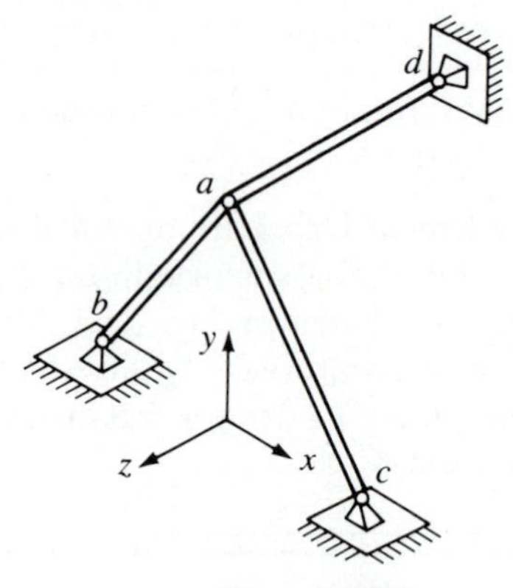

Figure P5.2

5.3 The three legs of the tripod in Fig. P5.3 are continuous from point d to the supports. The coordinates of the joints (m) are: $a_2(0.0, 0.0, 0.0)$; $b_2(3.2, 0.0, 2.4)$;

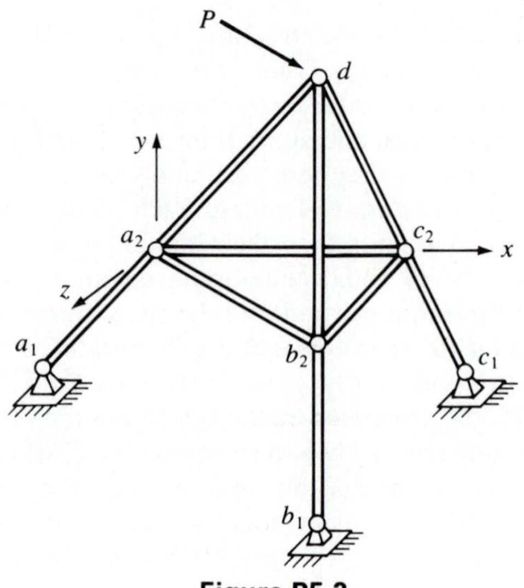

Figure P5.3

$c_2(5.0, 0.0, 0.0)$; and $d(3.2, 3.2, 0.0)$. Calculate the nodal displacements and element forces if $AE/L = \text{constant}$ for all members and $\mathbf{P} = 6\mathbf{i} + 0\mathbf{j} + 8\mathbf{k}$ (magnitudes in kN).

5.4 The triangles *a-b-c* and *d-e-f* in Fig. P5.4 are equilateral with sides of length L, and the distance between these two parallel planes is L. Calculate nodal displacements and element forces if $AE = \text{constant}$ for all members.

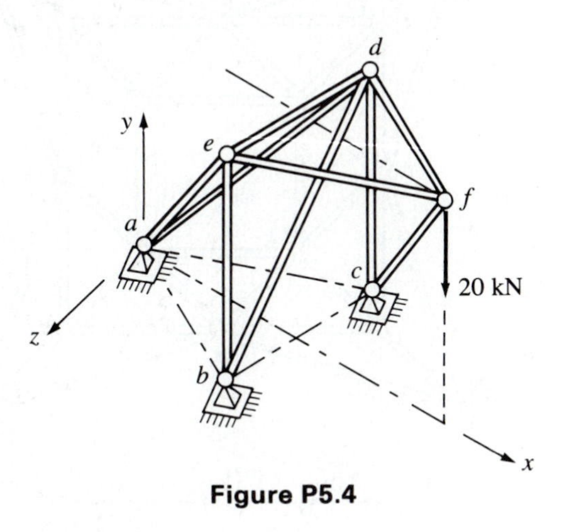

Figure P5.4

5.5 Calculate nodal displacements and element forces for the space truss in Fig. P5.5; $E = 29 \times 10^3$ ksi. Member cross-sectional areas are: 2 in.^{2} for verticals; 1.5 in.^{2} for horizontal edge members; 2.5 in.^{2} for vertical diagonals; and 2.83 in.^{2} for horizontal diagonals.

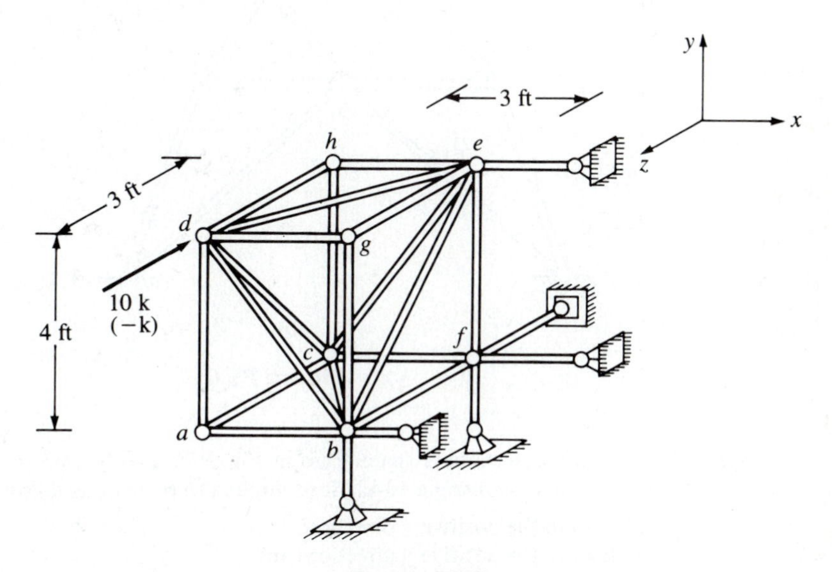

Figure P5.5

5.6 Calculate nodal displacements and element forces for the statically determinate space truss in Fig. P5.6. Note that a_i-c_i-e_i-g_i are squares in the horizontal plane and b_2-d_2-f_2-h_2 is also a square in the x-z plane.

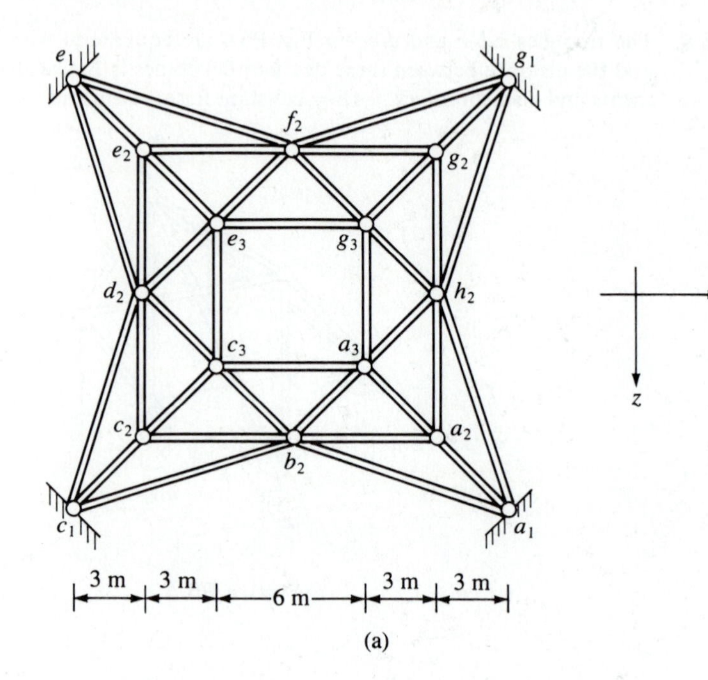

(a)

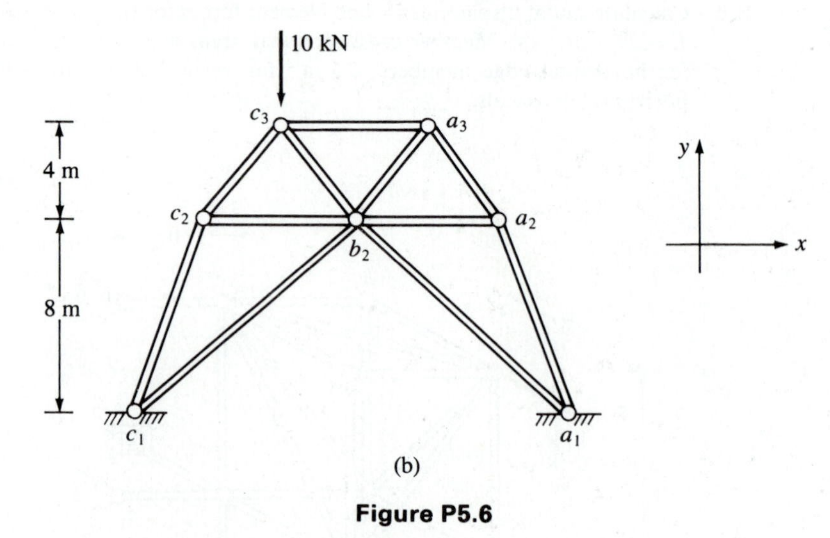

(b)

Figure P5.6

5.7 The statically determinate space truss in Fig. P5.7 is subjected to three separate load cases each involving a 10-kN load applied to node a_3 as follows:

LC1 load in the positive x direction;
LC2 load in the positive y direction; and
LC3 load in the positive z direction.

Calculate nodal displacements and element forces for the following configurations:

(a) The legs are all vertical and the square formed by the four joints at each level measures 5 m on a side.

(b) All legs have an equal and constant batter such that the top square $a_3 b_3 c_3 d_3$ is 5 m on each side and the bottom square $a_1 b_1 c_1 d_1$ is 25 m on each side.

(c) The sides of the squares at the top, middle, and bottom measure 5 m, 10 m, and 20 m, respectively (the tower is symmetrical about the vertical centerline, and all legs in a given level have the same batter).

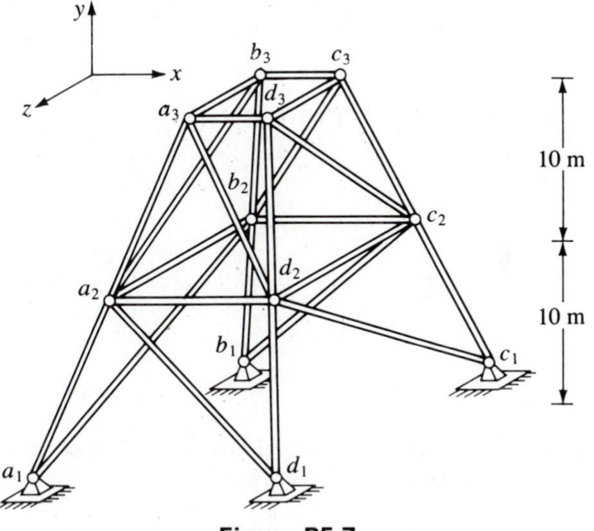

Figure P5.7

5.8 Calculate nodal displacements and element forces for the space truss in Fig. P5.8. Nodes 5, 6, 7, 12, 13, 14, 19, 20, and 21 are in the plane $z = 2.5$ ft.; all other nodes have $z = 0$. Nodes 1, 4, 22, and 25 are restrained in the x, y, and z directions. Nodes 2, 3, 8, 11, 15, 18, 23, and 24 have applied loads of $P_z = -2.5$ kips; nodes 9, 10, 16, and 17 have $P_z = -5.0$ kips. All elements have a cross-section of 1.0 in.^{2} and $E = 29 \times 10^3$ ksi.

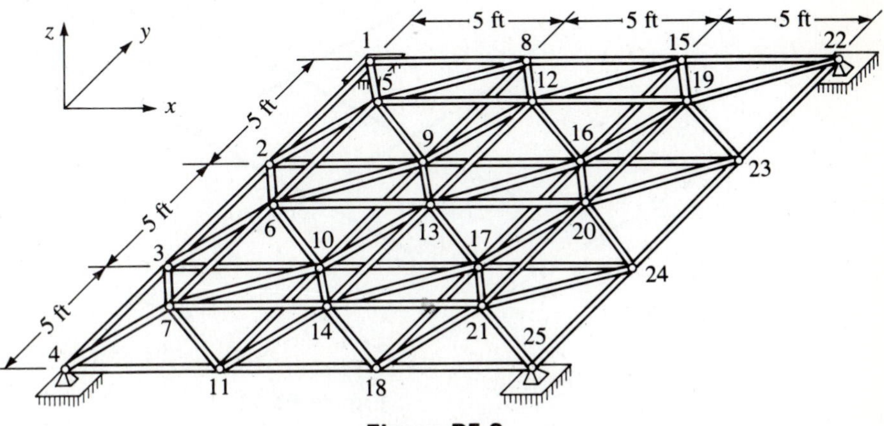

Figure P5.8

5.9 Test the stability of the trusses in Fig. P5.9 for the loadings shown; $AE/L =$ constant for all elements.

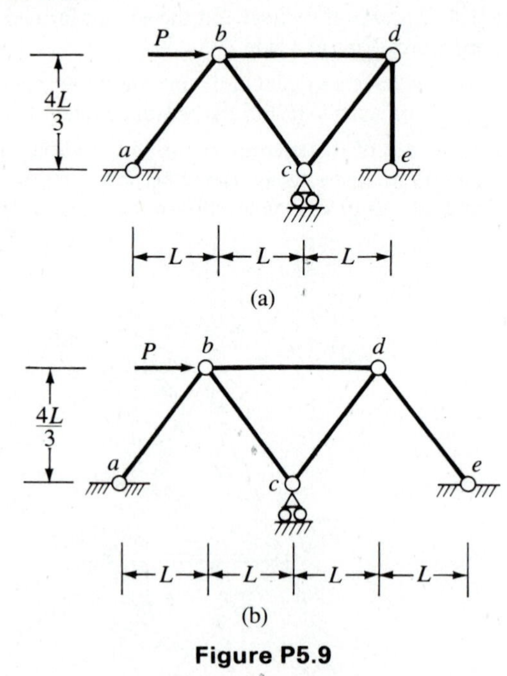

Figure P5.9

5.10 Contrast a matrix structural analysis computer program for space trusses with a similar program for grids. That is, what modules are the same and what ones are different (be specific)?

5.11 The grid in Fig. P5.11 has a single load of 10 kips applied at node h in the negative y direction. Calculate nodal displacements and element forces. All elements are steel ($E = 29 \times 10^3$ ksi) structural tubing: $10 \times 6 \times \frac{1}{4}$ in the x direction and $10 \times 4 \times \frac{3}{16}$ in the z direction.

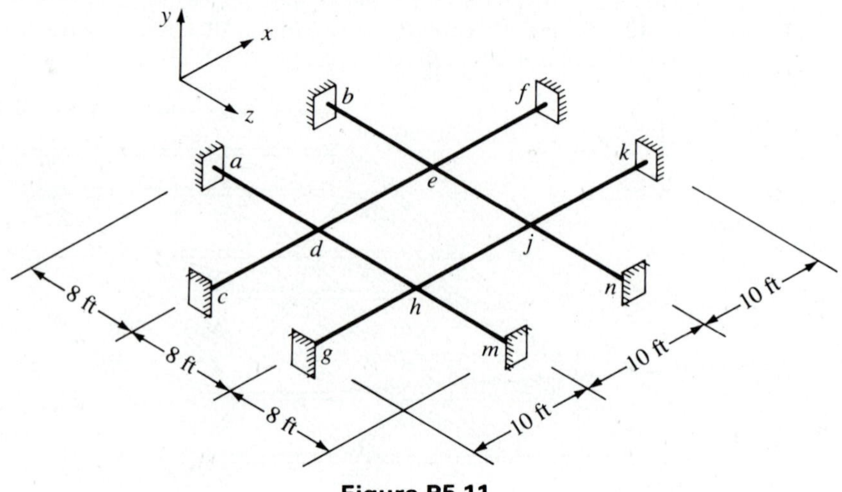

Figure P5.11

5.12 The grid of Prob. 5.11 is constructed so that the larger tubing runs in the z direction and the smaller in the x direction. Calculate nodal displacements and element forces for the downward load of 10 kips at node h.

5.13 Calculate nodal displacements and element forces for the rectangular grid in Fig. P5.13. All members have the illustrated cross section and $E = 29 \times 10^3$ ksi. Nodes 1, 4, 13, and 16 are restrained in the x, y, and z directions. Nodes 2, 3, 5, 8, 9, 12, 14, and 15 have $P_z = -2.5$ kips; nodes 6, 7, 10, and 11 have $P_z = -5.0$ kips.

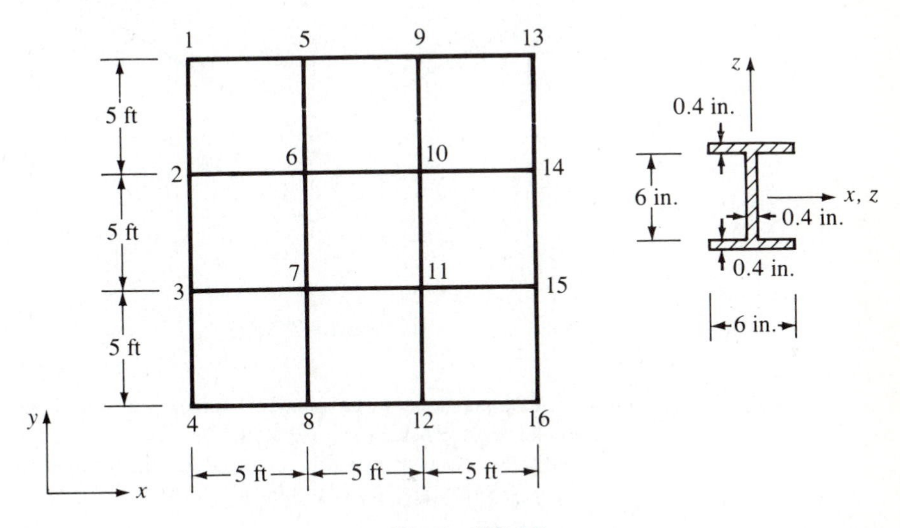

Figure P5.13

5.14 Contrast a matrix structural analysis computer program for grids with a similar program for space frames. That is, what modules are the same and what ones are different (be specific)? In general, how do the stiffness matrix bandwidths compare for these two types of structures?

5.15 Calculate the nodal displacements and element forces for the space frame in Fig. P5.15. For all elements $L = 10$ ft, $I_{\bar{z}\bar{z}} = 50$ in.^{4}, $I_{\bar{y}\bar{y}} = 10$ in.^{4}, $J = 60$ in.^{4}, $E = 30 \times 10^3$ ksi, and $G = 12 \times 10^3$ ksi. Elements ab, ac, and ad are oriented with the $\bar{y}$ axis in the z direction, and element ae has the $\bar{y}$ axis in the x direction

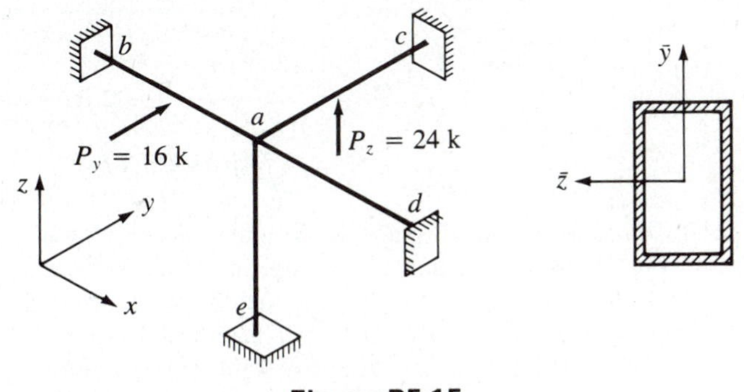

Figure P5.15

($\bar{z}$ is in the $-y$ direction). The forces on ab and ac are applied at midlength of the elements, and element ad has a linear temperature variation through the depth ($\bar{y}$ direction) with the top at 220°F and the bottom at 70°F [$\alpha = 6 \times 10^{-6}$ in./(in. · °F) and the element depth is 10 in.].

5.16 Calculate nodal displacements and element forces for the space frame in Fig. P5.16; E, G, I, and J are the same for both elements.

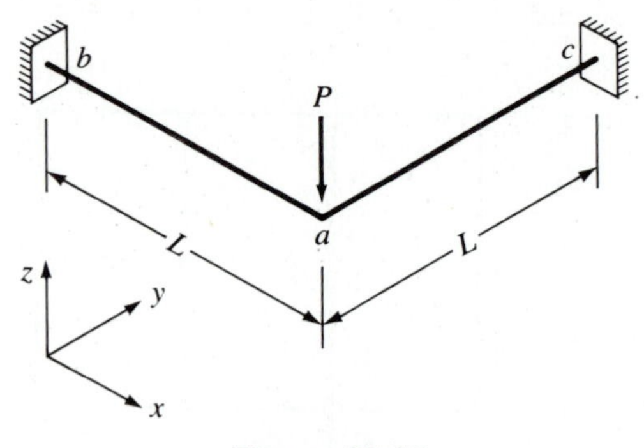

Figure P5.16

5.17 Calculate nodal displacements and element forces for the space frame in Fig. P5.17. Both elements are steel ($E = 29 \times 10^3$ ksi and $\nu = \frac{1}{3}$) structural tubing $12 \times 8 \times 0.50$ with the 12-in. dimension oriented in the y direction; therefore $A = 18.4$ in.^{2}, $I_{\bar{z}\bar{z}} = 353$ in.^{4}, and $I_{\bar{y}\bar{y}} = 188$ in.^{4}.

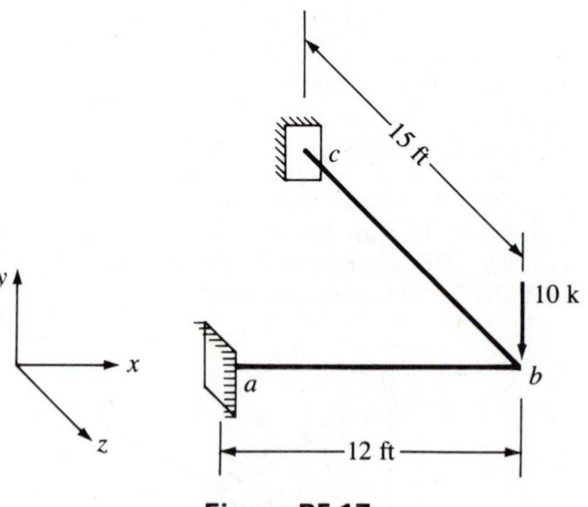

Figure P5.17

5.18 Calculate nodal displacements and element forces for the space frame in Fig. P5.18. In addition to the applied forces, element ac has a linear distribution of temperature through the depth with $T_u - T_l = 100$°F and $T_m = 0$°F [$\alpha = 6.5 \times 10^{-6}$ in./(in. · °F)]. All members have $L = 20$ ft and are made from steel ($E = 29 \times 10^3$ ksi and $\nu = 0.30$) pipe so that $I_{ab} = 160$ in.^{4}, $I_{ac} = 70$ in.^{4} (8-in. outside diameter), $I_{ad} = 280$ in.^{4}, and $I_{ae} = 70$ in.^{4}.

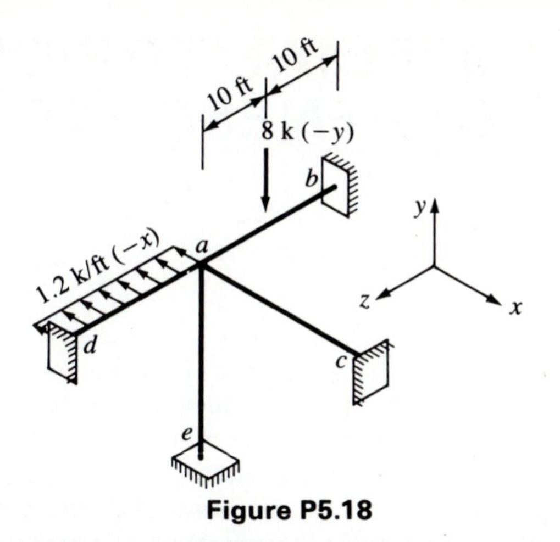

Figure P5.18

5.19 The rigid steel space frame in Fig. P5.19 is constructed from rectangular structural tubing 30 × 15 × 1 cm (30 cm dimension oriented in the y direction), with $L = 3$ m, $E = 200$ GPa, and $\nu = 0.30$. Calculate nodal displacements and element forces if: **(a)** a concentrated force of 10 kN ($-y$ direction) is applied at node c and **(b)** a uniformly distributed load of 2 kN/m ($-y$ direction) is applied to element bd.

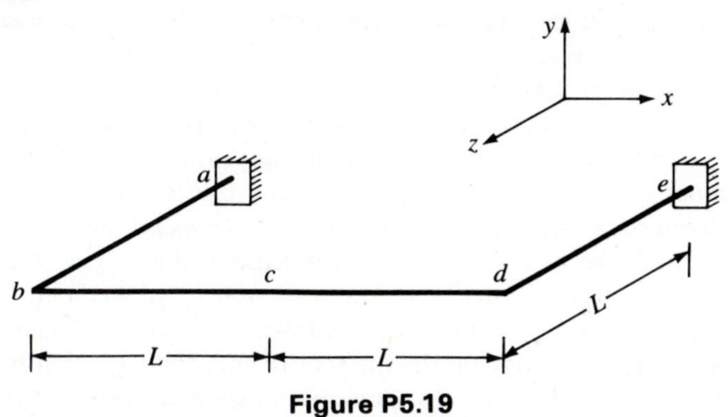

Figure P5.19

5.20 Calculate nodal displacements and element forces for the space frame in Fig. P5.20. Both members are made from steel ($E = 29 \times 10^3$ ksi and $\nu = 0.30$) structural tubing: ab is 18 × 6 × 0.375 and bc is 18 × 6 × 0.375 (dimensions in inches).

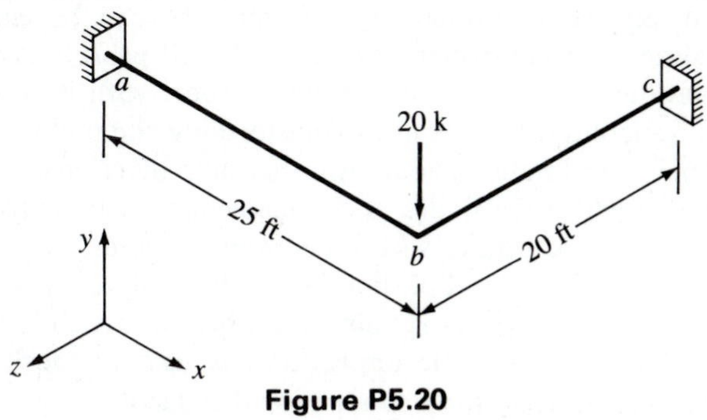

Figure P5.20

6 SPECIAL TOPICS FOR THE STIFFNESS METHOD

Various assumptions are embedded in the basic operations of the stiffness method as it is presented in the previous chapters. For example, we suppose that the materials are linearly elastic, the structure experiences small deformations, and the axial force and beam elements are prismatic. Frequently the structural engineer encounters systems of elements that violate these implicit fundamental tenets. For example, a distributed loading imposed on a beam appears to negate the assumption of constant shear and linearly varying moment of the prismatic beam element. In Ch. 4 we demonstrated how to approximate the structural behavior by replacing the actual loading with equivalent initial applied nodal forces. There are various other stratagems that can be used to circumvent apparent contradictions associated with the implicit assumptions of the stiffness method. In contrast, some unusual structural features can be dealt with more directly by incorporating special features into the framework of the stiffness method. These topics are described in this chapter.

A beam with an internal hinge appears to complicate the stiffness method since it results in a slope discontinuity in the displaced structure. A hinge is a common construction feature that can transmit shear but not moment and is an example of an *internal force release.* Other types of releases interrupt the transfer of axial force, shear, and torsion in an element. Force releases can be analyzed using a technique known as *matrix condensation*, which is an approach for reducing the number of equations to be solved. Also, large structures can be divided into a number of *substructures* and investigated efficiently using matrix condensation to interrelate the responses of the individual substructures.

Rotation of a large mat foundation and beam members that are axially rigid are two example in which several nodal degrees of freedom are dependent. Special coordinate transformations can be employed to describe and analyze such *constraint conditions. Nodal coordinates* are used to investigate situations such as that of a roller support with its support surface inclined to the horizontal; this requires transforming the nodal degrees of freedom to the global coordinate system. Another use of coordinate transformation is for cases in which the nodes of one element are offset from those of another. This occurs in rigid frames with large moment connections, and for roof systems in which the girders and purlines are to be analyzed as a single system.

Tapered elements are also described in this chapter. Using special-purpose elements for a specific application is usually more efficient than trying to devise a stratagem for a conventional element.

6.1 MATRIX CONDENSATION

Sometimes the identity of one or more degrees of freedom must be removed from the force-displacement equations without eliminating their influence on the response of the structure. This can be done using *matrix condensation* (also referred to as static condensation). First we consider the partition of the global force-displacement equations involving prescribed forces and unknown displacements, that is,

$$\mathbf{P}_f = \mathbf{K}_{ff}\,\mathbf{U}_f \tag{6.1}$$

and partitioning this matrix equation further as follows:

$$\underset{\mathbf{P}_f}{\begin{bmatrix} \mathbf{P}_\alpha \\ \hline \mathbf{P}_\beta \end{bmatrix}} = \underset{\mathbf{K}_{ff}}{\left[\begin{array}{c|c} \mathbf{K}_{\alpha\alpha} & \mathbf{K}_{\alpha\beta} \\ \hline \mathbf{K}_{\beta\alpha} & \mathbf{K}_{\beta\beta} \end{array}\right]} \underset{\mathbf{U}_f}{\begin{bmatrix} \mathbf{U}_\alpha \\ \hline \mathbf{U}_\beta \end{bmatrix}} \tag{6.2}$$

In general, $\mathbf{P}_\alpha$ and $\mathbf{P}_\beta$ are both column matrices of known forces, and $\mathbf{U}_\alpha$ and $\mathbf{U}_\beta$ are column matrices of unknown displacements. Expanding Eq. (6.2) gives the matrix equations

$$\begin{aligned} \mathbf{P}_\alpha &= \mathbf{K}_{\alpha\alpha}\,\mathbf{U}_\alpha + \mathbf{K}_{\alpha\beta}\,\mathbf{U}_\beta \\ \mathbf{P}_\beta &= \mathbf{K}_{\beta\alpha}\,\mathbf{U}_\alpha + \mathbf{K}_{\beta\beta}\,\mathbf{U}_\beta \end{aligned} \tag{6.3}$$

Solving the second of these yields

$$\mathbf{U}_\beta = \mathbf{K}_{\beta\beta}^{-1}(\mathbf{P}_\beta - \mathbf{K}_{\beta\alpha}\,\mathbf{U}_\alpha) \tag{6.4}$$

When this result is substituted into the first of Eqs. (6.3) and terms are rearranged, the combined equations become

$$\mathbf{P}_\alpha - \mathbf{K}_{\alpha\beta}\,\mathbf{K}_{\beta\beta}^{-1}\,\mathbf{P}_\beta = (\mathbf{K}_{\alpha\alpha} - \mathbf{K}_{\alpha\beta}\,\mathbf{K}_{\beta\beta}^{-1}\,\mathbf{K}_{\beta\alpha})\mathbf{U}_\alpha \tag{6.5}$$

that is, the so-called *condensed* force-displacement equations are

$$\mathbf{P}_c = \mathbf{K}_{cc}\,\mathbf{U}_c \tag{6.6}$$

where $\mathbf{U}_c = \mathbf{U}_\alpha$ and the *condensed force* and *stiffness matrices* are, respectively,

$$\mathbf{P}_c = \mathbf{P}_\alpha - \mathbf{K}_{\alpha\beta}\,\mathbf{K}_{\beta\beta}^{-1}\,\mathbf{P}_\beta \tag{6.7}$$

and

$$\mathbf{K}_{cc} = \mathbf{K}_{\alpha\alpha} - \mathbf{K}_{\alpha\beta}\,\mathbf{K}_{\beta\beta}^{-1}\,\mathbf{K}_{\beta\alpha} \tag{6.8}$$

This process of *matrix condensation* or *static condensation* can be very useful when it is desirable to reduce the order of the stiffness matrix. That is, since the order of $\mathbf{K}_{cc}$ is less than that of $\mathbf{K}_{ff}$, the solution is obtained by inverting the two smaller matrices $\mathbf{K}_{cc}$ and $\mathbf{K}_{\beta\beta}$. This approach can be used whether these matrices are explicitly inverted or the solution is found by a method such as Gauss elimination. Handling these reduced-order matrices can be extremely useful when investigating large problems or in working with a computer with limited core capacity. With this approach we solve smaller sets of equations, but we must also perform a number of matrix multiplications that can consume a great amount of computer time; the trade-off must be carefully considered.

Note that matrix condensation is analogous to solving a pair of simultaneous equations by eliminating one unknown. The matrix $\mathbf{K}_{cc}$ contains the implicit effect of the generalized displacements in $\mathbf{U}_\beta$ even though their stiffness coefficients do not appear explicitly. Thus, by condensing $\mathbf{K}_{ff}$, solving Eq. (6.6), and substituting back into Eq. (6.4) all the generalized displacements are calculated. This is quite different from the solution obtained by solving the equations $\mathbf{P}_\alpha = \mathbf{K}_{\alpha\alpha}\mathbf{U}_\alpha$ [see Eqs. (6.3)] since this formulation implies that $\mathbf{U}_\beta$ is null (i.e., that all the generalized displacements contained therein are constrained).

If the force matrix $\mathbf{P}_f$ is partitioned so that $\mathbf{P}_\beta$ is null, we note from Eq. (6.7) that $\mathbf{P}_c = \mathbf{P}_\alpha$, and Eq. (6.5) gives the condensed force-displacement equations as

$$\mathbf{P}_\alpha = \mathbf{K}_{cc}\,\mathbf{U}_\alpha \tag{6.9}$$

In this case, the force matrix does not have to be condensed, which reduces the number of manipulations required to obtain the governing equations. Example 6.1 illustrates the application of matrix condensation in the form of Eq. (6.9) to reduce the computational work required to calculate the generalized displacements.

EXAMPLE 6.1

Calculate the generalized displacement for Example 4.5 by matrix condensation.

Solution

Partitioning so that $\mathbf{U}_\alpha = v_b$ and $\mathbf{U}_\beta = \{\Theta_a \quad \Theta_b \quad \Theta_c\}$ we have

$$\mathbf{K}_{\alpha\alpha} = 16000$$

$$\mathbf{K}_{\alpha\beta} = [-8000 \quad 12000 \quad 20000]$$

$$\mathbf{K}_{\beta\alpha} = \begin{bmatrix} -8000 \\ 12000 \\ 20000 \end{bmatrix}$$

$$\mathbf{K}_{\beta\beta} = \begin{bmatrix} 32000 & 16000 & 0000 \\ 16000 & 72000 & 20000 \\ 0000 & 20000 & 40000 \end{bmatrix}$$

From Eq. (6.8) we have

$$\mathbf{K}_{cc} = \mathbf{K}_{\alpha\alpha} - \mathbf{K}_{\alpha\beta}\mathbf{K}_{\beta\beta}^{-1}\mathbf{K}_{\beta\alpha} = 3333.33$$

Since $\mathbf{P}_\alpha = -75$ kN and $\mathbf{P}_\beta = \{0 \quad 0 \quad 0\}$, we have

$$\mathbf{U}_\alpha = \mathbf{U}_c = -0.0225 \text{ m}$$

Equation (6.4) gives

$$\mathbf{U}_\beta = \mathbf{K}_{\beta\beta}^{-1}(\mathbf{P}_\beta - \mathbf{K}_{\beta\alpha}\mathbf{U}_\alpha) = 10^{-3}\{-6.875 \quad 2.500 \quad 10.000\} \text{ rad}$$

Discussion The generalized displacements and forces from Example 4.5 have been rearranged in the equations so that the moments and the associated rotations can be eliminated. Since $M_a = M_b = M_c = 0$, Eq. (6.9) was used and the solution requires the inverse of the 3 by 3 matrix $\mathbf{K}_{\beta\beta}$ and the scalar $\mathbf{K}_{cc}$, instead of that of the larger $\mathbf{K}_{ff}$.

6.1.1 RELEASE OF GENERALIZED ELEMENT END FORCES An internal hinge in a beam introduces a discontinuity in the slope of the elastic curve and supports zero bending moment. This behavior is an example of a broad class of conditions in which one or more of the generalized element forces at a node is specified as zero; this is sometimes referred to as a release of the generalized force. Construction details can be designed so that at a particular point a member transmits zero axial force, shear, bending moments, and/or torsion. These special conditions can be treated by starting with the usual element stiffness matrix and eliminating the known zero generalized force using Eq. (6.9). This yields a modified (condensed) element stiffness matrix with the appropriate generalized force identical to zero and the corresponding displacement eliminated.

For example, consider the beam element in Fig. 6.1a with an internal hinge

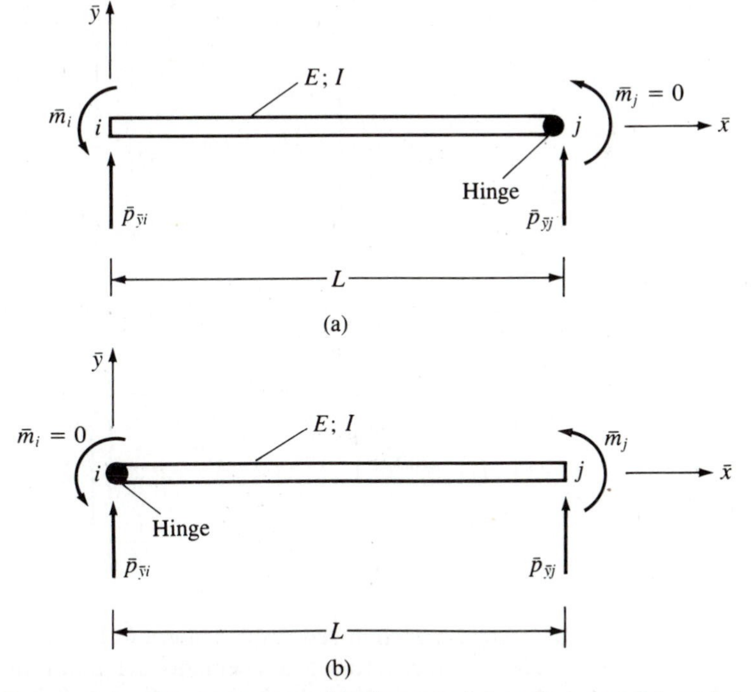

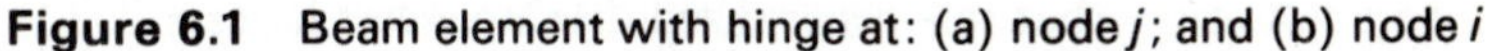

Figure 6.1 Beam element with hinge at: (a) node *j*; and (b) node *i*

at node j. To obtain the stiffness matrix incorporating this condition the element stiffness matrix from Eq. (4.19) is first partitioned as follows:

$$\bar{\mathbf{k}} = \frac{EI}{L}\left[\begin{array}{ccc|c} \dfrac{12}{L^2} & \dfrac{6}{L} & -\dfrac{12}{L^2} & \dfrac{6}{L} \\ \dfrac{6}{L} & 4 & -\dfrac{6}{L} & 2 \\ -\dfrac{12}{L^2} & -\dfrac{6}{L} & \dfrac{12}{L^2} & -\dfrac{6}{L} \\ \hline \dfrac{6}{L} & 2 & -\dfrac{6}{L} & 4 \end{array}\right]$$

The condensed stiffness matrix is calculated using Eq. (6.8), that is,

$$\bar{\mathbf{k}}_{cc} = \frac{EI}{L}\begin{bmatrix} \dfrac{12}{L^2} & \dfrac{6}{L} & -\dfrac{12}{L^2} \\ \dfrac{6}{L} & 4 & -\dfrac{6}{L} \\ -\dfrac{12}{L^2} & -\dfrac{6}{L} & \dfrac{12}{L^2} \end{bmatrix} - \frac{EI}{L}\begin{bmatrix} \dfrac{6}{L} \\ 2 \\ \dfrac{6}{L} \end{bmatrix}\frac{1}{4}\begin{bmatrix} \dfrac{6}{L} & 2 & -\dfrac{6}{L} \end{bmatrix}$$

Thus

$$\bar{\mathbf{k}}_{cc} = \frac{3EI}{L}\begin{bmatrix} \dfrac{1}{L^2} & \dfrac{1}{L} & -\dfrac{1}{L^2} \\ \dfrac{1}{L} & 1 & -\dfrac{1}{L} \\ -\dfrac{1}{L^2} & -\dfrac{1}{L} & \dfrac{1}{L^2} \end{bmatrix} \tag{6.10}$$

and the force-displacement matrix equations with the *hinge at node j* are

$$\begin{bmatrix} \bar{p}_{\bar{y}i} \\ \bar{m}_i \\ \bar{p}_{\bar{y}j} \end{bmatrix} = \frac{3EI}{L}\begin{bmatrix} \dfrac{1}{L^2} & \dfrac{1}{L} & -\dfrac{1}{L^2} \\ \dfrac{1}{L} & 1 & -\dfrac{1}{L} \\ -\dfrac{1}{L^2} & -\dfrac{1}{L} & \dfrac{1}{L^2} \end{bmatrix}\begin{bmatrix} \bar{v}_i \\ \bar{\theta}_i \\ \bar{v}_j \end{bmatrix} \tag{6.11}$$

Note that in this case the generalized displacement $\bar{\theta}_j$ has been eliminated and will not be calculated when nodal displacements are computed, but this fact does not imply that this generalized displacement is zero.

Similarly, the condensed force-displacement relations for a beam element with a *hinge at node i* (Fig. 6.1b) is obtained by eliminating the zero moment $\bar{m}_i$ and its corresponding rotation, to give

$$\begin{bmatrix} \bar{p}_{\bar{y}i} \\ \bar{p}_{\bar{y}j} \\ \bar{m}_j \end{bmatrix} = \frac{3EI}{L} \begin{bmatrix} \dfrac{1}{L^2} & -\dfrac{1}{L^2} & \dfrac{1}{L} \\ -\dfrac{1}{L^2} & \dfrac{1}{L^2} & -\dfrac{1}{L} \\ \dfrac{1}{L} & -\dfrac{1}{L} & 1 \end{bmatrix} \begin{bmatrix} \bar{v}_i \\ \bar{v}_j \\ \bar{\theta}_j \end{bmatrix} \tag{6.12}$$

Both forms of the hinged beam element in Eqs. (6.11) and (6.12) have a corresponding element force-transformation matrix **e** that must be used. Equations (6.11) and (6.12) can also be obtained by analyzing the beams using the basic equations. As an alternative to treating internal hinges with Eqs. (6.11) and (6.12), it is possible to derive special element stiffness matrices with internal moment releases between nodal points. The matrix equations for beams with internal shear releases can also be obtained by matrix condensation or by using basic beam theory. We urge the reader to experiment with these various possibilities as exercises. Example 6.2 illustrates the use of the special element stiffness matrices of Eqs. (6.11) and (6.12).

EXAMPLE 6.2

Calculate the generalized displacements at b and the element forces for the uniform beam in Fig. E6.2 with an internal hinge; EI = constant. Work the problem in three different ways: (a) consider the hinge as part of element ab;

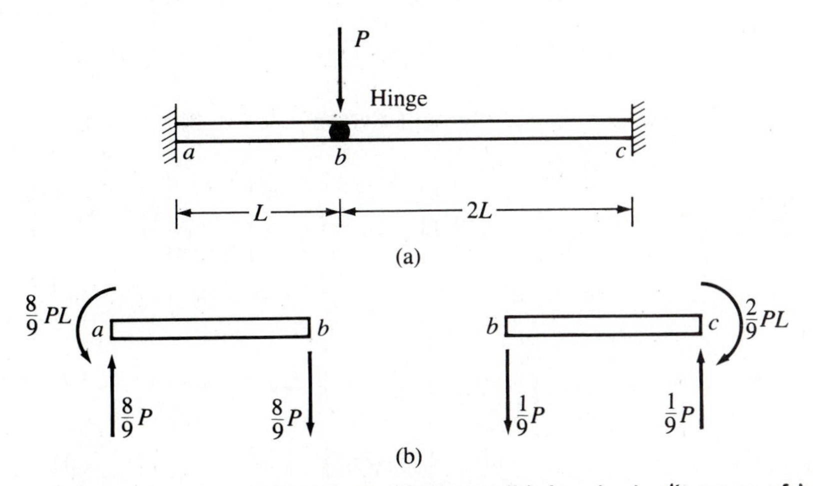

Figure E6.2 (a) Beam with an internal hinge; (b) free-body diagrams of beam elements

(b) consider the hinge as part of element bc; and (c) consider the hinge on both elements.

Solution

(a) For the hinge on the left element use Eq. (6.10) for element ab; this gives

$$\mathbf{K}_{ff} = \frac{EI}{L}\begin{bmatrix} \frac{3}{L^2} + \frac{12}{8L^2} & \frac{6}{4L} \\ \frac{6}{4L} & 2 \end{bmatrix} = \frac{EI}{L}\begin{bmatrix} \frac{9}{2L^2} & \frac{3}{2L} \\ \frac{3}{2L} & 2 \end{bmatrix}$$

$$\begin{bmatrix} V_b \\ \Theta_{b+} \end{bmatrix} = \frac{4L^3}{27EI}\begin{bmatrix} 2 & -\frac{3}{2L} \\ -\frac{3}{2L} & \frac{9}{2L^2} \end{bmatrix}\begin{bmatrix} -P \\ 0 \end{bmatrix} = \frac{4PL^3}{27EI}\begin{bmatrix} -2 \\ \frac{3}{2L} \end{bmatrix}$$

where Θ_{b+} = rotation to the right of b. The element forces are as follows:

$$\begin{bmatrix} V \\ M_a \end{bmatrix}^{ab} = \frac{3EI}{L}\begin{bmatrix} \frac{1}{L^2} & \frac{1}{L} & -\frac{1}{L^2} \\ \frac{1}{L} & 1 & -\frac{1}{L} \end{bmatrix}\frac{4PL^3}{27EI}\begin{bmatrix} 0 \\ 0 \\ -2 \end{bmatrix} = \frac{8PL^2}{9}\begin{bmatrix} \frac{1}{L^2} \\ \frac{1}{L} \end{bmatrix}$$

$$\begin{bmatrix} V \\ M_b \\ M_c \end{bmatrix}^{bc} = \frac{EI}{2L}\begin{bmatrix} \frac{12}{4L^2} & \frac{6}{2L} & -\frac{12}{4L^2} & \frac{6}{2L} \\ \frac{6}{2L} & 4 & -\frac{6}{2L} & 2 \\ \frac{6}{2L} & 2 & -\frac{6}{2L} & 4 \end{bmatrix}\frac{4PL^3}{27EI}\begin{bmatrix} -2 \\ \frac{3}{2L} \\ 0 \\ 0 \end{bmatrix} = \frac{2PL}{9}\begin{bmatrix} -\frac{1}{2L} \\ 0 \\ -1 \end{bmatrix}$$

(b) With a hinge on the right element use Eq. (6.12) for bc:

$$\mathbf{K}_{ff} = \frac{EI}{L}\begin{bmatrix} \frac{12}{L^2} + \frac{3}{8L^2} & -\frac{6}{L} \\ -\frac{6}{L} & 4 \end{bmatrix} = \frac{EI}{L}\begin{bmatrix} \frac{99}{8L^2} & -\frac{6}{L} \\ \frac{6}{L} & 4 \end{bmatrix}$$

$$\begin{bmatrix} V_b \\ \Theta_{b-} \end{bmatrix} = \frac{2L^3}{27EI}\begin{bmatrix} 4 & \frac{6}{L} \\ \frac{6}{L} & \frac{99}{8L^2} \end{bmatrix}\begin{bmatrix} -P \\ 0 \end{bmatrix} = \frac{4PL^3}{27EI}\begin{bmatrix} -2 \\ -\frac{3}{L} \end{bmatrix}$$

where Θ_{b-} = rotation to the left of b.

(c) With a hinge on both elements:

$$\mathbf{K} = \frac{3EI}{L^3} + \frac{3EI}{8L^3} = \frac{27EI}{8L^3}$$

$$V_b = -\frac{8L^3}{27EI} P$$

Discussion This structure can be analyzed using three different combinations of conventional and hinged beam elements, and all have been demonstrated for computing the generalized displacements. In using two hinged elements, only V_b is obtained, but in each of the other approaches the rotation at point b that is associated with the conventional beam element is obtained from the calculations.

Only the computation of element forces has been demonstrated for a hinged element between nodes a and b with a conventional beam element between nodes b and c (Fig. E6.2b). This required the **e** matrix corresponding to the stiffness matrix in each case. We suggest that the reader calculate the element forces for the other two idealizations as an exercise.

6.1.2 SUBSTRUCTURAL ANALYSIS Some structures are too large to be analyzed as a single system, and they can be investigated only after division into a number of smaller units or *substructures*. For example, in the case of an airframe as shown in Fig. 6.2, it is not unusual to require thousands of nodal points to completely describe the response of the entire structure. Each of the substructures defined is assigned to an individual engineering team, and the analysis of these separate units can proceed in parallel. With the smaller problem sizes, the engineers can make more convenient and reliable data checks. Also, the equations can be reasonably and economically manipulated on limited computing hardware. The response of the total airframe system is

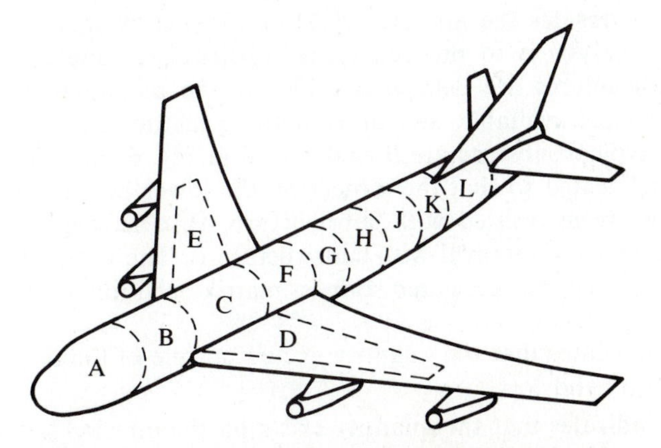

Figure 6.2 Typical substructures for an airframe

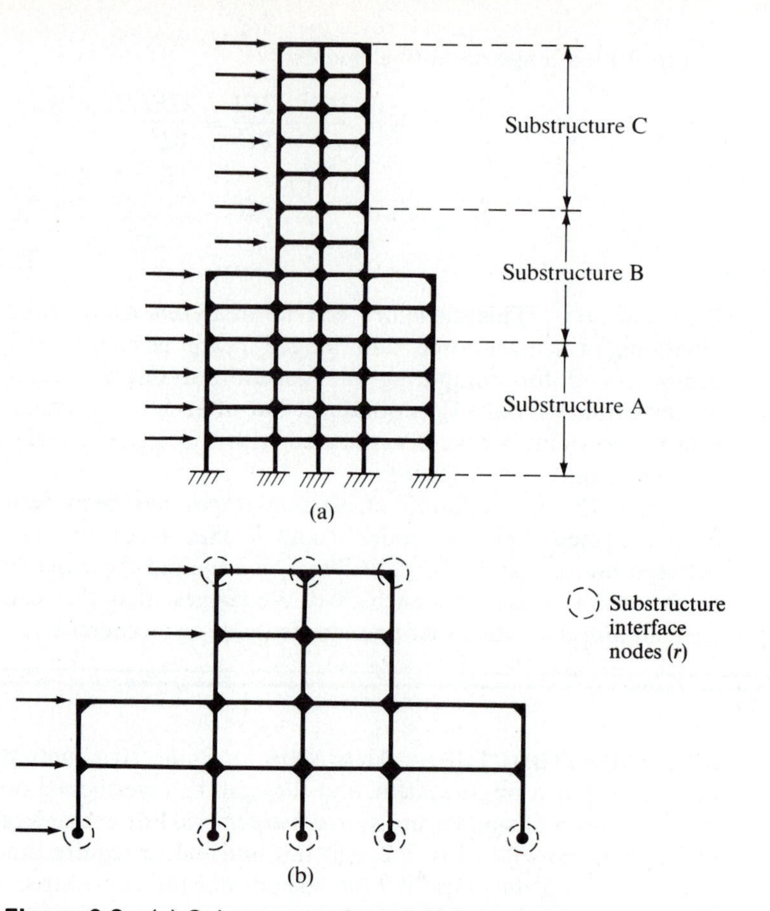

Figure 6.3 (a) Substructures for a rigid frame; (b) substructure B

established by synthesizing the individual substructural responses in a systematic fashion.

Consider the midrise rigid frame structure in Fig. 6.3a and assume that the analysis is to proceed using substructural analysis. First, the individual substructures are defined considering the maximum tractable problem size, personnel available, and other limiting factors. In this case, we will examine the typical substructure B as depicted in Fig. 6.3b. Note that this substructure is delineated so that the girders on the lower boundary are not included (i.e., they are associated with substructure A), whereas the girders on the upper interface are retained with substructure B. The generalized displacements and forces (and the associated stiffness matrix partitions) are subscripted so that

r indicates that the quantity occurs on one of the interfaces of the substructure and

e indicates that the quantity exists on the interior of the substructure (i.e., it is not on an interface).

The force-displacement equations for this substructure can be partitioned as follows:

$$\begin{bmatrix} \mathbf{P}_r^{\mathrm{B}} \\ \hline \mathbf{P}_e^{\mathrm{B}} \end{bmatrix} = \left[\begin{array}{c|c} \mathbf{K}_{rr}^{\mathrm{B}} & \mathbf{K}_{re}^{\mathrm{B}} \\ \hline \mathbf{K}_{er}^{\mathrm{B}} & \mathbf{K}_{ee}^{\mathrm{B}} \end{array}\right] \begin{bmatrix} \mathbf{U}_r^{\mathrm{B}} \\ \hline \mathbf{U}_e^{\mathrm{B}} \end{bmatrix} \tag{6.13}$$

where the superscript B denotes the substructure.

Using the method of matrix condensation described in Sec. 6.1, the interior quantities (e) are eliminated and the interface degrees of freedom (r) are retained. This yields

$$\mathbf{P}_r^{\mathrm{B}} - \bar{\mathbf{P}}_r^{\mathrm{B}} = \hat{\mathbf{K}}_{ee}^{\mathrm{B}} \mathbf{U}_e^{\mathrm{B}} \tag{6.14}$$

where the condensed column of forces is

$$\bar{\mathbf{P}}_r^{\mathrm{B}} = \mathbf{K}_{re}^{\mathrm{B}} [\mathbf{K}_{ee}^{\mathrm{B}}]^{-1} \mathbf{P}_e^{\mathrm{B}} \tag{6.15}$$

and the condensed stiffness matrix is

$$\hat{\mathbf{K}}_{rr}^{\mathrm{B}} = \mathbf{K}_{rr}^{\mathrm{B}} - \mathbf{K}_{re}^{\mathrm{B}} [\mathbf{K}_{ee}^{\mathrm{B}}]^{-1} \mathbf{K}_{er}^{\mathrm{B}} \tag{6.16}$$

The response of this substructure can be described exclusively by the force-displacement relations expressed in terms of the degrees of freedom at the interfaces; these contain the implicit effect of the interior structure. Thus the substructure can be envisaged as a large element (i.e., a *superelement*) that is composed of a collection of individual elements, and the entire rigid frame consists of only superelements A, B, and C connected at the interface nodal points.

The response of the entire structure can now be obtained using the usual equations of the stiffness method. For example, imposing equilibrium at the interface nodes gives

$$\mathbf{P}_r - \bar{\mathbf{P}}_r = \mathbf{K}_{rr} \mathbf{U}_r \tag{6.17}$$

In the case of the structure in Fig. 6.3a, these assembled equations are expressed in terms of the eight interface nodes (since the generalized displacements at the foundation are all zero). Note that this has considerably reduced the order of the equations that would have been obtained if the entire rigid frame were formulated in a single set of equations.

The solution of Eq. (6.17) gives the generalized displacements at these few interface nodes, and those within each substructure must be calculated by returning to the force-displacement equations for each substructure. For example, from Eq. (6.13) we note that the internal generalized displacements for substructure B are

$$\mathbf{U}_e^{\mathrm{B}} = -[\mathbf{K}_{ee}^{\mathrm{B}}]^{-1} \mathbf{K}_{er}^{\mathrm{B}} \mathbf{U}_r^{\mathrm{B}} + [\mathbf{K}_{ee}^{\mathrm{B}}]^{-1} \mathbf{P}_e^{\mathrm{B}} \tag{6.18}$$

Subsequent to this calculation, the element forces are obtained in the usual manner for the stiffness method. Example 6.3 illustrates the principle of substructural analysis for a relatively simple structure.

EXAMPLE 6.3

Analyze the beam in Fig. E6.3 using substructural analysis; $I = 1200$ in.^{4} and $E = 29 \times 10^3$ kips/in.^{2}.

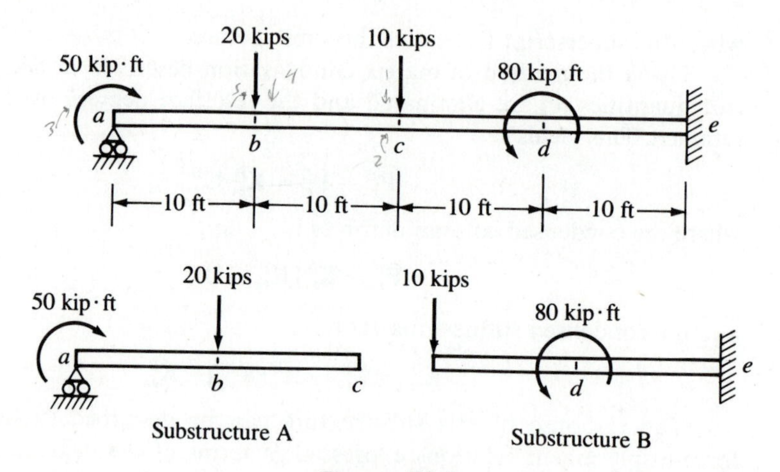

Figure E6.3

Solution

The stiffness matrix for each beam element is

$$\mathbf{k}^{ij} = 10^2 \begin{bmatrix} 2.4167 & 145.00 & -2.4167 & 145.00 \\ 145.00 & 11600. & -145.00 & 5800.0 \\ -2.4167 & -145.00 & 2.4167 & -145.00 \\ 145.00 & 5800.0 & -145.00 & 11600. \end{bmatrix}$$

where

$$\mathbf{u}^{ij} = \{v_i \quad \Theta_i \quad v_j \quad \Theta_j\}$$

For substructure A, Eq. (6.13) is

$$\begin{bmatrix} 0 \\ 0 \\ \hline -600 \\ -20 \\ 0 \end{bmatrix} = 10^2 \left[\begin{array}{cc|ccc} 2.4167 & -145.00 & 00000. & -2.4167 & -145.00 \\ -145.00 & 11600. & 00000. & 145.00 & 5800.0 \\ \hline 00000. & 00000. & 11600. & -145.00 & 5800.0 \\ -2.4167 & 145.00 & -145.00 & 4.8333 & 00000. \\ -145.00 & 5800.0 & 5800.0 & 00000. & 23200. \end{array}\right] \begin{bmatrix} V_c \\ \Theta_c \\ \hline \Theta_a \\ V_b \\ \Theta_b \end{bmatrix}$$

Using Eqs. (6.14), (6.15), and (6.16) gives the following condensed force-displacement relationships for substructure A:

$$\begin{bmatrix} -10.0 \\ 600. \end{bmatrix} = \begin{bmatrix} 7.5521 & -1812.5 \\ -1812.5 & 435000 \end{bmatrix} \begin{bmatrix} V_c \\ \Theta_c \end{bmatrix}$$

For substructure B, Eq. (6.13) is

$$\left[\begin{array}{r} -10 \\ 0 \\ \hline 0 \\ -960 \end{array}\right] = 10^2 \left[\begin{array}{rr|rr} 2.4167 & 145.00 & -2.4167 & 145.00 \\ 145.00 & 11600. & -145.00 & 5800.0 \\ \hline -2.4167 & -145.00 & 4.8333 & 0000.0 \\ 145.00 & 5800.0 & 0000.0 & 23200. \end{array}\right] \left[\begin{array}{c} V_c \\ \Theta_c \\ \hline V_d \\ \Theta_d \end{array}\right]$$

Using Eqs. (6.14), (6.15), and (6.16) gives the following condensed force-displacement relationships for substructure B:

$$\begin{bmatrix} -4.0 \\ 240. \end{bmatrix} = \begin{bmatrix} 30.208 & 3625.0 \\ 3625.0 & 580000 \end{bmatrix} \begin{bmatrix} V_c \\ \Theta_c \end{bmatrix}$$

The equations of nodal equilibrium at the interface degrees of freedom are

$$\begin{bmatrix} -14.0 \\ 840 \end{bmatrix} = \begin{bmatrix} 37.760 & 1812.5 \\ 1812.5 & 1015000 \end{bmatrix} \begin{bmatrix} V_c \\ \Theta_c \end{bmatrix}$$

Solving these equations gives

$$\{V_c \quad \Theta_c\} = \{-0.4490 \text{ in.} \quad 0.00163 \text{ rad}\}$$

and Eq. (6.18) yields the displacements of the eliminated degrees of freedom as follows:

$$\{\Theta_a \quad V_b \quad \Theta_b\}^{A} = \{-0.00362 \text{ rad} \quad -0.4234 \text{ in.} \quad -0.00231 \text{ rad}\}$$
$$\{V_d \quad \Theta_d\}^{B} = \{-0.1756 \text{ in.} \quad 0.00198 \text{ rad}\}$$

Discussion The applied load at node c could have been assigned to either substructure or a fraction of it associated with each. With all of the load at c related to substructure B, the matrix $\mathbf{P}_r^{A}$ is null, while $\mathbf{P}_r^{B} = \{-10 \quad 0\}$. Upon calculating the generalized displacements, the interior nodal displacements for each substructure can be computed using Eq. (6.18).

This example is not representative of the size of problems for which substructural analysis is typically performed; usually this structure would be investigated without the use of this powerful procedure. Generally, substructural analysis is only used when the number of degrees of freedom is very large.

6.2 COORDINATE TRANSFORMATIONS

In the stiffness method, we transform the element stiffness matrices from local to global coordinates while forming the structural stiffness matrix. Initial force matrices must also be transformed from local to global coordinates during formation of the force-displacement equations for the structure. Global-to-local transformations are subsequently carried out while calculating element forces. The transformation matrices are square if the matrices of variables in the two coordinates are of the same order, that is, $\mathbf{T}$ is n by n if $\bar{\mathbf{u}}$ and $\mathbf{u}$ are both n by 1, and

$$\bar{\mathbf{u}} = \mathbf{T}\mathbf{u}$$

or

$$\mathbf{u} = \mathbf{T}^T\bar{\mathbf{u}}$$

since $\mathbf{T}$ is orthogonal in this case. The same type of transformations apply to the force matrices, that is,

$$\bar{\mathbf{p}} = \mathbf{T}\mathbf{p} \quad \text{and} \quad \mathbf{p} = \mathbf{T}^T\bar{\mathbf{p}}$$

If the number of degrees of freedom of an element differ in local and global coordinates (e.g., $\bar{\mathbf{u}}$ and $\mathbf{u}$ are different orders), the theorem of virtual work can be applied to obtain the following contragredient transformations:

$$\mathbf{u} = \mathbf{T}^T\bar{\mathbf{u}}$$

and

$$\bar{\mathbf{p}} = \mathbf{T}\mathbf{p}$$

In this case $\mathbf{T}$ is a rectangular matrix.

Transformations can be used in other ways to deal with a number of special situations. For example, if several degrees of freedom in a structure are dependent, we can apply the theory of transformations to impose these *constraints* on the structural force-displacement equations. Structural supports that are not aligned with the global coordinate system also impose unique demands on the force-displacement equations. Such skew supports can be treated using a set of *nodal coordinates* located at the subject node; the affected degrees of freedom are transformed into the nodal coordinates and the prescribed skew boundary conditions are imposed on the transformed equations. Coordinate transformations can also be applied to handle *offset nodes*; for example, if a structural member has a cross section that is not doubly symmetric and the centroidal axis and shear center do not coincide. The flexure and torsion are coupled; the stiffnesses at nodes must be transferred to a common nodal point.

6.2.1 CONSTRAINT EQUATIONS A *constraint* indicates any type of relationship between generalized displacements. An unyielding support is a constraint that is described by putting the corresponding degree of freedom into the partition $\mathbf{U}_s$, and its effect does not appear in the partition $\mathbf{K}_{ff}$. Another type of constraint occurs if there is a dependency between several generalized displacements; this condition must be imposed on the structural force-displacement equations before solution. For example, the girders of the plane rigid frame in Fig. 6.4a are axially rigid and it is required to indicate this fact by noting that all the horizontal nodal displacements at a given floor level are equal. Another type of constraint occurs for the structure in Fig. 6.4b, which has experienced a support settlement; here the relationships for the vertical displacements of the foundation are prescribed. Constraint

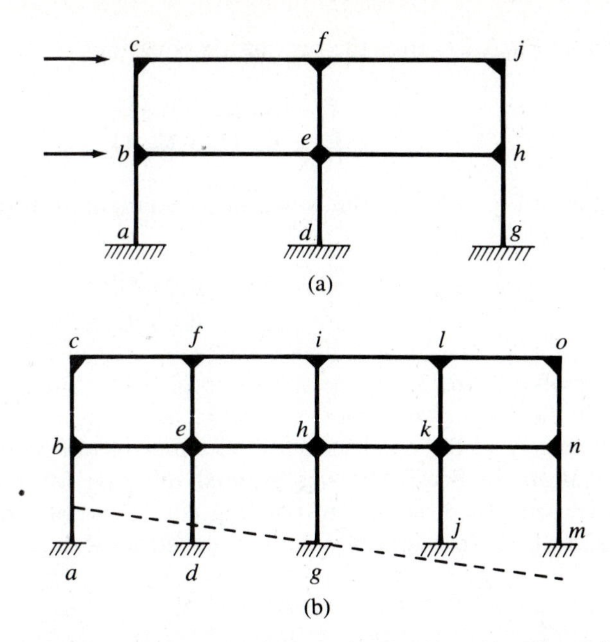

Figure 6.4 (a) Rigid frame with lateral loads, $U_c = U_f = U_j$ and $U_b = U_e = U_h$; (b) rigid frame with foundation settlement, $V_a = -V_m$, $V_d = -V_j$, $V_d = 0.5V_a$

conditions can be introduced into the structural force-displacement relations if we partition the displacement column matrix as follows:

$$\mathbf{U}_f = \begin{bmatrix} \mathbf{U}_r \\ \hline \mathbf{U}_d \end{bmatrix} \tag{6.19}$$

where U_r contains the generalized displacements to be retained and $\mathbf{U}_d$ contains those that are constrained (i.e., they have a prescribed dependency). The constraint conditions are expressed in the form of the following transformation:

$$\mathbf{U}_d = \mathbf{T}\mathbf{U}_r \tag{6.20}$$

For example, for the structure in Fig. 6.4b,

$$\begin{bmatrix} V_d \\ V_g \\ V_j \\ V_m \end{bmatrix} = \begin{bmatrix} 0.5 & 0 & 0 & 0 & \dots & 0 \\ 0 & 0 & 0 & 0 & \dots & 0 \\ -0.5 & 0 & 0 & 0 & \dots & 0 \\ -1 & 0 & 0 & 0 & \dots & 0 \end{bmatrix} \begin{bmatrix} V_a \\ U_b \\ V_b \\ \vdots \\ U_o \\ V_o \end{bmatrix}$$

The force-displacement equations are partitioned as implied by Eq. (6.19) to yield

$$\begin{bmatrix} \mathbf{P}_r \\ \hline \mathbf{P}_d \end{bmatrix} = \left[\begin{array}{c|c} \mathbf{K}_{rr} & \mathbf{K}_{rd} \\ \hline \mathbf{K}_{dr} & \mathbf{K}_{dd} \end{array}\right] \begin{bmatrix} U_r \\ \hline \mathbf{U}_d \end{bmatrix} \tag{6.21}$$

Expanding Eq. (6.21) gives the two matrix equations

$$\begin{aligned}\mathbf{P}_r &= \mathbf{K}_{rr}\mathbf{U}_r + \mathbf{K}_{rd}\mathbf{U}_d \\ \mathbf{P}_d &= \mathbf{K}_{dr}\mathbf{U}_r + \mathbf{K}_{dd}\mathbf{U}_d\end{aligned} \tag{6.22}$$

Substituting Eq. (6.20) for the constrained generalized displacements into Eqs. (6.22) yields

$$\begin{aligned}\mathbf{P}_r &= (\mathbf{K}_{rr} + \mathbf{K}_{rd}\mathbf{T})\mathbf{U}_r \\ \mathbf{P}_d &= (\mathbf{K}_{dr} + \mathbf{K}_{dd}\mathbf{T})\mathbf{U}_r\end{aligned} \tag{6.23}$$

The first of Eqs. (6.23) can be solved to give $\mathbf{U}_r$, but note that the stiffness matrix in this form is not generally symmetric.

Symmetry of the condensed stiffness matrix can be preserved by further manipulation. In Sec. 3.3 it was pointed out that for a given displacement transformation there is a corresponding contragredient force transformation. That is, the force transformation for the constrained degrees of freedom is

$$\bar{\mathbf{P}}_r = \mathbf{T}^T\mathbf{P}_d \tag{6.24}$$

where $\bar{\mathbf{P}}_r$ denotes that the forces are transformed to the r generalized displacements; we shall observe that these augment $\mathbf{P}_r$. For the structure in Fig. 6.4b this contragredient force transformation is

$$\begin{bmatrix} \bar{P}_{ya} \\ \bar{P}_{xb} \\ \bar{P}_{yb} \\ \vdots \\ \bar{P}_{xo} \\ \bar{P}_{yo} \end{bmatrix} = \begin{bmatrix} 0.5 & 0 & -0.5 & -1 \\ 0 & 0 & 0 & 0 \\ 0 & 0 & 0 & 0 \\ \vdots & \vdots & \vdots & \vdots \\ 0 & 0 & 0 & 0 \\ 0 & 0 & 0 & 0 \end{bmatrix} \begin{bmatrix} P_{yd} \\ P_{yg} \\ P_{yj} \\ P_{ym} \end{bmatrix}$$

Multiplying the second matrix equation in Eq. (6.23) by $\mathbf{T}^T$ and adding the two equations gives

$$\mathbf{P}_r + \mathbf{T}^T\mathbf{P}_d = (\mathbf{K}_{rr} + \mathbf{K}_{rd}\mathbf{T} + \mathbf{T}^T\mathbf{K}_{dr} + \mathbf{T}^T\mathbf{K}_{dd}\mathbf{T})\mathbf{U}_r \tag{6.25}$$

or

$$\hat{\mathbf{P}}_r = \hat{\mathbf{K}}_{rr}\mathbf{U}_r \tag{6.26}$$

where

$$\hat{\mathbf{P}}_r = \mathbf{P}_r + \mathbf{T}^T\mathbf{P}_d \tag{6.27}$$

and

$$\hat{\mathbf{K}}_{rr} = \mathbf{K}_{rr} + \mathbf{K}_{rd}\mathbf{T} + \mathbf{T}^T\mathbf{K}_{dr} + \mathbf{T}^T\mathbf{K}_{dd}\mathbf{T} \tag{6.28}$$

This gives the symmetric matrix $\hat{\mathbf{K}}_{rr}$, which can be used to obtain a solution with the usual computational efficiencies. Symmetry can be demonstrated by transposing $\hat{\mathbf{K}}_{rr}$ and noting that the matrix is equal to its transpose. Example 6.4 demonstrates this approach for the plane rigid frame of Example 4.13.

EXAMPLE 6.4

Solve for the generalized displacements of Example 4.13 using a constraint equation to impose the condition that $U_b = U_c$.

Solution

Partitioning the force-displacement equations according to Eq. (6.21) with $\mathbf{U}_r = \{\Theta_a \quad U_b \quad \Theta_b \quad \Theta_c\}$ and $\mathbf{U}_d = U_c$, Eq. (6.20) is

$$U_c = [0 \quad 1 \quad 0 \quad 0]\begin{bmatrix}\Theta_a \\ U_b \\ \Theta_b \\ \Theta_c\end{bmatrix}$$

Using Eq. (6.26) and noting that $\hat{\mathbf{P}}_r = \mathbf{P}_r$ since $\mathbf{P}_d = 0$ yields

$$\begin{bmatrix}0 \\ 80 \\ 0 \\ 0\end{bmatrix} = 10^2\begin{bmatrix}4833.3 & 50.347 & 2416.7 & 0000.0 \\ 50.347 & 1.3985 & 50.347 & 50.347 \\ 2416.7 & 50.347 & 10633. & 2900.0 \\ 0000.0 & 50.347 & 2900.0 & 10633.\end{bmatrix}\begin{bmatrix}\Theta_a \\ U_b \\ \Theta_b \\ \Theta_c\end{bmatrix}$$

Solving gives

$$\{\Theta_a \quad U_b \quad \Theta_b \quad \Theta_c\}$$
$$= \{-0.01258 \text{ rad} \quad 1.2891 \text{ in.} \quad -0.00171 \text{ rad} \quad -0.00564 \text{ rad}\}$$

Also we know that $U_c = 1.2891$ in. because of the constraint condition.

Discussion The transformation matrix $\mathbf{T} = [0 \quad 1 \quad 0 \quad 0]$ relates the partitions $\mathbf{U}_d = U_c$ and $\mathbf{U}_r = \{\Theta_a \quad U_b \quad \Theta_b \quad \Theta_c\}$ and affirms the fact that member *bc* is axially inextensible. Note that the conditions $V_b = V_c = 0$ could have been included in $\mathbf{T}$, but these degrees of freedom were assigned to $\mathbf{U}_s$ and as such did not appear in $\mathbf{K}_{ff}$. The matrix $\mathbf{K}_{ff}$ from Example 4.13 is partitioned according to Eq. (6.21), and components of $\hat{\mathbf{K}}_{rr}$, as prescribed by Eq. (6.28), are computed. In this case $\hat{\mathbf{P}}_r = \mathbf{P}_r$ since $\mathbf{P}_d = 0$. Note that the solution of $\mathbf{P}_r = \hat{\mathbf{K}}_{rr}\mathbf{U}_r$ yields the same generalized displacements (within the precision of the computer) obtained in Example 4.13 where the elements are idealized as axially rigid by adopting large values for their cross-sectional areas.

6.2.2 NODAL COORDINATES Sometimes it is necessary to express the forces, displacements, and stiffness coefficients at a particular node in terms of a unique nodal coordinate system that is different from the global coordinate system. For example, since the support surface at point *c* for the truss in

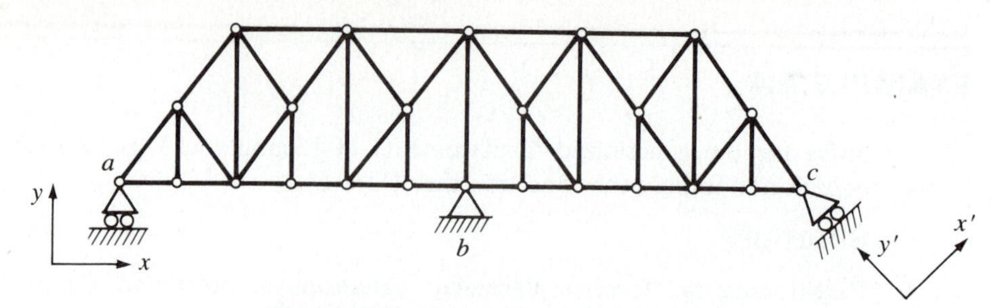

Figure 6.5 Truss with a skew support

Fig. 6.5 is aligned with neither the x nor y global coordinate direction, the conventional method of enforcing zero support displacements cannot be used. If the degrees of freedom at c are transformed into the special x'-y' coordinate system fixed to that node, we can enforce the condition that $V'_c = 0$ in the usual fashion. In this case it would be pointless to transform all nodal degrees of freedom to the x'-y' coordinates since the zero support displacement at node a could not be enforced in this system; furthermore, it is preferable to express the majority of the nodal displacements and forces in terms of the x-y global coordinate directions. Thus, there is a need for expressing the structural response of this truss in terms of both the global coordinates and a single nodal coordinate system. In summary, the truss can be analyzed by: (a) transforming only the displacements at node c into a special nodal coordinate system while maintaining all others in the x-y global system; (b) enforcing the zero displacement support conditions in the force-displacement equations; and (c) proceeding with the solution in the usual fashion for the stiffness method.

Nodal coordinates may also be mandated in some other situations. Figure 6.6 shows the end of a truss bridge with the usual portal bracing spanning perpendicular to the two primary trusses. This portal bracing is composed of a truss that has zero stiffness perpendicular to its plane, and the displacements in the global x direction must be retained in the overall problem since the primary trusses can display deflections in this direction. It is possible to treat this system as three distinct planar trusses, but we must assume that the total structure is idealized as a space truss for purposes of studying the interaction of the main trusses and the portal bracing. Since the stiffnesses of the individual primary trusses and the portal bracing are planar, it is necessary to constrain the displacements normal to the plane of the portal truss for nodes c, d, e, and f. This can be accomplished using the coordinate system x'-y'-z' for these four nodes and enforcing the condition that W' (displacement in the z' direction) be zero for each of them. These constraints perpendicular to the portal bracing are required to prevent a major numerical difficulty introduced by a singular stiffness matrix. Note that nodes in the primary trusses other than those stabilized by the portal bracing must also be constrained against movement perpendicular to the plane of the truss; this is done by stipulating zero displacements in the global z direction.

The displacements in the nodal coordinates, $\mathbf{U}'$, can be expressed in terms of those described in the global coordinates, $\mathbf{U}$, using a transformation

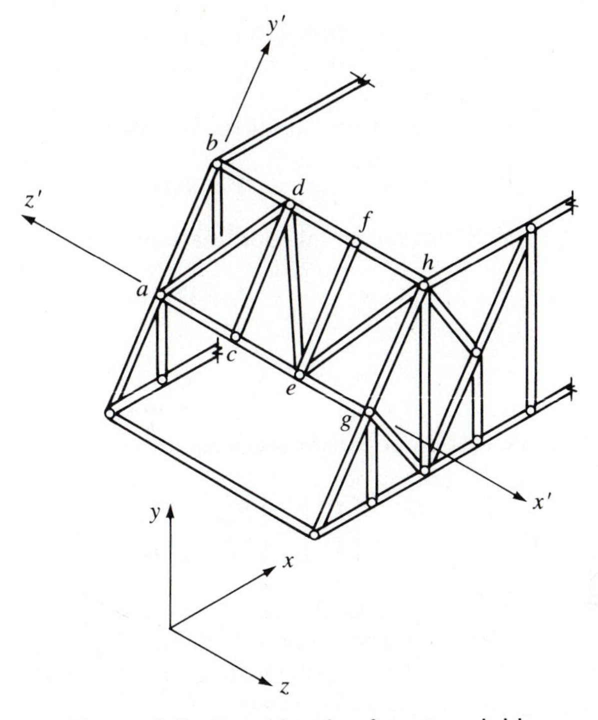

Figure 6.6 Portal bracing for a truss bridge

expressed in the usual manner, that is,

$$\mathbf{U} = \mathbf{T}\mathbf{U}' \tag{6.29}$$

and there is a corresponding transformation for the forces so that

$$\mathbf{P} = \mathbf{T}\mathbf{P}' \tag{6.30}$$

Since the column matrices of displacements and forces expressed in both coordinate systems contain all nodal quantities of the structure, $\mathbf{T}$ is a square matrix that is also orthogonal (i.e., $\mathbf{T}^T = \mathbf{T}^{-1}$). Usually, only one or two nodal coordinate systems need be used for a given structure; therefore, $\mathbf{T}$ will generally resemble the identity matrix except for those rows and colums that are associated with a nodal coordinate system. For these degrees of freedom the transformation from nodal to global coordinates is expressed in terms of the appropriate direction cosines. For example, for the truss of Fig. 6.5,

$$\mathbf{T} = \begin{bmatrix} 1 & 0 & 0 & \dots & 0 & 0 \\ 0 & 1 & 0 & \dots & 0 & 0 \\ 0 & 0 & 1 & \dots & 0 & 0 \\ \vdots & \vdots & \vdots & \vdots & \vdots & \vdots \\ 0 & 0 & 0 & \dots & \lambda_{x'} & \mu_{x'} \\ 0 & 0 & 0 & \dots & \lambda_{y'} & \mu_{y'} \end{bmatrix}$$

where $\lambda_{x'}$ and $\mu_{x'}$ are the direction cosines of the global x axis with respect to the x'-y' nodal axis at c and $\lambda_{y'}$ and $\mu_{y'}$ are the direction cosines of the global y axis.

The force-displacement equations in global coordinates,

$$\mathbf{P} = \mathbf{K}\mathbf{U} \tag{6.31}$$

are transformed into the nodal coordinate system to give

$$\mathbf{P}' = \mathbf{K}'\mathbf{U}' \tag{6.32}$$

where

$$\mathbf{K}' = \mathbf{T}^T\mathbf{K}\mathbf{T} \tag{6.33}$$

which can be written in partitioned form as

$$\left[\begin{array}{c} \mathbf{P}'_g \\ \hline \mathbf{P}'_n \end{array}\right] = \left[\begin{array}{c|c} \mathbf{K}'_{gg} & \mathbf{K}'_{gn} \\ \hline \mathbf{K}'_{ng} & \mathbf{K}'_{nn} \end{array}\right] \left[\begin{array}{c} \mathbf{U}'_g \\ \hline \mathbf{U}'_n \end{array}\right] \tag{6.34}$$

Quantities with a subscript g are expressed in global coordinates and those with an n are in nodal coordinates. Note that $\mathbf{P}'_g = \mathbf{P}_g$ and $\mathbf{U}'_g = \mathbf{U}_g$, where $\mathbf{P}_g$ and $\mathbf{U}_g$ are the corresponding partitions of $\mathbf{P}$ and $\mathbf{U}$, respectively. With this partitioning scheme Eq. (6.33) becomes

$$\begin{aligned} \mathbf{K}' &= \left[\begin{array}{c|c} \mathbf{K}'_{gg} & \mathbf{K}'_{gn} \\ \hline \mathbf{K}'_{ng} & \mathbf{K}'_{nn} \end{array}\right] \\ &= \left[\begin{array}{c|c} \mathbf{I} & \mathbf{0} \\ \hline \mathbf{0}^T & \mathbf{\Gamma}^T \end{array}\right] \left[\begin{array}{c|c} \mathbf{K}_{gg} & \mathbf{K}_{gn} \\ \hline \mathbf{K}_{ng} & \mathbf{K}_{nn} \end{array}\right] \left[\begin{array}{c|c} \mathbf{I} & \mathbf{0}^T \\ \hline \mathbf{0} & \mathbf{\Gamma} \end{array}\right] \\ &= \left[\begin{array}{c|c} \mathbf{K}_{gg} & \mathbf{K}_{gn}\mathbf{\Gamma} \\ \hline \mathbf{\Gamma}^T\mathbf{K}_{ng} & \mathbf{\Gamma}^T\mathbf{K}_{nn}\mathbf{\Gamma} \end{array}\right] \end{aligned} \tag{6.35}$$

where $\mathbf{I}$ = identity matrix

$\mathbf{0}$ = null matrix

$\mathbf{\Gamma}$ = transformation matrix containing appropriate direction cosines

For example, for the truss of Fig. 6.5,

$$\mathbf{\Gamma} = \begin{bmatrix} \lambda_{x'} & \mu_{x'} \\ \lambda_{y'} & \mu_{y'} \end{bmatrix}$$

From Eq. (6.35) we can observe that $\mathbf{K}'$ is obtained with only a few multiplications, and it is not necessary to carry out the complete triple product $\mathbf{T}^T\mathbf{K}\mathbf{T}$ to transform $\mathbf{K}'$ since a considerable portion of $\mathbf{K}$ is completely unaltered if only a limited number of nodal coordinate systems are used. That is, for the truss of Fig. 6.5, $\mathbf{K}'_{gn}$ is a matrix with only two columns, $\mathbf{K}'_{ng}$ is the transpose of $\mathbf{K}'_{gn}$, and $\mathbf{K}'_{nn}$ is merely a 2 by 2 matrix.

Upon calculating $\mathbf{K}'$ the condition of the zero nodal support displacements can be invoked to give $\mathbf{K}'_{ff}$ (the partition of the merged stiffness matrix containing the effect of the unconstrained degrees of freedom), and the solution for the generalized displacements can be obtained by the stiffness method. This procedure is illustrated in Example 6.5 for a simple truss with one skew support condition.

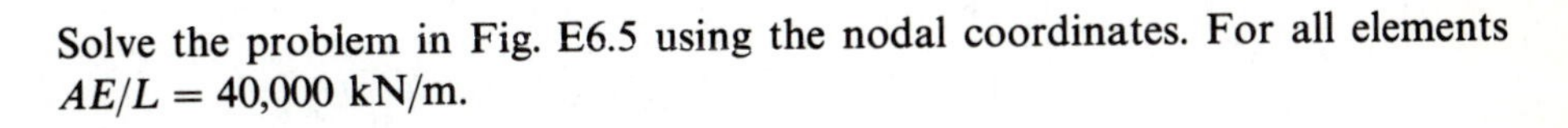

EXAMPLE 6.5

Solve the problem in Fig. E6.5 using the nodal coordinates. For all elements $AE/L = 40{,}000$ kN/m.

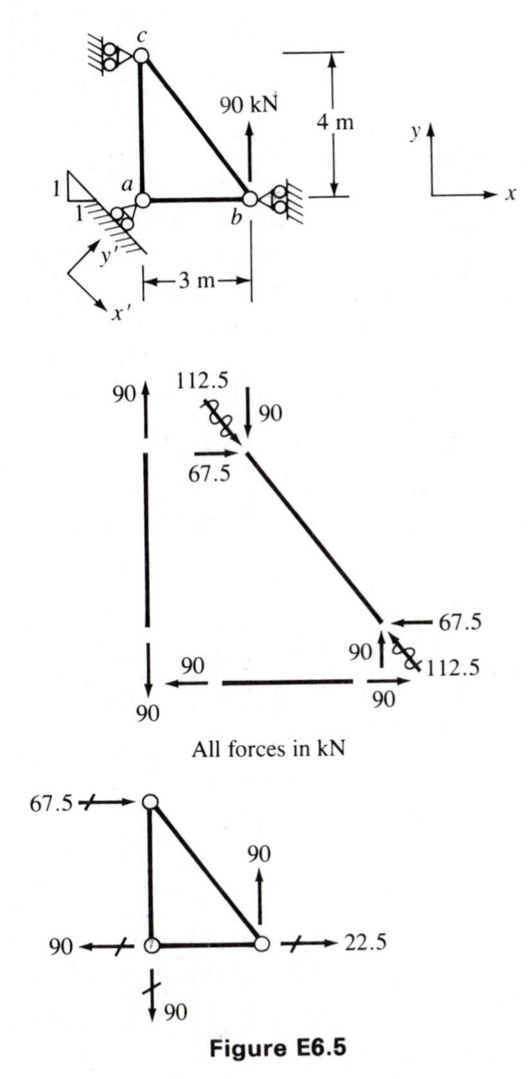

Figure E6.5

Solution

The 6 by 6 stiffness matrix for the assemblage is given in Example 2.3. With the boundary conditions at b and c enforced,

$$\begin{bmatrix} P_{xa} \\ P_{ya} \\ P_{yb} \\ P_{yc} \end{bmatrix} = (4 \times 10^4) \begin{bmatrix} 1.00 & 0.00 & 0.00 & 0.00 \\ 0.00 & 1.00 & 0.00 & -1.00 \\ 0.00 & 0.00 & 0.64 & -0.64 \\ 0.00 & -1.00 & -0.64 & 1.64 \end{bmatrix} \begin{bmatrix} U_a \\ V_a \\ V_b \\ V_c \end{bmatrix}$$

For this structure the transformation of Eq. (6.29) is

$$\begin{bmatrix} U_a \\ V_a \\ V_b \\ V_c \end{bmatrix} = \begin{bmatrix} 0.707 & 0.707 & 0 & 0 \\ -0.707 & 0.707 & 0 & 0 \\ 0 & 0 & 1 & 0 \\ 0 & 0 & 0 & 1 \end{bmatrix} \begin{bmatrix} U'_a \\ V'_a \\ V_b \\ V_c \end{bmatrix}$$

From Eq. (6.33),

$$\begin{bmatrix} P'_{xa} \\ P'_{ya} \\ P_{yb} \\ P_{yc} \end{bmatrix} = (4 \times 10^4) \begin{bmatrix} 1.000 & 0.000 & 0.000 & 0.707 \\ 0.000 & 1.000 & 0.000 & -0.707 \\ 0.000 & 0.000 & 0.640 & -0.640 \\ 0.707 & -0.707 & -0.640 & 1.640 \end{bmatrix} \begin{bmatrix} U'_a \\ V'_a \\ V_b \\ V_c \end{bmatrix}$$

Imposing the condition that $V'_a = 0$ yields

$$\mathbf{K}_{ff} = (4 \times 10^4) \begin{bmatrix} 1.000 & 0.000 & 0.707 \\ 0.000 & 0.640 & -0.640 \\ 0.707 & -0.640 & 1.640 \end{bmatrix}$$

Hence

$$\begin{bmatrix} U'_a \\ V_b \\ V_c \end{bmatrix} = 10^{-4} \begin{bmatrix} 0.5000 & -0.3535 & -0.3535 \\ -0.3535 & 0.8905 & 0.5000 \\ -0.3535 & 0.5000 & 0.5000 \end{bmatrix} \begin{bmatrix} 0 \\ 90 \\ 0 \end{bmatrix}$$

$$= 10^{-4} \begin{bmatrix} -3.18 \\ 8.02 \\ 4.50 \end{bmatrix} \text{m}$$

From Eq. (6.29),

$$\begin{bmatrix} U_a \\ V_a \\ V_b \\ V_c \end{bmatrix} = \begin{bmatrix} 0.707 & 0.707 & 0 & 0 \\ -0.707 & 0.707 & 0 & 0 \\ 0 & 0 & 1 & 0 \\ 0 & 0 & 0 & 1 \end{bmatrix} 10^{-3} \begin{bmatrix} -3.18 \\ 0.00 \\ 8.02 \\ 4.50 \end{bmatrix} = 10^{-3} \begin{bmatrix} -2.25 \\ 2.25 \\ 8.02 \\ 4.50 \end{bmatrix} \text{m}$$

The element forces are computed as usual to give

$$s^{ab} = \mathbf{eu}^{ab} = 90 \text{ kN (t)}$$
$$s^{bc} = \mathbf{eu}^{bc} = -112.5 \text{ kN (c)}$$
$$s^{ac} = \mathbf{eu}^{ac} = 90 \text{ kN (t)}$$

Discussion The assembled stiffness matrix from Example 2.3 is used, and the support conditions $U_b = U_c = 0$ have been enforced to give the $\mathbf{K}$ matrix. In this case the entire $\mathbf{T}$ matrix is used in calculating $\mathbf{K}'$ since the multiplication is straightforward. For the equation

$$\mathbf{P}' = \mathbf{K}'\mathbf{U}' \qquad \text{with } \mathbf{U}' = \{U'_a \quad V'_a \quad V_b \quad V_c\},$$

the condition $V'_a = 0$ is enforced to give the 3 by 3 stiffness matrix. The solution yields $\mathbf{U}'$, and all the element forces are presented for completeness.

Another method of treating skew supports is to use a special boundary element consisting of a fictitious spring, as illustrated in Fig. 6.7. This approach should be used with discretion since in order to accomplish the desired zero displacement (along the axis of the spring) it is necessary to use an extremely stiff element. If this stiffness is many orders of magnitude greater than the other nodal stiffness values for the truss, a stiffness matrix, ill conditioned with respect to solution, may result.

6.2.3 OFFSET NODES The analytical model of a structure is an idealized configuration yielding displacements and element forces that approximate the behavior of the true structure. Sometimes we specify nodal locations that may not be points where the loads are physically applied. The wind and snow loads are transmitted to the cladding in Fig. 6.8. Therefore, if the roof beam is to be analyzed independently of the remainder of the roof system the wind and snow loads must be transferred to the centerline of the roof beam. Other situ-

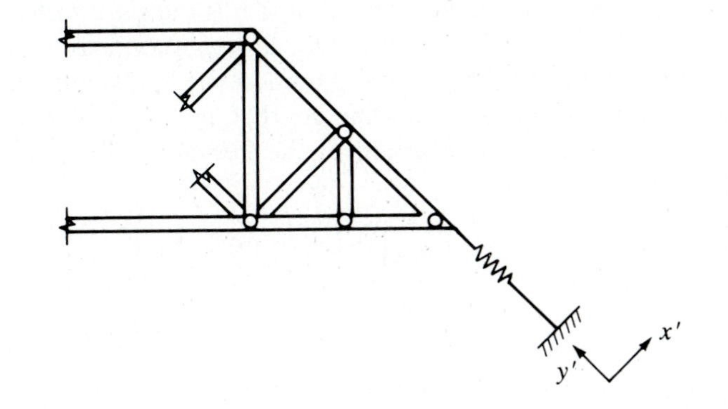

Figure 6.7 Alternate approach for analyzing skew supports using a substitute spring

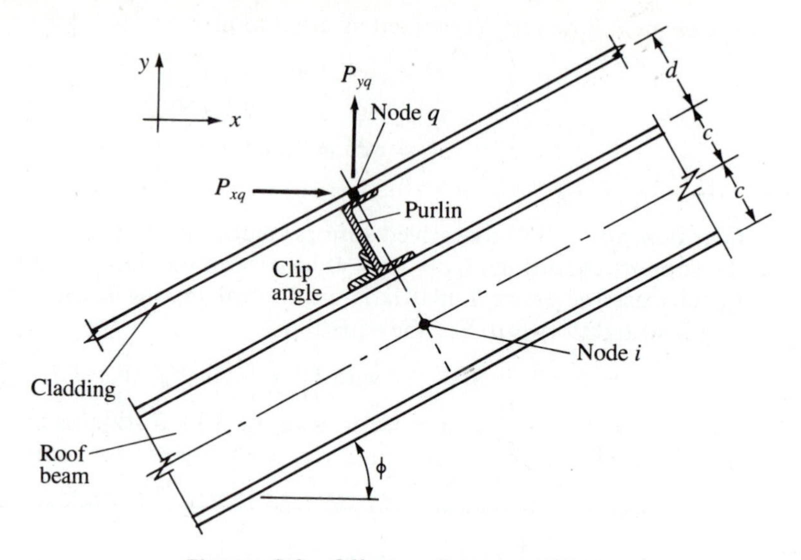

Figure 6.8 Offset node on a roof beam

ations require that displacements and stiffness matrices be transformed to nodal points. Large moment-resisting connections mandate that nodal quantities be transformed to the connection centerline. We observe an additional application of offset nodes if the purlin in Fig. 6.8 is analyzed. For a cross section that is not doubly symmetric the centroidal axis does not correspond to the shear center (i.e., transverse loads applied to unsymmetrical beams must pass through the shear center of the cross section for the beam to bend without twisting). Since the loads are not applied to the shear center of the purlin cross section (the channel-angle combination), the member will deflect both normal and tangent to the roof beam centerline. This behavior can be investigated by transforming either: (a) the element stiffness matrix to the centroidal axis or (b) the applied loads to the shear center. In Fig. 6.8 node q is an *offset node* attached to the element node i by a *rigid link iq*. Node i is sometimes referred to as the *master node* and q is the *slave node*.

The beam element in Fig. 6.9a is a generalization of the roof beam in Fig. 6.8. That is, in the physical situation the loads applied to the cladding are offset from the beam centerline so that $a_i = a_j = -(c + d) \sin \phi$, and $b_i = b_j = (c + d) \cos \phi$. Equilibrium of the free-body diagram of the rigid link in Fig. 6.9b yields

$$\begin{bmatrix} p_{xq} \\ p_{yq} \\ m_q \end{bmatrix} = \begin{bmatrix} 1 & 0 & 0 \\ 0 & 1 & 0 \\ b_i & -a_i & 1 \end{bmatrix} \begin{bmatrix} p_{xi} \\ p_{yi} \\ m_i \end{bmatrix} \tag{6.36}$$

$$\mathbf{p}^q = \boldsymbol{\Gamma}_o^i \quad \mathbf{p}^i$$

We note by inspection that the transformation matrix $\boldsymbol{\Gamma}_o^i$ is not orthogonal (i.e., $\boldsymbol{\Gamma}_o^T \neq \boldsymbol{\Gamma}_o^{-1}$). This can be verified by writing the equilibrium equations for the rigid link with the $\mathbf{p}^i$ as the independent variables. A transformation

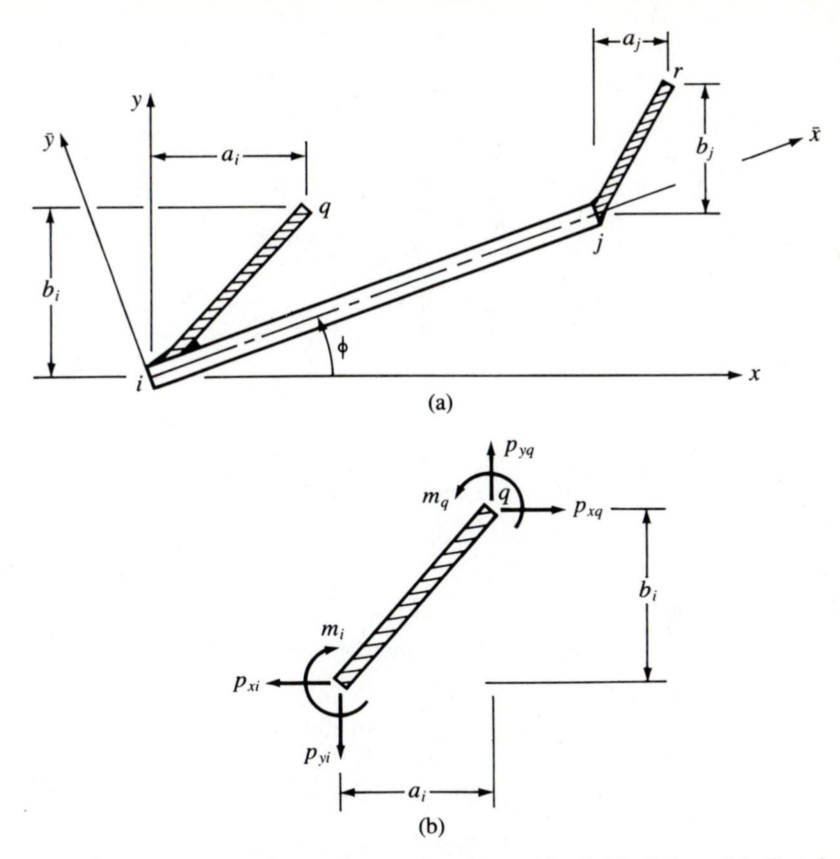

Figure 6.9 (a) Elastic plane frame element with rigid links; (b) free-body diagram of rigid link iq

similar to that in Eq. (6.36) relates the generalized forces at r and j as follows:

$$\begin{bmatrix} p_{xr} \\ p_{yr} \\ m_r \end{bmatrix} = \begin{bmatrix} 1 & 0 & 0 \\ 0 & 1 & 0 \\ b_j & -a_j & 1 \end{bmatrix} \begin{bmatrix} p_{xj} \\ p_{yj} \\ m_j \end{bmatrix} \tag{6.37}$$

$$\mathbf{p}^r = \mathbf{\Gamma}_o^j \quad \mathbf{p}^j$$

Combining Eqs. (6.36) and (6.37) gives the transformation for the elastic element and its rigid links, that is,

$$\begin{bmatrix} p_{xq} \\ p_{yq} \\ m_q \\ \hline p_{xr} \\ p_{yr} \\ m_r \end{bmatrix} = \left[\begin{array}{ccc|ccc} 1 & 0 & 0 & 0 & 0 & 0 \\ 0 & 1 & 0 & 0 & 0 & 0 \\ b_i & -a_i & 1 & 0 & 0 & 0 \\ \hline 0 & 0 & 0 & 1 & 0 & 0 \\ 0 & 0 & 0 & 0 & 1 & 0 \\ 0 & 0 & 0 & b_j & -a_j & 1 \end{array}\right] \begin{bmatrix} p_{xi} \\ p_{yi} \\ m_i \\ \hline p_{xj} \\ p_{yj} \\ m_j \end{bmatrix} \tag{6.38}$$

$$\mathbf{p}^{qr} = \mathbf{T}_o \quad \mathbf{p}^{ij}$$

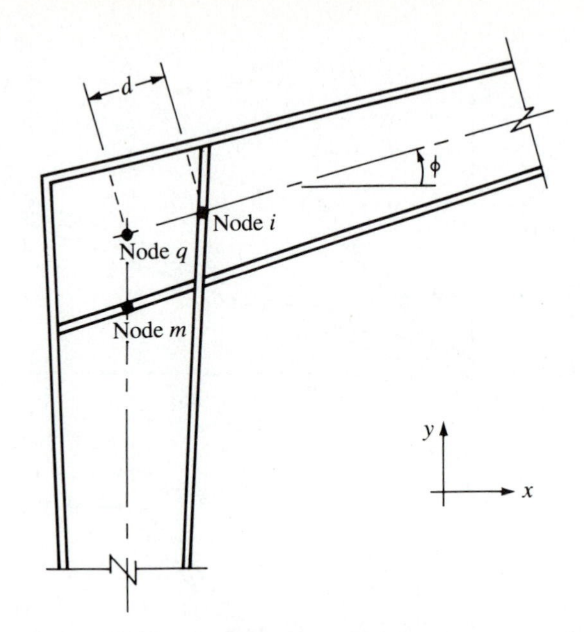

Figure 6.10 A moment-resisting connection

The transformation matrix $\mathbf{T}_o$ is also not orthogonal. From the principle of virtual work (see Sec. 3.3), we know that there is a corresponding contragredient transformation for the displacements:

$$\mathbf{u}^{ij} = \mathbf{T}_o^T \mathbf{u}^{qr} \tag{6.39}$$

where $\mathbf{u}^{ij} = \{u_i \quad v_i \quad \theta_i \quad u_j \quad v_j \quad \theta_j\}$

and $\mathbf{u}^{qr} = \{u_q \quad v_q \quad \theta_q \quad u_r \quad v_r \quad \theta_r\}$

The moment-resisting connection at the beam-column intersection of Fig. 6.10 is a very stiff component of the structure; therefore, the analysis would be in error if the ends of the column and beam elements were located at point q. It is preferable to terminate the column element at node m, have the beam element start at node i, and attach the two elements to point q with rigid links. Thus, we require that the stiffness matrices be transformed from nodes m and i to node q. The roof beam of Fig. 6.10 has the typical force-displacement equations referred to the element nodal points i and j, that is,

$$\mathbf{p}^{ij} = \mathbf{k}^{ij}\mathbf{u}^{ij} \tag{6.40}$$

Substituting Eq. (6.39) in Eq. (6.40) yields

$$\mathbf{p}^{ij} = \mathbf{k}^{ij}\mathbf{T}_o^T \mathbf{u}^{qr} \tag{6.41}$$

Using Eq. (6.41) with Eq. (6.38) gives the force-displacement equations for the element with forces and displacements referenced to the ends of the rigid links, that is,

$$\mathbf{p}^{qr} = \mathbf{k}^{qr}\mathbf{u}^{qr} \tag{6.42}$$

where $\mathbf{k}^{qr}$, the stiffness matrix for the beam element with respect to the ends of the rigid links (see Fig. 6.9a), is

$$\mathbf{k}^{qr} = \mathbf{T}_o \mathbf{k}^{ij} \mathbf{T}_o^T \tag{6.43}$$

For the roof beam of Fig. 6.10, $a_i = -d \cos \phi$, $b_i = -d \sin \phi$, $a_j = b_j = 0$. Note that $\mathbf{\Gamma}_o^j = \mathbf{I}$ (i.e., the 3 by 3 identity matrix), and the element stiffness matrix at node j is unchanged.

Equation (6.43) can also be obtained using the principle of virtual work from Ch. 3. Since virtual work is invariant under a coordinate transformation, we have

$$\boldsymbol{\delta}\mathbf{u}^{ij^T} \mathbf{p}^{ij} = \boldsymbol{\delta}\mathbf{u}^{qr^T} \mathbf{p}^{qr} \tag{6.44}$$

where $\boldsymbol{\delta}\mathbf{u}^{ij}$ and $\boldsymbol{\delta}\mathbf{u}^{qr}$ are virtual displacements corresponding to $\mathbf{u}^{ij}$ and $\mathbf{u}^{qr}$, respectively; therefore the transformation of Eq. (6.39) is applicable, that is,

$$\boldsymbol{\delta}\mathbf{u}^{ij} = \mathbf{T}_o^T \, \boldsymbol{\delta}\mathbf{u}^{qr} \tag{6.45}$$

Substituting the force-displacement equations in Eq. (6.44) yields

$$\boldsymbol{\delta}\mathbf{u}^{ij^T} \mathbf{k}^{ij} \mathbf{u}^{ij} = \boldsymbol{\delta}\mathbf{u}^{qr^T} \mathbf{k}^{qr} \mathbf{u}^{qr} \tag{6.46}$$

Using Eqs. (6.39) and (6.45) with Eq. (6.46) we have

$$\boldsymbol{\delta}\mathbf{u}^{qr^T} \mathbf{T}_o \mathbf{k}^{ij} \mathbf{T}_o^T \mathbf{u}^{qr} = \boldsymbol{\delta}\mathbf{u}^{qr^T} \mathbf{k}^{qr} \mathbf{u}^{qr} \tag{6.47}$$

Noting that $\boldsymbol{\delta}\mathbf{u}^{qr}$ is arbitrary and nonzero gives the result of Eq. (6.43), that is,

$$\mathbf{k}^{qr} = \mathbf{T}_o \mathbf{k}^{ij} \mathbf{T}_o^T$$

The transformation matrices in Eqs. (6.36) and (6.38) are not orthogonal, but each has an inverse. Therefore, by using the inverse relationship of Eq. (6.39) (i.e., by inverting $\mathbf{T}_o$), it is possible to predict the displacements of the top of the purlin in Fig. 6.8 after the displacements of the roof beam have been calculated. All transformation matrices do not have an inverse. If the axial deformations of the beam element in Fig. 6.11a are to be ignored, there are two degrees of freedom at nodes i and j (i.e., v and θ), whereas all applied forces at nodes q and r must be considered. Therefore, the force transformation for the rigid link iq (see Fig. 6.11b) is

$$\begin{bmatrix} p_{yi} \\ m_i \end{bmatrix} = \begin{bmatrix} 0 & 1 & 0 \\ -b_i & a_i & 1 \end{bmatrix} \begin{bmatrix} p_{xq} \\ p_{yq} \\ m_q \end{bmatrix} \tag{6.48}$$
$$\mathbf{p}^i = \mathbf{\Gamma}_o^{iT} \quad \mathbf{p}^q$$

A similar transformation can be written for the rigid link jr. Note that these are noninvertible relationships. From the principle of virtual work we have a corresponding contragredient transformation for the displacements, that is,

$$\begin{bmatrix} u_q \\ v_q \\ \theta_q \end{bmatrix} = \begin{bmatrix} 0 & -b_i \\ 1 & a_i \\ 0 & 1 \end{bmatrix} \begin{bmatrix} v_i \\ \theta_i \end{bmatrix} \tag{6.49}$$
$$\mathbf{u}^q = \mathbf{\Gamma}_o^i \quad \mathbf{u}^i$$

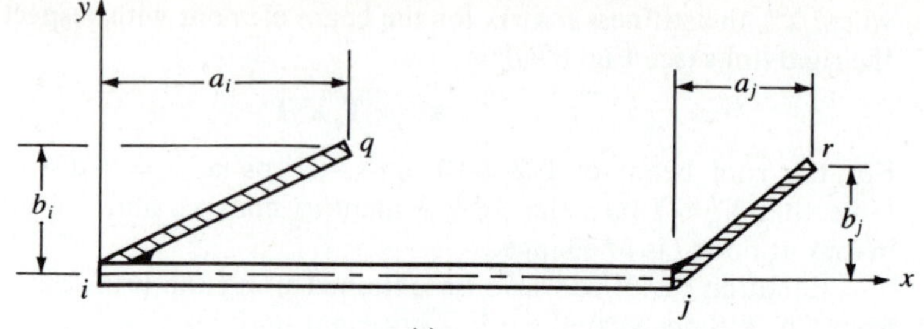

(a)

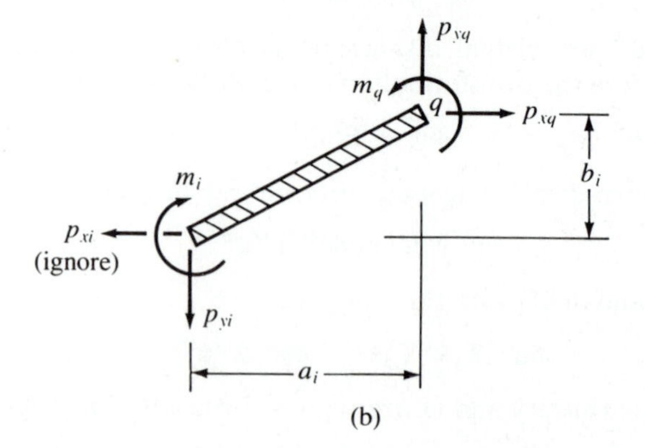

(b)

Figure 6.11 (a) Elastic beam element with rigid links; (b) free-body diagram of rigid link *iq*

Using the fact that virtual work is invariant under a coordinate transformation yields the following transformation law for the stiffness matrices:

$$\mathbf{k}^{ij} = \mathbf{T}_o^T \mathbf{k}^{qr} \mathbf{T}_o \tag{6.50}$$

where $\mathbf{T}_o$ is the transformation matrix for the combination of the elastic element and the rigid links [similar to the juxtaposition shown in Eq. (6.38)], $\mathbf{k}^{qr}$ is the stiffness matrix referenced to the ends of the rigid links, and $\mathbf{k}^{ij}$ is the conventional beam element stiffness matrix [see Eq. (4.19)].

The beam of Fig. 6.12 is a component of a rigid planar frame, and the element stiffness is to be transformed to the centerline of the columns at points q and r; the element has $b_i = b_j = 0$, $a_i = -c_1$, and $a_j = c_2$. Assuming that the beam is to be considered axially rigid so that the local u degrees of freedom for the beam need not be considered, we can use $\mathbf{k}^{ij}$ from Eq. (4.19) with

$$\mathbf{\Gamma}_o^i = \begin{bmatrix} 1 & 0 \\ c_1 & 0 \end{bmatrix} \quad \text{and} \quad \mathbf{\Gamma}_o^j = \begin{bmatrix} 1 & 0 \\ -c_2 & 1 \end{bmatrix}$$

Constructing the 4 by 4 matrix $\mathbf{T}_o$ from $\mathbf{\Gamma}_o^i$ and $\mathbf{\Gamma}_o^j$ and carrying out the matrix triple product of Eq. (6.43) yields the required beam stiffness matrix referenced

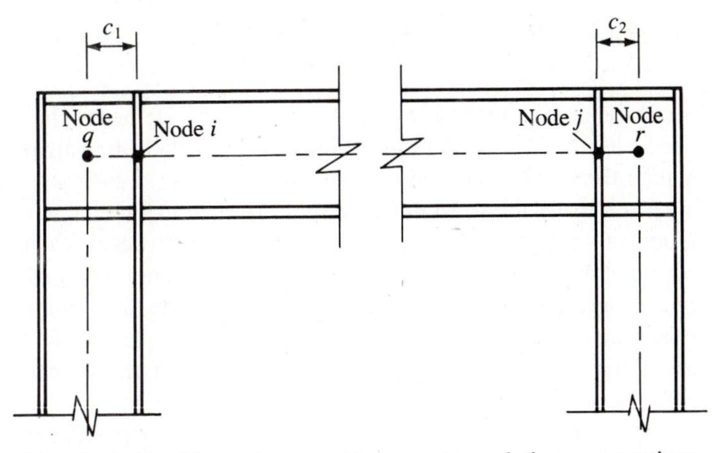

Figure 6.12 Planar frame with moment-resisting connections

to nodes q and r:

$$\mathbf{k}^{qr} = \frac{EI}{L^3}\begin{bmatrix} 12 & 6L + 12c_1 & -12 & 6L + 12c_2 \\ & 4L^2 + 12Lc_1 + 12c_1^2 & -6L - 12c_1 & L^2 + 6Lc_1 + 6Lc_2 + 12c_1c_2 \\ & & 12 & -6L - 12c_2 \\ & \text{Symmetrical} & & 4L^2 + 6Lc_2 + 12c_2^2 \end{bmatrix}$$

where

$$\mathbf{u}^{qr} = \{v_q \quad \theta_q \quad v_r \quad \theta_r\}$$

and the corresponding force matrix is obtained from Eq. (6.38) which gives

$$\mathbf{p}^{qr} = \{p_{yi} \quad c_1 p_{yi} + m_i \quad p_{yj} \quad -c_2 p_{yj} + m_j\}$$

The transformations described in this section relate forces and displacements that are expressed in global coordinates. In addition, the usual relationships between local and global quantities must be invoked in many situations. For the roof beam of Fig. 6.8 it is necessary to transform the stiffness matrix for the beam from local to global coordinates before the transformations associated with the offset nodes are used. Therefore the stiffness matrix in global coordinates is obtained by first using Eq. (4.88), followed by the transformation of Eq. (6.43). This gives the following:

$$\mathbf{k}^{qr} = \mathbf{T}_o(\mathbf{T}^T\bar{\mathbf{k}}\mathbf{T})\mathbf{T}_o^T \tag{6.51}$$

where $\mathbf{T}$ is the transformation from local to global coordinates shown in Eq. (4.83). Note that after nodal displacements have been calculated it is necessary to carry out the transformation of displacements to the ends of the element (in global coordinates) using Eq. (6.39), and these displacements, in turn, must be transformed to local coordinates with Eq. (4.85), giving

$$\bar{\mathbf{u}} = \mathbf{T}(\mathbf{T}_o^T\,\mathbf{u}^{qr}) \tag{6.52}$$

6.3 SPECIAL-PURPOSE ELEMENTS

For some structures it is most efficient to use elements that are different from those discussed in previous chapters. Members requiring careful scrutiny include those with: curved or tapered geometry; significant shear deformations or coupling of flexure and axial behavior; and/or elastic end restraints. The structural engineer who routinely investigates structures with unusual features will typically adopt an appropriate special-purpose element. Most production-level programs generally contain only the standard elements; therefore, if possible, the user must incorporate additional elements. The approach for describing these special members is discussed in the subsequent sections. The stiffness matrix is not given in closed form in all cases; the reader should realize that sometimes it is computationally more efficient to program the equations and numerically construct the basic matrices for each individual element. An alternative approach for obtaining stiffness matrices for special-purpose elements is described in Sec. 7.7.

6.3.1 THE TAPERED AXIAL FORCE ELEMENT The cross-sectional area of the axial force element in Fig. 6.13 varies linearly and the change in area (m) can take on values between zero (prismatic member) and unity (member terminates in a point). The differential equation of equilibrium for this member is

$$\frac{d}{dx}\left(AE\,\frac{du}{dx}\right) = 0 \tag{6.53}$$

Using $A = A_o(1 - mx/L)$, integrating twice, and using the boundary conditions $u(x = 0) = u_i$ and $u(x = L) = u_j$ yields

$$u = \left[1 - \frac{\ln(1 - mx/L)}{\ln(1 - m)}\right]u_i + \left[\frac{\ln(1 - mx/L)}{\ln(1 - m)}\right]u_j \tag{6.54}$$

or

$$u = N_1 u_i + N_2 u_j$$

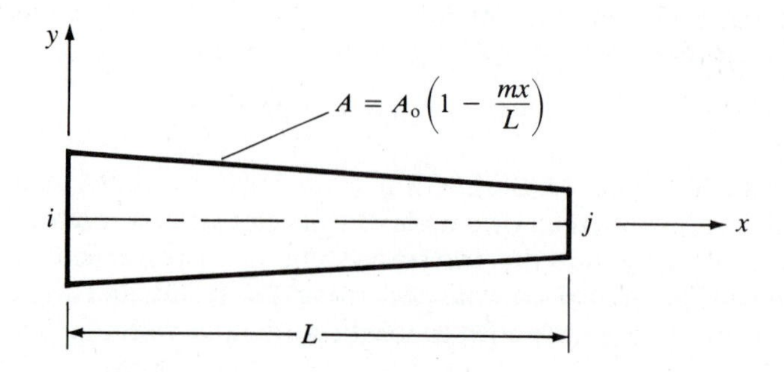

Figure 6.13 The tapered axial force element

Therefore

$$B_1 = \frac{m/L}{\ln(1-m)[1-mx/L]}$$

and

$$B_2 = -\frac{m/L}{\ln(1-m)[1-mx/L]} \tag{6.55}$$

Using Eq. (3.49) to obtain the stiffness matrix yields

$$\mathbf{k} = -\frac{A_o E}{L}\frac{m}{\ln(1-m)}\begin{bmatrix} 1 & -1 \\ -1 & 1 \end{bmatrix} \tag{6.56}$$

If $m = \frac{2}{3}$ and the element has an applied axial load P applied at node j in the positive x direction, while node i is constrained we have $u_j = 1.6479PL/A_o E$. This is an exact solution, and it is the same result as that obtained in Sec. 3.1.3.

Consider an approximate solution to this problem. For example, if we use the shape functions for a prismatic axial force element in conjunction with Eq. (3.49) the stiffness matrix is

$$\mathbf{k} = \int_0^L \mathbf{B}^T AE\,\mathbf{B}\,dx = \frac{A_o E}{L^2}\int_0^L \begin{bmatrix} -1 \\ 1 \end{bmatrix}\left(1 - \frac{2x}{3L}\right)[-1 \quad 1]\,dx$$

$$= \frac{2A_o E}{3L}\begin{bmatrix} 1 & -1 \\ -1 & 1 \end{bmatrix} \tag{6.57}$$

Solving for the tip displacement if $u_i = 0$ and an axial load P is applied at node j yields $u_j = 1.50PL/A_o E$.

As a final alternative we consider using the shape functions for a three-node element with the third node k located at the middle (see Fig. E3.3), that is,

$$\mathbf{N} = \left[\left(1 - 3\frac{x}{L} + 2\frac{x^2}{L^2}\right) \quad \left(-\frac{x}{L} + 2\frac{x^2}{L^2}\right) \quad \left(4\frac{x}{L} - 4\frac{x^2}{L^2}\right)\right]$$

If $m = \frac{2}{3}$ the stiffness matrix for the three-node representation is

$$\mathbf{k} = \frac{2A_o E}{9L}\begin{bmatrix} 9 & 1 & -10 \\ 1 & 5 & -6 \\ -10 & -6 & 16 \end{bmatrix} \tag{6.58}$$

where $\mathbf{u} = \{u_i \quad u_j \quad u_k\}$ with node k at the middle of the element. For purposes of comparison with the previous two stiffness matrices [i.e., Eqs. (6.56) and (6.57)] we condense node k from the stiffness matrix to yield

$$\mathbf{k} = \frac{11A_o E}{18L}\begin{bmatrix} 1 & -1 \\ -1 & 1 \end{bmatrix} \tag{6.59}$$

For the case with $u_i = 0$ and an applied axial force P at node j, the stiffness matrix in Eq. (6.59) yields $u_i = 1.6364PL/A_o E$. We note that the linear approximation of the displacements gives a tip displacement that is approximately 9.0% in error, while the displacement has an error of 0.7% using the quadratic approximation. These approximate solutions violate the differential equation of equilibrium [Eq. (6.53)], but analyses incorporating elements with approximate displacement fields are valid if certain conditions are met. We can observe monotonic convergence for the tapered axial force problem of Fig. 6.13 with the left end constrained by performing a series of analyses with an increasing number of linear elements. Recall that with one element the tip displacement is 9% less than the actual answer. Using more and more linear elements we would observe that the solution would become larger and larger, but would never exceed the actual displacement. We will attain *monotonic convergence if the elements are complete and compatible.*

Completeness implies that the displacement functions for the element can represent the *rigid-body displacements* and characterize the state of *constant strain.* For example, in the linear displacement function

$$u = \left(1 - \frac{x}{L}\right)u_i + \frac{x}{L}\,u_j$$

the constant term u_i is the rigid-body displacement and since $\varepsilon = du/dx$ we note that $(u_j - u_i)/L$ is a description of constant strain. Therefore the linear displacement function satisfies the condition of completeness. The reader is urged to identify the rigid-body and constant-strain terms in the quadratic displacement function.

Beam and frame elements have both translational and rotational rigid-body modes of displacements. The plane frame element must be capable of translating in two directions and rotating about an axis normal to the plane of the member while remaining stress free. The beam elements *ef* and *fg* at the tip of the loaded cantilever beam in Fig. 6.14 displace significantly, but they must remain stress free because static equilibrium indicates that the cantilever has zero shear and bending moment to the right of the applied load.

The requirement of constant strain is apparent if we envisage increasing the number of elements used to idealize a structure. For example, consider the

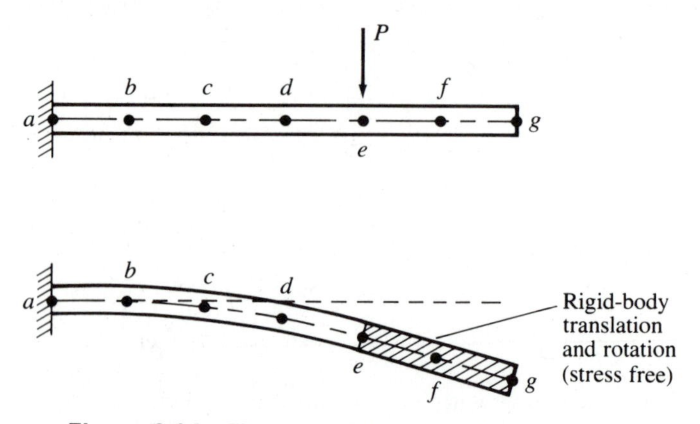

Figure 6.14 Elements with only rigid-body modes

analysis of a tapered member using linear axial force elements. In the limit as each element approaches a very small length, the strain in each element must approach a constant value.

Compatibility means that the displacements within elements and across nodes connecting elements must be continuous. The axial force element contains only translational degrees of freedom; therefore, we can conveniently observe compatibility for both the linear and quadratic axial force elements. Beam and frame elements have rotational degrees of freedom that are obtained by taking the first derivative of the displacement function (i.e., v and/or w depending upon whether two- or three-dimensional behavior is displayed). Thus it is also necessary to have element continuity in the appropriate first displacement derivatives. This stems from the assumption that displacements normal to the axis of bending, as well as rotations, are continuous over the depth of the beam; continuity of the displacements and their derivatives along the respective beam edges assures continuity of the corresponding displacements throughout the depth of adjoining elements.

6.3.2 THE TAPERED BEAM The tapered beam in Fig. 6.15 has a constant width b, and the depth tapers linearly with the rate of change of depth equal to m. The differential equation of equilibrium for a flexural member is

$$\frac{d^2}{dx^2}\left(EI\,\frac{d^2v}{dx^2}\right) = 0 \tag{6.60}$$

Using $I = bh^3/12$, $h = 2h_o(1 - mx/L)$, with $m = \frac{1}{2}$, and integrating twice gives

$$\frac{d^2v}{d^2x} = C_1\,\frac{x}{(1 - x/L)^3} + C_2\,\frac{1}{(1 - x/L)^3}$$

where C_1 and C_2 are constants of integration. Integration produces

$$\frac{dv}{dx} = C_1\left[\frac{xL}{(1 - x/L)^2} - \frac{2L^2}{(1 - x/L)}\right] + C_2\,\frac{L}{(1 - x/L)^2} + C_3 \tag{6.61}$$

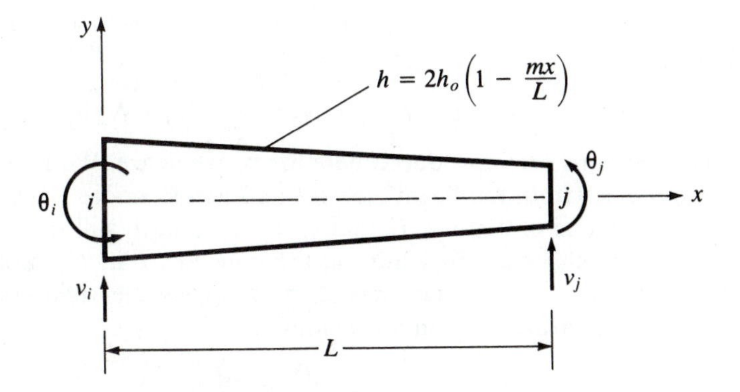

Figure 6.15 The tapered beam element

and
$$v = C_1\left[2xL^2\left(1 - \frac{x}{L}\right)^{-1} + 8L^3 \ln\left(1 - \frac{x}{L}\right)\right]$$
$$+2C_2 L^2\left(1 - \frac{x}{L}\right)^{-1} + C_3 x + C_4 \tag{6.62}$$

The constants of integration C_1, C_2, C_3, and C_4 are evaluated using the boundary conditions

$$\begin{aligned} v &= v_i & &\text{at } x = 0 \\ \frac{dv}{dx} &= \theta_i & &\text{at } x = 0 \\ v &= v_j & &\text{at } x = L \\ \frac{dv}{dx} &= \theta_j & &\text{at } x = L \end{aligned}$$

which gives the matrix equation

$$\begin{bmatrix} v_i \\ \theta_i \\ v_j \\ \theta_j \end{bmatrix} = \begin{bmatrix} 0 & 2L^2 & 0 & 1 \\ -2L^2 & L & 1 & 0 \\ -1.5452L^3 & 4L^2 & L & 1 \\ 0 & 4L & 1 & 0 \end{bmatrix} \begin{bmatrix} C_1 \\ C_2 \\ C_3 \\ C_4 \end{bmatrix} \tag{6.63}$$

Solving for the C_i's and substituting into Eq. (6.62) produces the shape functions for the tapered beam element.

The prismatic beam element utilized the following shape functions [see Eqs. (4.22) and 4.23)]:

$$\begin{aligned} N_1 &= 2\left(\frac{x}{L}\right)^3 - 3\left(\frac{x}{L}\right)^2 + 1 \\ N_2 &= \frac{x^3}{L^2} - 2\frac{x^2}{L} + x \\ N_3 &= -2\frac{x}{L} + 3\left(\frac{x}{L}\right)^2 \\ N_4 &= \frac{x^3}{L^2} - \frac{x^2}{L} \end{aligned} \tag{6.64}$$

where

$$v = N_1 v_i + N_2 \theta_i + N_3 v_j + N_4 \theta_j \tag{6.65}$$

The equations are reproduced here for convenience. We note that v and θ are continuous within the beam element and across nodes i and j. That is, the N_i are continuous in $0 \leq x \leq L$, and v_i, v_j, θ_i, and θ_j will match up with contiguous beam elements. In addition, the constant 1 in N_1 represents the rigid-body translation, and the term x in N_2 gives the rigid-body rotation. The moment-curvature relation for bending is

$$\frac{d^2v}{dx^2} = \frac{M}{EI} \tag{6.66}$$

Combining Eqs. (6.64) and (6.65) with Eq. (6.66) we note that these shape functions can represent a state of constant strain. Therefore monotonic convergence for a tapered beam element is guaranteed by using the shape functions for a prismatic element (see the discussion on convergence in Sec. 6.3.1). The stiffness matrix for the tapered beam element can be obtained by integration of

$$\mathbf{k} = \frac{8Ebh_o^3}{12} \int_0^L \mathbf{B}^T \left(1 - m\frac{x}{L}\right)^3 \mathbf{B}\, dx \tag{6.67}$$

where

$$\mathbf{B} = -y \frac{d^2}{dx^2} \mathbf{N}$$

See Eqs. (4.24) through (4.27) for the development of this relationship. Calculating the terms of the stiffness matrix for the tapered beam using either the results from Eq. (6.62) or those from Eq. (6.67) may be accomplished either in closed form or numerically.

6.4 Problems

6.1 Using matrix condensation calculate $\mathbf{K}_{cc}$ and the displacements at node b by eliminating the degrees of freedom at node c in the stiffness matrix for the truss in Fig. P6.1. Calculate the displacements at node c from those obtained for node b. Elements ab, bc, and cd have cross-sectional areas A, while diagonals ac and bd have areas of $A\sqrt{2}$.

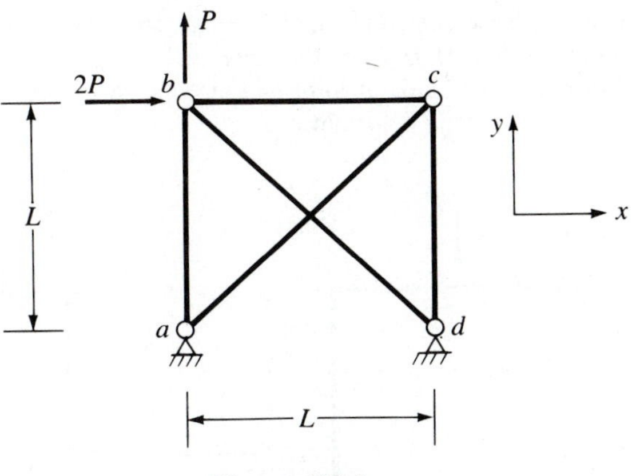

Figure P6.1

6.2 Using matrix condensation, calculate U_c and V_c if $U_b = 4.828(PL/AE)$ and $V_b = 2.000(PL/AE)$ for the truss of Fig. P6.1 if AE/L = constant for all elements.

6.3 The rigid frame of Prob. 4.27 is modified by placing a hinge at point b. Calculate the displacements and element forces idealizing the structure as a two-element, three-node structure in which member axial deformations are ignored.

6.4 Calculate nodal displacements and element forces for the steel beam in Fig. P6.4; $E = 29 \times 10^3$ ksi and $I = 600$ in.^{4}.

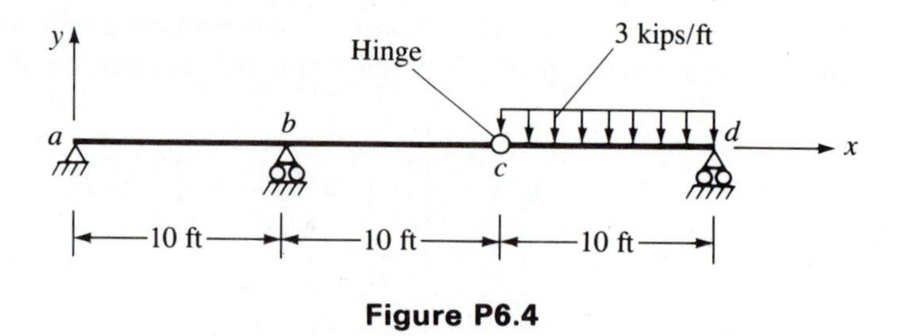

Figure P6.4

6.5 Calculate nodal displacements and element forces using the beam element with a hinge for the beam in Fig. P6.5; $E = 30 \times 10^3$ ksi and $I = 300$ in.^{4}.

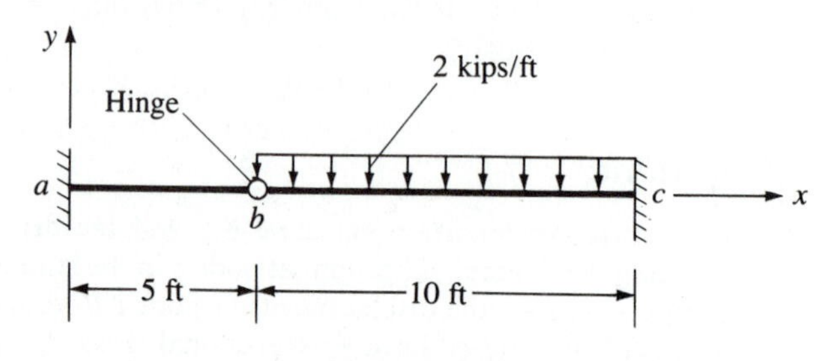

Figure P6.5

6.6 Using the beam element with a hinge, show that the analysis of the rigid frames shown in Fig. P6.6 yields the same results; $I_{bc} = 2I_{ab} = 2I$ and $E =$ constant. The hinge is constructed at joint b; Fig. P6.6 shows two interpretations of this structure. Ignore axial deformations.

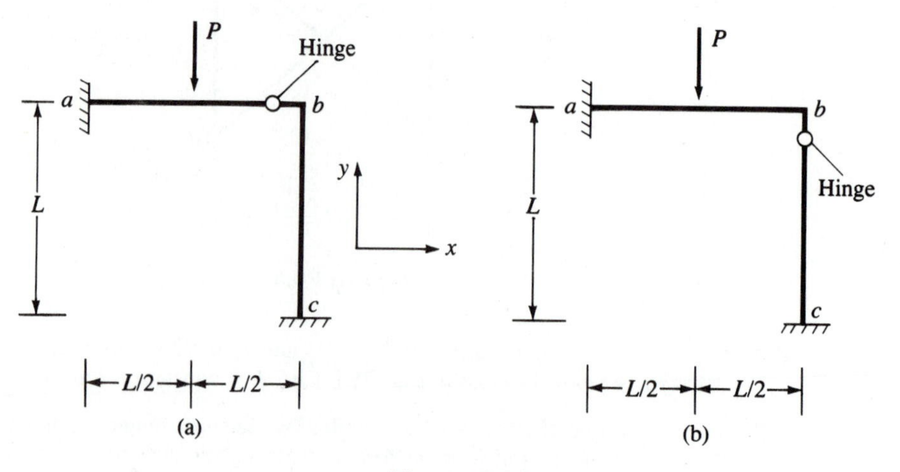

Figure P6.6

6.7 Calculate nodal displacements and element forces for the rigid frame in Fig. P6.7 using the beam element with a hinge; $I_{ab} = I_{cd} = 2I_{bc} = 2I$ and $E = \text{constant}$. Use constraint equations to impose the condition of axial inextensibility of all elements.

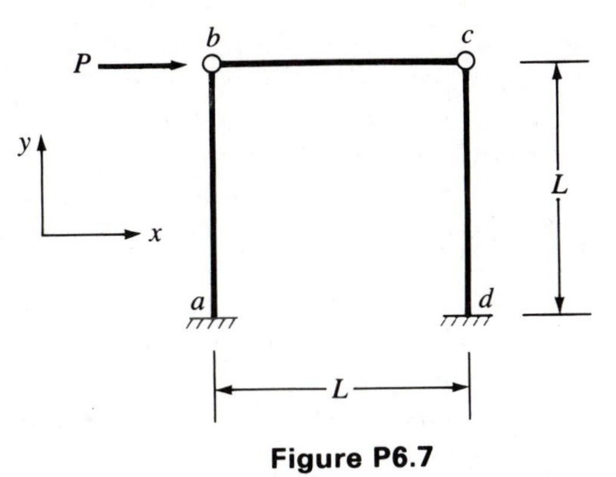

Figure P6.7

In Problems 6.8 through 6.10 use constraint equations to impose the condition of axial inextensibility of all elements of the original problem. Solve for nodal displacements.

6.8 Problem 4.26.

6.9 Problem 4.27.

6.10 Problem 4.29(a).

6.11 Use substructural analysis to obtain nodal displacements and element forces for the truss shown in Fig. P6.11. Cross-sectional areas of horizontal and vertical elements are 1.00 in.^{2} and the diagonal elements have areas of 0.707 in.^{2}; $E = 10^7$ psi.

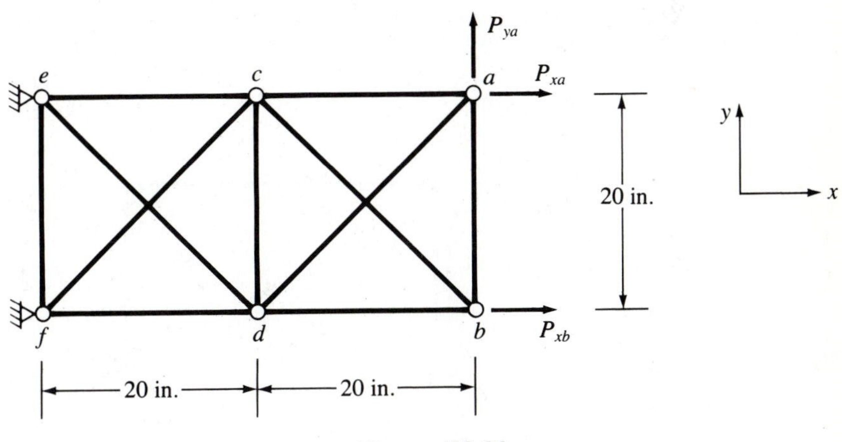

Figure P6.11

In Problems 6.12 through 6.16 use substructural analysis to obtain nodal displacements and element forces for the original problem.

6.12 Problem 2.12.

6.13 Problem 2.13.

6.14 Problem 2.24.

6.15 Problem 2.26.

6.16 Problem 2.32.

6.17 Examine the truss in Fig. 6.17 for its suitability to support the illustrated loading; $AE/L = 6 \times 10^4$ kN/m for all elements. Use nodal coordinates and then investigate the consistency of the set of equations as described in Sec. 5.1.

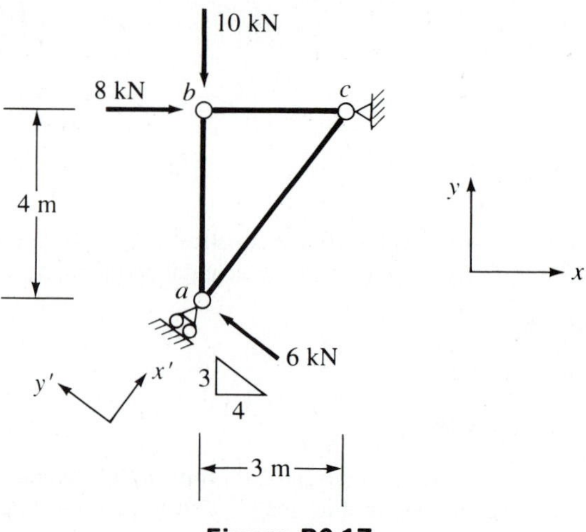

Figure P6.17

6.18 Calculate nodal displacements and element forces in Fig. P6.18 using nodal coordinates; $E = 29 \times 10^3$ ksi, while $I = 600$ in.^{4} for the columns and $I = 1500$ in.^{4} for the beam. Use constraint equations to impose the condition of axial inextensibility of all elements.

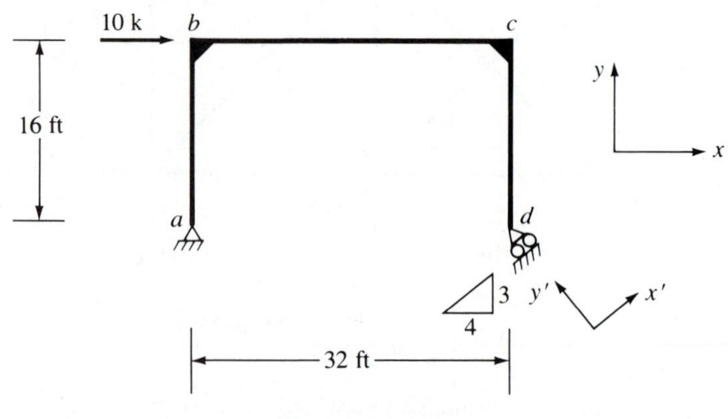

Figure P6.18

6.19 Calculate nodal displacements for the truss in Fig. P6.19; $E = 29 \times 10^3$ ksi. Cross-sectional areas are as follows: verticals $A = 5$ in.^{2}; horizontals $A = 10$ in^2; and diagonals $A = 12.5$ in.^{2}.

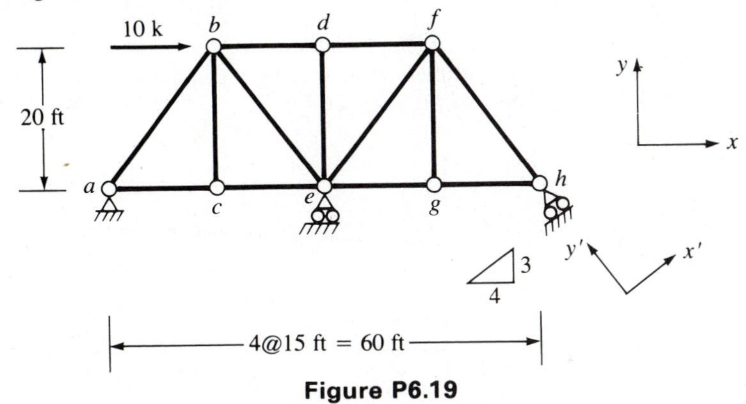

Figure P6.19

6.20 (a) Calculate nodal displacements and element forces for the truss in Fig. P6.20 using nodal coordinates; $AE/L = 20{,}000$ kN/m for all elements.

(b) Attempt to obtain nodal displacements using constraint equations (see Sec. 6.2.1) with $U_a = \frac{4}{3}V_a$. Explain the inadequacy of constraint equations in dealing with skew support conditions.

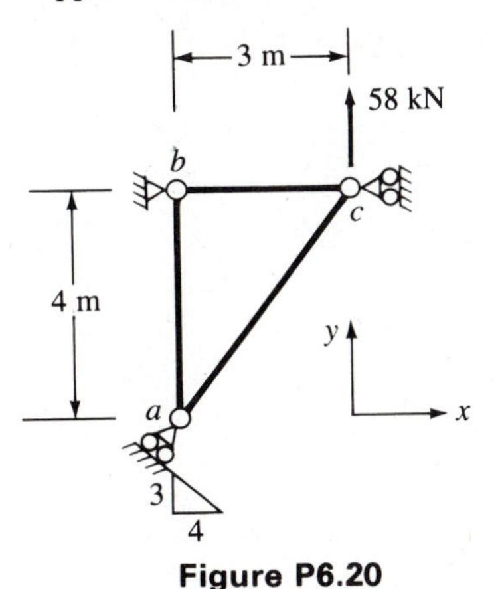

Figure P6.20

6.21 Obtain the stiffness matrix for the tapered beam element in Fig. P6.21 using the shape functions for a uniform beam as given in Eq. (4.23).

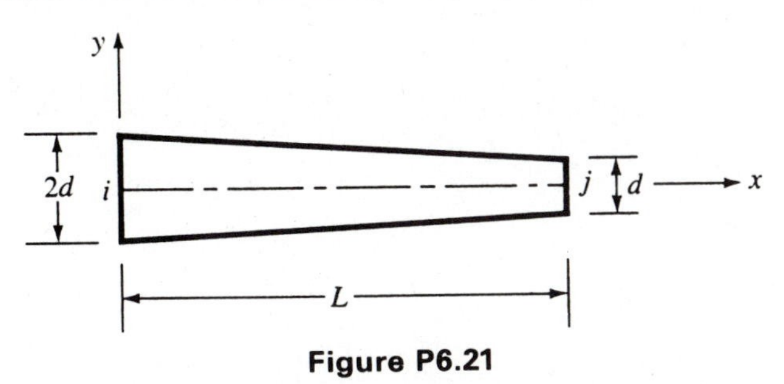

Figure P6.21

6.22 The beam element in Fig. P6.22 has a linearly varying moment of inertia, that is,

$$I(x) = I_0\left(1 + n\frac{x}{L}\right)$$

(a) Obtain the exact shape functions.
(b) Obtain the stiffness matrix using the shape functions for a uniform beam as given in Eq. (4.23).

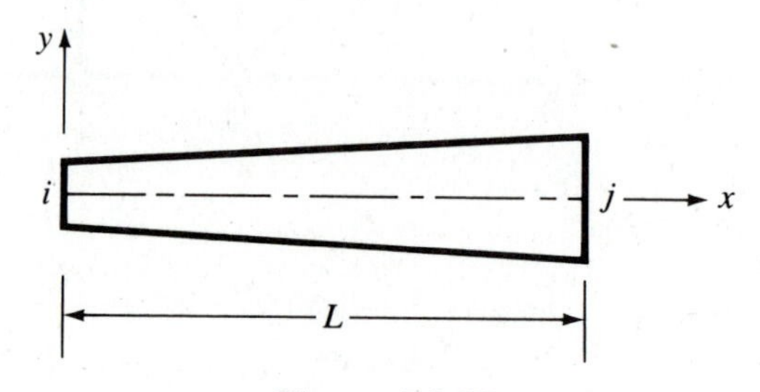

Figure P6.22

7 THE FLEXIBILITY METHOD

Compatibility methods provide an alternative to equilibrium techniques for investigating structures. For example, the methods of consistent displacements and the three-moment equation invoke compatibility conditions throughout the structure by superposing a set of partial solutions, all of which satisfy equilibrium, force-deformation relations, and boundary conditions. The compatibility approach, broadly categorized as the force or flexibility method, gives a set of equations with the forces as the unknowns and the flexibility quantities as the coefficients. The principle of complementary virtual work also yields the flexibility method.

This chapter derives the general flexibility method from the basic equations of mechanics and illustrates the approach for axial force elements in series, trusses, beams, and rigid frames. We also illustrate how the principle of complementary virtual work yields the flexibility method. In addition, some useful relationships between the stiffness and flexibility methods are developed.

While Secs. 7.1, 7.2, and 7.3 contain the essence of the flexibility method, the reader may wish to follow the discussion of the principle of complementary virtual work in Sec. 7.4 and see this energy principle used to develop the flexibility method in Sec. 7.5. An understanding of initial strains in Sec. 7.8 is required for a thorough understanding of the flexibility method, but Secs. 7.7 and 7.9 will probably be of interest only to those implementing the flexibility method. Section 7.6 explains relationships between the stiffness and flexibility approaches; we urge the reader to study it carefully.

7.1 THE FLEXIBILITY METHOD USING THE BASIC EQUATIONS

The assemblage of two axial force elements in series, investigated in Ch. 2, provides a simple means for illustrating the flexibility method. Each weightless prismatic axial force element in Fig. 7.1a has a cross-sectional area A_i, a length L_i, and is made from a homogeneous, isotropic, linearly elastic material with a modulus of elasticity E_i. Let $L_i/A_i E_i = f_i$ and $f_i = 1/k_i$ for convenience. The assemblage, constrained at nodes a and c, has one statical indeterminacy. We regard the support reaction at a as the redundant force and designate it as X (see Fig. 7.1b). Equilibrium of the assemblage yields

$$\begin{bmatrix} s^{ab} \\ s^{bc} \end{bmatrix} = \begin{bmatrix} 0 \\ -1 \end{bmatrix} P_{xb} + \begin{bmatrix} -1 \\ -1 \end{bmatrix} X \tag{7.1}$$

$$\mathbf{s} = \boldsymbol{b}_0 \boldsymbol{P} + \mathbf{b}_1 \mathbf{X}$$

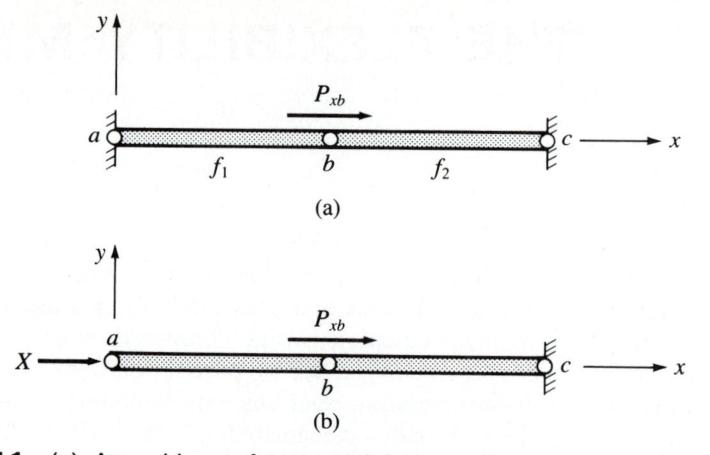

Figure 7.1 (a) Assemblage of two axial force elements; (b) primary structure showing redundant force

where s^{ij} is the axial force in element ij; in general

$$\mathbf{s} = [\mathbf{b}_0 \quad \mathbf{b}_1]\begin{bmatrix}\mathbf{P}\\ \mathbf{X}\end{bmatrix} \tag{7.1a}$$

Designating the deformation of each element as d^{ij}, Hooke's law gives

$$\underset{\mathbf{d}}{\begin{bmatrix}d^{ab}\\ d^{bc}\end{bmatrix}} = \underset{\mathbf{f}}{\begin{bmatrix}f_1 & 0\\ 0 & f_2\end{bmatrix}}\underset{\mathbf{s}}{\begin{bmatrix}s^{ab}\\ s^{bc}\end{bmatrix}} \tag{7.2}$$

where **f** is the flexibility matrix for all elements. The axial force s^{ij} and deformation d^{ij} for a single element are shown in Fig. 7.2. The displacement of node a (i.e., the redundant d_X) is the combination of the deformations of the two elements, as follows:

$$d_X = -d^{ab} - d^{bc} = [-1 \quad -1]\begin{bmatrix}d^{ab}\\ d^{bc}\end{bmatrix} \tag{7.3}$$

Deformations are positive if the element is in tension; therefore, positive deformations will yield a displacement of support a in the negative coordinate direction (see Fig. 7.1b). Recalling the form of $\mathbf{b}_1$ from Eq. (7.1), we note from Eq. (7.3) that

$$d_X = \mathbf{b}_1^T \mathbf{d} \tag{7.4}$$

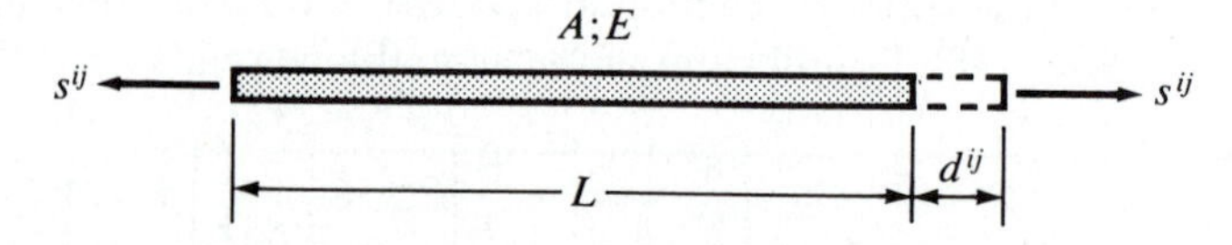

Figure 7.2 Axial force element showing element force and deformation

While not proved here, Eq. (7.4) is true in general; the proof is furnished in the discussion on complementary virtual work. Substituting Eqs. (7.1) and (7.2) into Eq. (7.4), and noting that $d_X = 0$ yields

$$0 = \mathbf{b}_1^T \mathbf{f}\mathbf{s} = \mathbf{b}_1^T\mathbf{f}\mathbf{b}_0 \mathbf{P} + \mathbf{b}_1^T\mathbf{f}\mathbf{b}_1\mathbf{X} \tag{7.5}$$

Solution for the redundants gives

$$\mathbf{X} = -\mathbf{F}_{11}^{-1}\mathbf{F}_{10}\mathbf{P} \tag{7.6}$$

where

$$\mathbf{F}_{11} = \mathbf{b}_1^T\mathbf{f}\mathbf{b}_1$$

and

$$\mathbf{F}_{10} = \mathbf{b}_1^T\mathbf{f}\mathbf{b}_0 \tag{7.7}$$

Thus, from Eqs. (7.6) and (7.1) the element forces are

$$\mathbf{s} = \mathbf{b}_0 \mathbf{P} - \mathbf{b}_1\mathbf{F}_{11}^{-1}\mathbf{F}_{10}\mathbf{P} \tag{7.8}$$

We can express the nodal displacements in terms of the element deformations as follows:

$$U_b = [0 \quad -1]\begin{bmatrix} d^{ab} \\ d^{bc} \end{bmatrix} = -d^{bc} \tag{7.9}$$

that is,

$$\mathbf{U} = \mathbf{b}_0^T \mathbf{d} \tag{7.10}$$

Equation (7.10) is true in general; this is proved in the discussion on the principle of complementary virtual work. Substituting Eqs. (7.2) and (7.8) into Eq. (7.10) we have

$$\begin{aligned} \mathbf{U} &= \mathbf{b}_0^T \mathbf{f}\mathbf{s} = \mathbf{b}_0^T\mathbf{f}(\mathbf{b}_0 \mathbf{P} - \mathbf{b}_1\mathbf{F}_{11}^{-1}\mathbf{F}_{10}\mathbf{P}) \\ &= (\mathbf{b}_0^T\mathbf{f}\mathbf{b}_0 - \mathbf{b}_0^T\mathbf{f}\mathbf{b}_1\mathbf{F}_{11}^{-1}\mathbf{F}_{10})\mathbf{P} = (\mathbf{F}_{00} - \mathbf{F}_{01}\mathbf{F}_{11}^{-1}\mathbf{F}_{10})\mathbf{P} \\ &= \mathbf{F}\mathbf{P} \end{aligned} \tag{7.11}$$

where the structural flexibility matrix for the assemblage, $\mathbf{F}$, is

$$\mathbf{F} = \mathbf{F}_{00} - \mathbf{F}_{01}\mathbf{F}_{11}^{-1}\mathbf{F}_{10} \tag{7.12}$$

with

$$\mathbf{F}_{01} = \mathbf{F}_{10}^T \tag{7.12a}$$

and

$$\mathbf{F}_{00} = \mathbf{b}_0^T\mathbf{f}\mathbf{b}_0 \tag{7.12b}$$

For the assemblage in Fig. 7.1a, $\mathbf{F}_{00} = f_2$, $\mathbf{F}_{01} = f_2$, and $\mathbf{F}_{11} = (f_1 + f_2)$; therefore

$$\mathbf{F} = \frac{f_1 f_2}{f_1 + f_2}$$

and, from Eq. (7.11), $U_b = [f_1 f_2/(f_1 + f_2)]P_{xb}$. From Eq. (7.8), the forces in the elements are

$$\begin{bmatrix} s^{ab} \\ s^{bc} \end{bmatrix} = \left[\begin{bmatrix} 0 \\ -1 \end{bmatrix} - \begin{bmatrix} -1 \\ -1 \end{bmatrix} \frac{f_2}{f_1 + f_2}\right] P_{xb}$$
$$= \frac{P_{xb}}{f_1 + f_2} \begin{bmatrix} f_2 \\ -f_1 \end{bmatrix}$$

The governing equations are written for the constrained assemblage; therefore, in the context of the stiffness method, Eq. (7.11) refers to the governing equations with boundary conditions imposed. Comparing Eqs. (7.11) and (2.34) reveals that

$$\mathbf{F} = \mathbf{K}_{ff}^{-1} \tag{7.13}$$

Since $f_i = 1/k_i$, $\mathbf{F} = 1/(k_1 + k_2)$, and $\mathbf{K}_{ff} = (k_1 + k_2)$, Eq. (7.13) is identical to the results obtained using the stiffness method for this two element assemblage with nodes a and c constrained [see Eq. (2.24)].

7.2 THE FLEXIBILITY METHOD FOR TRUSSES

We will apply the flexibility method as displayed in the previous section to the truss of Example 2.7 (see Fig. 7.3a). Since element ab transmits zero force the truss could be considered to have one statical redundancy, but two redundants will be included to illustrate the effects of internal (i.e., element) and external (i.e., support) redundancies. For this truss $\mathbf{P} = \{P_{xc} \quad P_{yc} \quad P_{xd} \quad P_{yd}\}$ and $\mathbf{X} = \{X_1 \quad X_2\}$. A column of the $\mathbf{b}_0$ matrix contains the forces in the elements for a unit value of the associated external force; for example, the fourth column corresponds to element forces required to equilibrate a horizontal force at node d (see Fig. 7.3b). Similarly, a column of the matrix $\mathbf{X}$ is composed of the element forces that equilibrate a unit value of the associated redundant force (e.g., Fig. 7.3c). For this structure, Eq. (7.1) is

$$\begin{bmatrix} s^{ab} \\ s^{cd} \\ s^{ac} \\ s^{bd} \\ s^{ad} \\ s^{bc} \end{bmatrix} = \begin{bmatrix} 0 & 0 & 0 & 0 \\ -1 & 0 & 0 & 0 \\ 0 & 1 & 0 & 0 \\ -1 & 0 & -1 & 1 \\ \sqrt{2} & 0 & \sqrt{2} & 0 \\ 0 & 0 & 0 & 0 \end{bmatrix} \begin{bmatrix} P_{xc} \\ P_{yc} \\ P_{xd} \\ P_{yd} \end{bmatrix} + \begin{bmatrix} -\frac{\sqrt{2}}{2} & 1 \\ -\frac{\sqrt{2}}{2} & 0 \\ -\frac{\sqrt{2}}{2} & 0 \\ -\frac{\sqrt{2}}{2} & 0 \\ 1 & 0 \\ 1 & 0 \end{bmatrix} \begin{bmatrix} X_1 \\ X_2 \end{bmatrix} \tag{7.14}$$

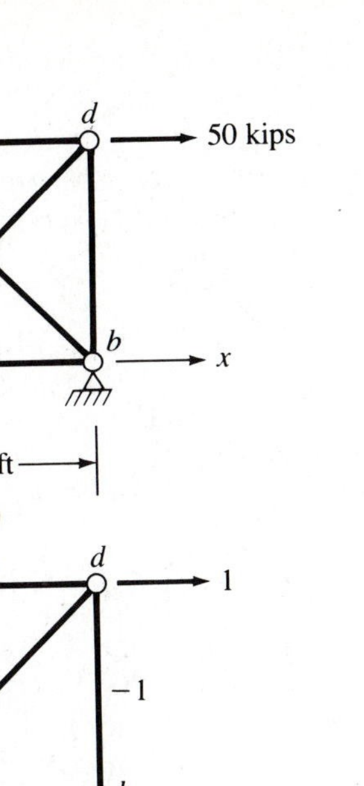
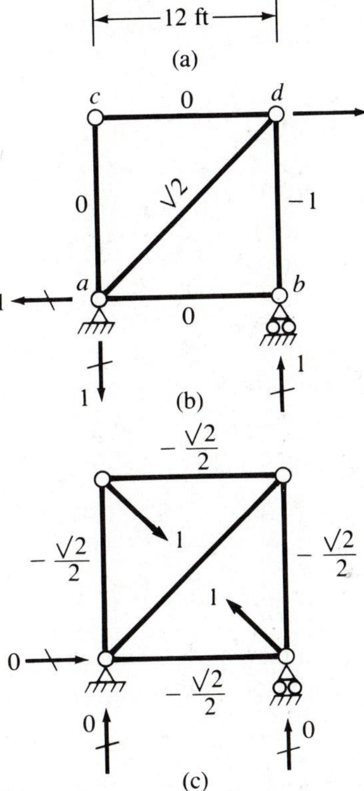

Figure 7.3 (a) Truss with two static redundants; (b) primary structure with applied unit force; (c) equilibrium solution for unit redundant X_1

From Example 2.7 we note that L/AE is 1/200 for horizontal and vertical elements and $\sqrt{2}/200$ for diagonal elements; therefore, the flexibility matrix of all elements is

$$\mathbf{f} = \frac{1}{200}\begin{bmatrix} 1 & 0 & 0 & 0 & 0 & 0 \\ 0 & 1 & 0 & 0 & 0 & 0 \\ 0 & 0 & 1 & 0 & 0 & 0 \\ 0 & 0 & 0 & 1 & 0 & 0 \\ 0 & 0 & 0 & 0 & \sqrt{2} & 0 \\ 0 & 0 & 0 & 0 & 0 & \sqrt{2} \end{bmatrix} \tag{7.15}$$

Substituting the basic matrices $\mathbf{b}_0$ and $\mathbf{b}_1$ from Eq. (7.14) and $\mathbf{f}$ from Eq. (7.15) into Eq. (7.12) yields the 4 by 4 flexibility matrix, $\mathbf{F} = \mathbf{K}_{ff}^{-1}$, identical to that shown in Example 2.7.

Alternatively, we could use the $\mathbf{b}_0$ matrix associated with P_{xd}: the only applied force with a nonzero value. That is, using only the third column of $\mathbf{b}_0$ from Eq. (7.14) yields

$$\mathbf{b}_0 = \{0 \quad 0 \quad 0 \quad -1 \quad \sqrt{2} \quad 0\} \tag{7.16}$$

Therefore,

$$\mathbf{F}_{00} = \frac{1}{200}[3.828]$$

$$\mathbf{F}_{01} = \mathbf{F}_{10}^T = \frac{1}{200}[2.707 \quad 0.000] \tag{7.17}$$

$$\mathbf{F}_{11} = \frac{1}{200}\begin{bmatrix} 4.828 & -0.707 \\ -0.707 & 1.000 \end{bmatrix}$$

and

$$\mathbf{F} = \frac{2.135}{200} \tag{7.18}$$

Since $P_{xd} = 50$ kips, Eq. (7.11) gives $U_d = 0.534$ in., which is identical to the result obtained from the stiffness method in Example 2.7.

We obtain the element forces from Eq. (7.8). Using $\mathbf{b}_0$ from Eq. (7.16) along with $\mathbf{b}_1$ from Eq. (7.14) and the matrices of Eq. (7.17) we have

$$\mathbf{b}_0 - \mathbf{b}_1\mathbf{F}_{11}^{-1}\mathbf{F}_{10} = \{0.000 \quad 0.442 \quad 0.442 \quad -0.558 \quad 0.789 \quad -0.625\}$$

and

$$\mathbf{s} = \{s^{ab} \quad s^{cd} \quad s^{ac} \quad s^{bd} \quad s^{ad} \quad s^{bc}\}$$
$$= \{0.0 \quad 22.1 \quad 22.1 \quad -27.9 \quad 39.4 \quad -31.2\} \text{ kips} \tag{7.19}$$

Positive and negative signs indicate elements in tension and compression, respectively. The two redundants can be calculated using Eq. (7.6), along with the results shown in Eq. (7.17); thus

$$\{X_1 \quad X_2\} = \{-0.625 \; 0.442\}50 = \{-31.2 \quad -22.1\} \text{ kips} \tag{7.20}$$

The negative sign associated with both redundants indicates that we assumed the directions incorrectly. That is, element bc, X_1, is in compression, and the horizontal reaction at node b, X_2, acts in the negative coordinate direction.

We will resolve the dilemma of the possible sizes of the $\mathbf{b}_0$ matrix after deriving the equations using the principle of complementary virtual work. For now it is sufficient to observe the outcome using the single column $\mathbf{b}_0$ matrix. Using the 4 by 4 stiffness matrix from Example 2.7 and condensing (see Sec. 6.1) all degrees of freedom with the exception of U_d, we obtain $\mathbf{K}_{cc} = 0.4687(200)$. Note that $\mathbf{K}_{cc}^{-1} = 2.134/200$. That is, in general we obtain the global flexibility matrix ($\mathbf{F}$) corresponding to the inverted condensed stiffness matrix associated with the degree(s) of freedom included in $\mathbf{b}_0$, as indicated by Eq. (7.13).

7.3 THE FLEXIBILITY METHOD FOR BEAMS

The general theory of the flexibility method can also be applied to beam structures, but we will require unique **f**, $\mathbf{b}_0$, and $\mathbf{b}_1$ matrices. The simply supported prismatic beam element in Fig. 7.4 has a moment of inertia I, a length L, and is made from homogeneous, isotropic, linearly elastic material with a modulus of elasticity E. Using the engineering theory of bending, with a classical approach for calculating displacements, we have the following relationships between the end moments m_i and m_j and the corresponding rotations θ_i and θ_j, respectively:

$$\begin{bmatrix} \theta_i \\ \theta_j \end{bmatrix} = \frac{L}{6EI} \begin{bmatrix} 2 & -1 \\ -1 & 2 \end{bmatrix} \begin{bmatrix} m_i \\ m_j \end{bmatrix} \qquad (7.21)$$

$$\mathbf{d} = \mathbf{f} \quad \mathbf{s}$$

The supports for the beam element need only be statically determinate and stable; therefore the element in Fig. 7.5 could also be used. In this case the forces p_{yj} and m_j relate to the associated displacements v_j and θ_j, respectively, as follows:

$$\begin{bmatrix} v_j \\ \theta_j \end{bmatrix} = \frac{L}{6EI} \begin{bmatrix} 2L^2 & 3L \\ 3L & 6 \end{bmatrix} \begin{bmatrix} p_{yj} \\ m_j \end{bmatrix} \qquad (7.22)$$

The form of the $\mathbf{b}_0$ and $\mathbf{b}_1$ matrices is dictated by the element flexibility matrices, but final results are unaffected by the choice of **f**.

The prismatic beam of Example 4.6 (see Fig. 7.6a), with I and E constant between nodes a and d, has two statical redundancies. We choose the primary structures to be two simply supported beams as shown in Fig. 7.6b and the

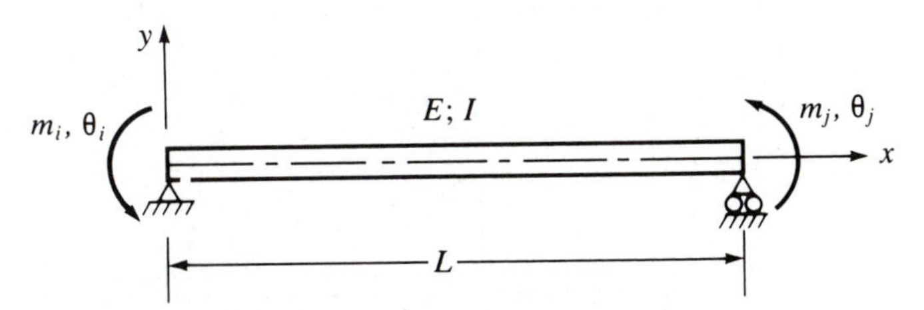

Figure 7.4 Simply supported prismatic beam element

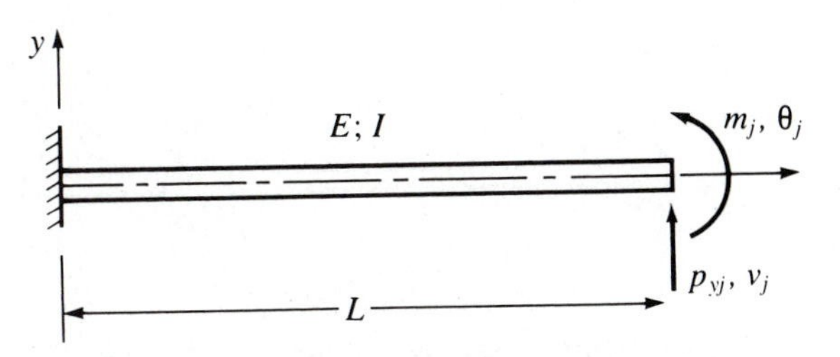

Figure 7.5 Cantilever prismatic beam element

redundants as the moments at nodes a and c (see Fig. 7.6c). Using Eq. (7.21), the flexibility matrix for all elements (i.e., ab, bc, and cd) is

$$\mathbf{f} = \frac{L}{6EI}\begin{bmatrix} 2 & -1 & 0 & 0 & 0 & 0 \\ -1 & 2 & 0 & 0 & 0 & 0 \\ 0 & 0 & 2 & -1 & 0 & 0 \\ 0 & 0 & -1 & 2 & 0 & 0 \\ 0 & 0 & 0 & 0 & 2 & -1 \\ 0 & 0 & 0 & 0 & -1 & 2 \end{bmatrix} \tag{7.23}$$

With this choice of flexibility matrix, Eq. (7.1) for the structure is

$$\begin{bmatrix} m_a^{ab} \\ m_{b-}^{ab} \\ m_{b+}^{bc} \\ m_{c-}^{bc} \\ m_{c+}^{cd} \\ m_d^{cd} \end{bmatrix} = \begin{bmatrix} 0 \\ -\frac{L}{2} \\ \frac{L}{2} \\ 0 \\ 0 \\ 0 \end{bmatrix} P_{yd} + \begin{bmatrix} -1 & 0 \\ \frac{1}{2} & \frac{1}{2} \\ -\frac{1}{2} & -\frac{1}{2} \\ 0 & 1 \\ 0 & -1 \\ 0 & 0 \end{bmatrix} \begin{bmatrix} X_a \\ X_c \end{bmatrix} \tag{7.24}$$

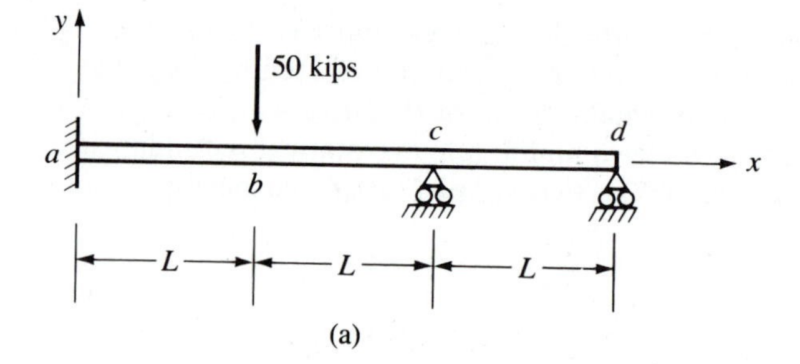

(a)

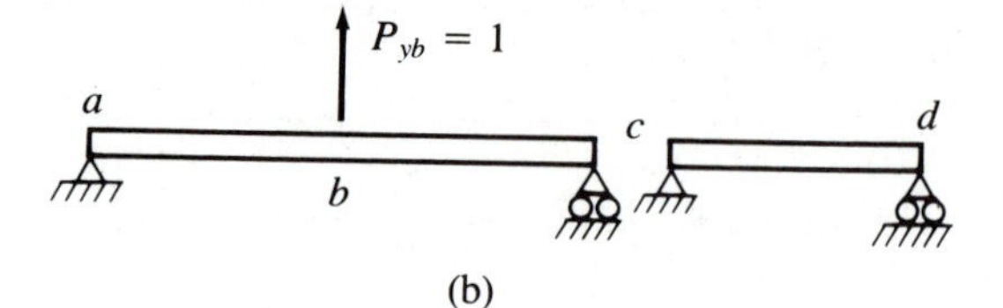

(b)

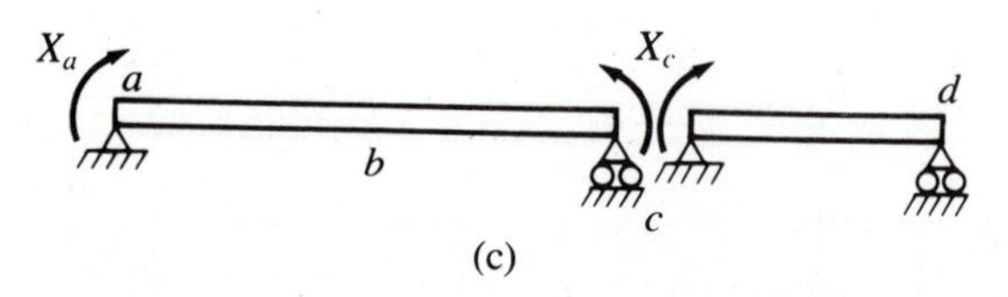

(c)

Figure 7.6 (a) Beam with two static redundants; (b) primary structure with applied unit force; (c) primary structure showing redundant moments

where m_{b-}^{ab} indicates the moment in element ab to the left of node b, and m_{b+}^{bc} is the moment in element bc to the right of node b, etc. Therefore,

$$\mathbf{F}_{00} = \frac{L^3}{6EI}$$

$$\mathbf{F}_{01} = \mathbf{F}_{10}^T = -\frac{L^2}{4EI}\begin{bmatrix}1 & 1\end{bmatrix} \tag{7.25}$$

$$\mathbf{F}_{11} = \frac{L}{3EI}\begin{bmatrix}2 & 1\\ 1 & 3\end{bmatrix}$$

and

$$\mathbf{F} = \frac{13L^3}{240EI} \tag{7.26}$$

We calculate the redundants using Eq. (7.6) and noting that $\mathbf{P} = P_{yb} = -50$ kips, $I = 1200$ in.^{4}, $L = 14.5$ ft, and $E = 29 \times 10^3$ ksi; thus

$$\begin{bmatrix}X_a\\ X_c\end{bmatrix} = -\mathbf{F}_{11}^{-1}\mathbf{F}_{10}\mathbf{P} = -15L\begin{bmatrix}1\\ \frac{1}{2}\end{bmatrix} = \begin{bmatrix}-2610\\ -1305\end{bmatrix} \text{kip}\cdot\text{in.} \tag{7.27}$$

Using Eq. (7.11), $\mathbf{U} = U_b = -0.40999$ in.; furthermore, substituting the values of the redundants X_a and X_c into Eq. (7.24) yields the following element forces:

$$\begin{bmatrix}m_a^{ab}\\ m_{b-}^{ab}\\ m_{b+}^{bc}\\ m_{c-}^{bc}\\ m_{c+}^{cd}\\ m_d^{cd}\end{bmatrix} = \begin{bmatrix}2610\\ 2392\\ -2392\\ -1305\\ 1305\\ 0\end{bmatrix} \text{kip}\cdot\text{in.} \tag{7.28}$$

We obtained these same results when analyzing the continuous beam by the stiffness method (see Example 4.6).

Alternatively, the beam of Fig. 7.5a could be analyzed using the flexibility matrix of Eq. (7.22), wherein the beam element is fixed at end i and the quantities explicitly included are those at end j (see Fig. 7.5). In this case

$$\mathbf{f} = \frac{L}{6EI}\begin{bmatrix}2L^2 & 3L & 0 & 0 & 0 & 0\\ 3L & 6 & 0 & 0 & 0 & 0\\ 0 & 0 & 2L^2 & 3L & 0 & 0\\ 0 & 0 & 3L & 6 & 0 & 0\\ 0 & 0 & 0 & 0 & 2L^2 & 3L\\ 0 & 0 & 0 & 0 & 3L & 6\end{bmatrix} \tag{7.29}$$

Choosing the moments at nodes a and c as the redundants (i.e., the same choice as in the previous analysis) and the appropriate element forces for the flexibility matrix of Eq. (7.29) we have

$$\begin{bmatrix} p_{yb}^{ab} \\ m_b^{ab} \\ p_{yc}^{bc} \\ m_c^{bc} \\ p_{yd}^{cd} \\ m_d^{cd} \end{bmatrix} = \begin{bmatrix} \frac{1}{2} \\ -\frac{L}{2} \\ -\frac{1}{2} \\ 0 \\ 0 \\ 0 \end{bmatrix} P_{yd} + \begin{bmatrix} \frac{1}{2L} & -\frac{1}{2L} \\ \frac{1}{2} & \frac{1}{2} \\ \frac{1}{2L} & -\frac{1}{2L} \\ 0 & 1 \\ 0 & \frac{1}{L} \\ 0 & 0 \end{bmatrix} \begin{bmatrix} X_a \\ X_c \end{bmatrix} \tag{7.30}$$

Using the basic matrices from Eqs. (7.29) and (7.30), we get the same results for $\mathbf{F}_{00}$, $\mathbf{F}_{10}$, $\mathbf{F}_{01}$, $\mathbf{F}_{11}$, and $\mathbf{F}$ as in the previous analysis [see Eqs. (7.25) and 7.26)], as well as the same nodal displacement at node b and values of the redundants X_a and X_c. In this case, the element forces calculated from Eq. (7.1) are

$$\begin{bmatrix} p_{yb}^{ab} \\ m_b^{ab} \\ p_{yc}^{bc} \\ m_c^{bc} \\ p_{yd}^{cd} \\ m_d^{cd} \end{bmatrix} = \begin{bmatrix} -28.8 \text{ kips} \\ 2392 \text{ kip} \cdot \text{in.} \\ 21.2 \text{ kips} \\ -1305 \text{ kip} \cdot \text{in.} \\ -7.5 \text{ kips} \\ 0 \text{ kip} \cdot \text{in.} \end{bmatrix} \tag{7.31}$$

Rigid frames can also be analyzed using the element flexibility matrices of Eqs. (7.21) and (7.22); the matrices $\mathbf{b}_0$ and $\mathbf{b}_1$ must be compatible with the flexibility matrices and contain entries appropriate for the structures analyzed. If degrees of freedom other than v and θ are of interest, we must obtain a unique flexibility matrix. For example, investigating flexure plus axial deformations requires combining the flexibility of the axial force element $(L_i/A_i E_i)$ with the beam flexibility matrix [i.e., Eq. (7.21) or (7.22)].

7.4 THE PRINCIPLE OF COMPLEMENTARY VIRTUAL WORK

Work- and energy-related principles, dating back to Archimedes' (287–212 B.C.) study of levers, provide an alternative to solving the three basic equations of mechanics. In Ch. 3 we presented the principle of virtual work, which is an equilibrium approach. The principle of complementary virtual work stipulates that a solution for the real system, which satisfies compatibility, should be accompanied by an equilibrated system of virtual forces. We can

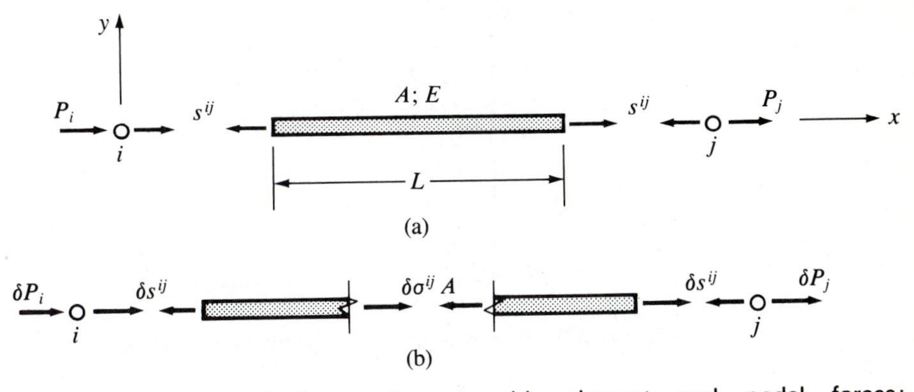

Figure 7.7 (a) Axial force element with element and nodal forces; (b) virtual force system

illustrate the principle of complementary virtual work (also known as the principle of virtual forces) by considering the prismatic axial force element in Fig. 7.7, with cross section A, length L, and linear elastic material property E. The tensile force s^{ij} produces a compatible displacement state, u, along the element, wherein u_i and u_j are the displacements at points i and j, respectively. Nodal forces P_i and P_j are directed in the positive x coordinate direction, and the constant element force, s^{ij}, is applied to the nodal points in conformance with Newton's third law (see Fig. 7.7a). Figure 7.8 shows the force-displacement graph of node i and Fig. 7.9 portrays the element force-deformation plot. The area above the force-displacement line in Fig. 7.8 is the *complementary external work*, W_e^*, and the *complementary work of the internal forces* (or *complementary strain energy*), W_i^*, is the area above the element force-deformation plot in Fig. 7.9. In contrast, the areas below these respective plots are the external work, W_e, and the strain energy, W_i, respectively (see Figs. 3.4 and 3.5, respectively).

We give the nodal forces virtual changes, δP_i amd δP_j, that result in a virtual change to the element force, δs^{ij}, and a corresponding virtual change in

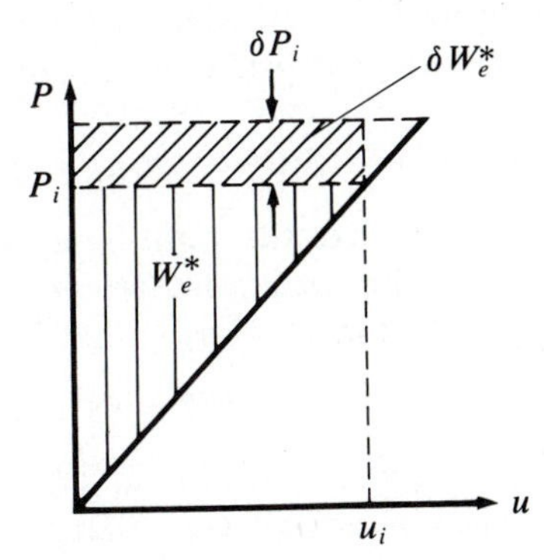

Figure 7.8 Force-displacement graph for a linear element

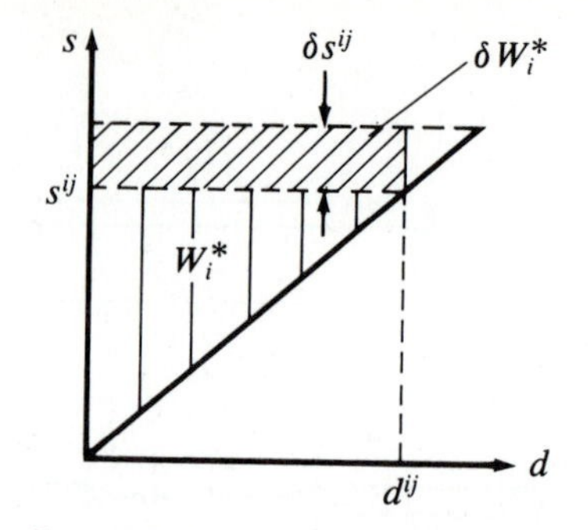

Figure 7.9 Force-deformation graph for a linear element

the element stress $\delta\sigma^{ij}$. The virtual force system must be in equilibrium; therefore, for the element (see Fig. 7.7b)

$$\delta s^{ij} - \delta\sigma^{ij} A = 0 \tag{7.32}$$

and at nodes i and j, respectively,

$$\delta P_i + \delta s^{ij} = \delta P_i + \delta\sigma^{ij} A = 0 \tag{7.33}$$

$$\delta P_j - \delta s^{ij} = \delta P_j - \delta\sigma^{ij} A = 0 \tag{7.34}$$

Multiplying Eqs. (7.33) and (7.34) by u_i and u_j, respectively, and adding the two equations yields

$$\delta P_i u_i + \delta P_j u_j - \delta\sigma^{ij} A(u_j - u_i) = 0 \tag{7.35}$$

Noting that the deformation of the element, d^{ij}, is

$$d^{ij} = u_j - u_i \tag{7.36}$$

Eq. (7.35) becomes

$$\delta P_i u_i + \delta P_j u_j = \delta\sigma^{ij} A d^{ij} \tag{7.37}$$

$$= \delta s^{ij} d^{ij} \tag{7.37a}$$

$$= \delta\sigma^{ij} \varepsilon^{ij} AL \tag{7.37b}$$

Note from Fig. 7.8 that the first term on the left-hand side of Eq. (7.37) is the cross-hatched area, which is the complementary external virtual work at node i. Therefore, the entire left-hand side of Eq. (7.37) is the total *complementary external virtual work* for the system, δW_e^*, subjected to the virtual forces, that is,

$$\delta W_e^* = \delta P_i u_i + \delta P_j u_j \tag{7.38}$$

From Fig. 7.9 we note that the right-hand side of Eq. (7.37a) is the cross-hatched area, which is the *complementary virtual work of the internal forces* or the *complementary virtual strain energy*; that is,

$$\delta W_i^* = \delta s^{ij} d^{ij} \tag{7.39}$$

Substituting Eqs. (7.38) and (7.39) in Eq. (7.37) we have the statement of the *principle of complementary virtual work*, that is,

$$\delta W_e^* = \delta W_i^* \tag{7.40}$$

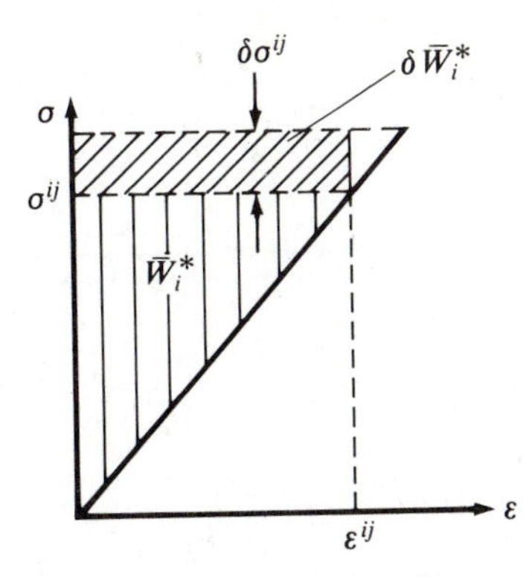

Figure 7.10 Stress-strain graph for a linear elastic element

That is,

> **If a linearly elastic structure is in a compatible state of deformation under a system of forces and temperature distribution, then for any virtual stresses and forces away from the equilibrium state of stress the complementary external virtual work is equal to the complementary virtual strain energy of the virtual forces.**

Figure 7.10 shows the stress-strain diagram for the single element of Fig. 7.7a. The area above the line, indicated as $\bar{W}_i^*$, is the *complementary strain energy density*, and the cross-hatched area denoted as $\delta\bar{W}_i^*$ is the *complementary virtual strain energy density*. Therefore,

$$\delta W_i^* = \int_{\text{vol}} \delta\sigma^{ij}\varepsilon^{ij}\, dV \tag{7.41}$$

For the prismatic axial force element with area A and length L, $\delta\sigma^{ij}$ and ε^{ij} are constant and can be factored outside the integral; thus,

$$\delta W_i^* = \delta\sigma^{ij}\varepsilon^{ij} \int_{\text{vol}} dV = \delta\sigma^{ij}\varepsilon^{ij} AL \tag{7.42}$$

which is the right-hand side of Eq. (7.37b).

Note that in computing all complementary virtual work quantities we have neglected the second-order terms. That is, in Figs. 7.8, 7.9, and 7.10 the small triangles to the right of the cross-hatched areas are not included. For example, in Fig. 7.8 the virtual force δP_i would produce a change in u_i (i.e., δu_i), and δW_i^* should include the quantity $\delta P_i\, \delta u_i/2$. This second-order quantity is not important for linear theory; therefore, we have ignored these terms in all calculations.

Applying the principle of complementary virtual work yields equations cast in the mold of the flexibility method. Consider the two-element assemblage investigated in Sec. 7.1, which is reproduced in Fig. 7.11a; element ab has flexibility f_1, element bc has flexibility f_2, and all forces are applied in the x direction only. For the primary structure (i.e., statically determinate

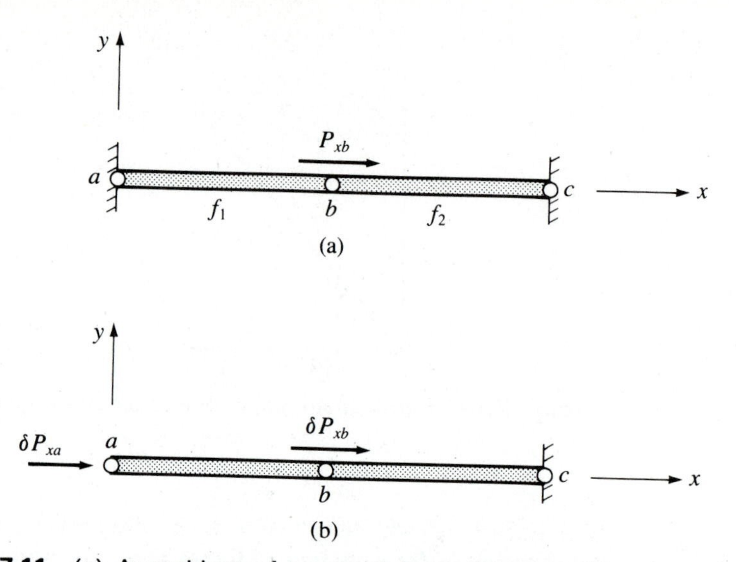

Figure 7.11 (a) Assemblage of two axial force elements; (b) primary structure with virtual force system

assemblage), with the virtual force system in Fig. 7.11b, the complementary external virtual work is

$$\delta W_e^* = \delta P_{xa} u_a + \delta P_{xb} u_b \tag{7.43}$$

where δP_{xa} and δP_{xb} are virtual forces and u_a and u_b are actual displacements. Similarly, the complementary strain energy for the assemblage of Fig. 7.11a is

$$\delta W_i^* = \delta s^{ab} d^{ab} + \delta s^{bc} d^{bc} \tag{7.44}$$

where δs^{ab} and δs^{bc} are the virtual element forces produced by δP_i and δP_j and d^{ab} and d^{bc} are actual element deformations. Applying the principle of complementary virtual work [i.e., Eq. 7.40)] yields

$$\delta P_{xa} u_a + \delta P_{xb} u_b = \delta s^{ab} d^{ab} + \delta s^{bc} d^{bc} \tag{7.45}$$

The force-deformation relations give

$$\begin{aligned} d^{ab} &= f_1 s^{ab} \\ d^{bc} &= f_2 s^{bc} \end{aligned} \tag{7.46}$$

Equilibrium for the virtual force system requires (see Fig. 7.11b)

$$\begin{aligned} \delta s^{ab} &= \delta P_{xa} \\ \text{and} \qquad \delta s^{bc} &= \delta P_{xa} + \delta P_{xb} \end{aligned} \tag{7.47}$$

Similarly, for the actual force system

$$\begin{aligned} s^{ab} &= P_{xa} \\ \text{and} \qquad s^{bc} &= P_{xa} + P_{xb} \end{aligned} \tag{7.48}$$

Substituting Eqs. (7.46), (7.47), and (7.48) in Eq. (7.45) and rearranging terms results in

$$\delta P_{xa}[u_a - f_1 P_{xa} + f_2(P_{xa} + P_{xb})] + \delta P_{xb}[u_b - f_2 P_{xa} + f_2 P_{xb}] = 0 \quad (7.49)$$

Since δP_{xa} and δP_{xb} are arbitrary, independent, and nonzero, Eq. (7.49) can only be satisfied if the expressions in the square brackets are each zero; thus

$$u_a = (f_1 + f_2)P_{xa} + f_2 P_{xb} \quad (7.50a)$$

and

$$u_b = f_2 P_{xa} + f_2 P_{xb} \quad (7.50b)$$

But compatibility requires that $u_a = 0$; therefore Eq. (7.50a) gives

$$P_{xa} = -\frac{f_2}{f_1 + f_2} P_{xb} \quad (7.51)$$

Substituting Eq. (7.51) into Eq. (7.50b) we have

$$u_b = \frac{f_1 f_2}{f_1 + f_2} P_{xb} \quad (7.52)$$

Note that Eqs. (7.50a) and (7.50b) are in the format of flexibility equations. That is, the flexibilities are the coefficients and the forces are cast as the unknowns. The coefficient of P_{xb} in Eq. (7.52) is $\mathbf{F}$ [i.e., Eq. (7.12), which was derived using the basic equations]. Assigning the values $f_1 = 0.20$ in./kip, $f_2 = 0.10$ in./kip, and $P_{xb} = 9$ kips defines the identical assemblage investigated by the stiffness method in Example 3.2. Substituting these values in Eqs. (7.51) and (7.52) we have $P_{xa} = -3$ kips and $u_b = 0.60$ in., which are identical to the results calculated by the stiffness method.

Thus, in general, we formulate the principle of complementary virtual work by visualizing that a structure is perturbed from an equilibrium position by an equilibrated system of imposed virtual forces. The *complementary external virtual work*, δW_e^*, is the product of the virtual forces, $\boldsymbol{\delta}\mathbf{P}$, and the associated real displacements, $\mathbf{U}$; that is,

$$\delta W_e^* = \boldsymbol{\delta}\mathbf{P}^T\mathbf{U} \quad (7.53)$$

The virtual force system induces virtual stresses, $\boldsymbol{\delta\sigma}$, within each element, and the *complementary virtual strain energy*, δW_i^*, for each element is

$$\delta W_i^* = \int_{\text{vol}} \boldsymbol{\delta\sigma}^T \boldsymbol{\varepsilon} \, dV \quad (7.54)$$

where $\boldsymbol{\varepsilon}$ contains the real element strains. The total complementary virtual strain energy, which is the sum of the complementary internal virtual work for all elements, equals the complementary external virtual work accomplished by the virtual force system. The virtual force system must be equilibrated by structural elements (i.e., not necessarily by the aggregate of all the real elements); therefore, the resulting virtual stress system need not correspond to that produced by the actual structural system.

7.5 THE FLEXIBILITY METHOD USING COMPLEMENTARY VIRTUAL WORK

We can obtain element flexibility matrices, as well as formulate the flexibility equations for an entire structure, using the principle of complementary virtual work.

The *axial force element* in Fig. 7.2, with cross-sectional area A, length L, and modulus of elasticity E, is equilibrated by equal opposing axial forces s; the corresponding deformation is d. Note that in this discussion no ambiguity is introduced by omitting the superscript ij on s and d. The virtual force δs results in an axial virtual stress of $\delta s/A$ throughout the element. The real strain in the element is s/AE and the external complementary virtual work is $\delta s\, d$. Applying the principle of complementary virtual work to this element yields

$$\delta s\, d = \int_{\text{vol}} \left(\frac{\delta s}{A}\right)\left(\frac{s}{AE}\right) dV = \delta s \frac{s}{A^2 E} \int_{\text{vol}} dV = \delta s \frac{s}{A^2 E} AL$$

Therefore

$$\delta s\left(d - \frac{L}{AE} s\right) = 0$$

Since δs is nonzero and arbitrary, the equation can only be satisfied if

$$d = \frac{L}{AE} s \tag{7.55}$$

This relationship between the element deformation and force for an axial element is an expression of Hooke's law; it gives the following element flexibility:

$$f = \frac{L}{AE} \tag{7.56}$$

The *prismatic simply supported beam element* in Fig. 7.4 with moment of inertia I, length L, modulus of elasticity E, and applied end moments m_i and m_j is in equilibrium so that

$$M(x) = \begin{bmatrix} \left(\frac{x}{L} - 1\right) & \frac{x}{L} \end{bmatrix} \begin{bmatrix} m_i \\ m_j \end{bmatrix} \tag{7.57}$$

For convenience we introduce a virtual force system similar to the real force system; that is,

$$\delta M(x) = \begin{bmatrix} \left(\frac{x}{L} - 1\right) & \frac{x}{L} \end{bmatrix} \begin{bmatrix} \delta m_i \\ \delta m_j \end{bmatrix} \tag{7.58}$$

According to the engineering theory of bending,

$$\varepsilon = -\frac{My}{EI} \quad \text{and} \quad \delta\sigma = -\frac{\delta My}{I} \tag{7.59}$$

Thus the theorem of complementary virtual work yields

$$\boldsymbol{\delta}\mathbf{s}\ \mathbf{d} = \int_{\text{vol}} \boldsymbol{\delta}\mathbf{M}^T \frac{y}{I} \mathbf{M} \frac{y}{EI}\, dV = \frac{1}{EI}\int_0^L \boldsymbol{\delta}\mathbf{M}^T\mathbf{M}\, dx \tag{7.60}$$

Note that $\mathbf{s} = \{m_i \quad m_j\}$, $\boldsymbol{\delta}\mathbf{s} = \{\delta m_i \quad \delta m_j\}$, $\mathbf{d} = \{\theta_i \quad \theta_j\}$, and

$$I = \int_{\text{area}} y^2\, dA$$

Substituting Eqs. (7.57) and (7.58) into Eq. (7.60) yields

$$\boldsymbol{\delta}\mathbf{s}^T\, \mathbf{d} = \boldsymbol{\delta}\mathbf{s}^T \frac{1}{EI}\int_0^L \begin{bmatrix} \frac{x}{L} - 1 \\ \frac{x}{L} \end{bmatrix} \left[\left(\frac{x}{L} - 1\right) \quad \frac{x}{L}\right] dx \tag{7.61}$$

Noting that $\boldsymbol{\delta}\mathbf{s}$ is arbitrary and nonzero, we have

$$\begin{bmatrix} \theta_i \\ \theta_j \end{bmatrix} = \frac{L}{6EI}\begin{bmatrix} 2 & -1 \\ -1 & 2 \end{bmatrix}\begin{bmatrix} m_i \\ m_j \end{bmatrix} \tag{7.62}$$

$$\mathbf{d} \quad = \quad \mathbf{f} \quad \mathbf{s}$$

which is the same result obtained using the basic equations [see Eq. (7.21)].

To satisfy equilibrium, the *prismatic cantilever beam element* in Fig. 7.5 with moment of inertia I, length L, modulus of elasticity E, and applied end forces p_{yj} and m_j must have the following internal bending moment:

$$M(x) = [(L - x) \quad 1]\begin{bmatrix} p_{yj} \\ m_j \end{bmatrix} \tag{7.63}$$

By using a virtual force system similar to that in Eq. (7.63) and applying the theorem of complementary virtual work, as in the simply supported beam element, we obtain the element flexibility matrix **f** shown in Eq. (7.22), which relates the nodal deformations **d** and the element forces **s**.

We can also obtain the governing equations of the flexibility method for the total structure by applying the theorem of complementary virtual work. That is, for the assemblage

$$\delta W_e^* = \delta W_i^* \tag{7.64}$$

The virtual force system consists of $\boldsymbol{\delta}\mathbf{P}$ at the nodes plus the virtual forces at the redundants $\boldsymbol{\delta}\mathbf{X}$; therefore

$$\delta W_e^* = \boldsymbol{\delta}\mathbf{P}^T\mathbf{U} + \boldsymbol{\delta}\mathbf{X}^T\, \mathbf{d}_x \tag{7.65}$$

where $\mathbf{d}_x$ contains the displacements of the redundants. Furthermore,

$$\begin{aligned}\delta W_i^* &= \boldsymbol{\delta}\mathbf{s}^T\, \mathbf{d} = [\mathbf{b}_0\, \boldsymbol{\delta}\mathbf{P} + \mathbf{b}_1\, \boldsymbol{\delta}\mathbf{X}]^T\, \mathbf{d} \\ &= \boldsymbol{\delta}\mathbf{P}^T\mathbf{b}_0^T\, \mathbf{d} + \boldsymbol{\delta}\mathbf{X}^T\mathbf{b}_1^T\, \mathbf{d}\end{aligned} \tag{7.66}$$

Substituting Eqs. (7.65) and (7.66) into Eq. (7.64) yields

$$\delta\mathbf{P}^T\mathbf{U} + \delta\mathbf{X}^T\,\mathbf{d}_x = \delta\mathbf{P}^T\mathbf{b}_0^T\,\mathbf{d} + \delta\mathbf{X}^T\mathbf{b}_1^T\,\mathbf{d} \tag{7.67}$$

Thus

$$\delta\mathbf{P}^T[\mathbf{U} - \mathbf{b}_0^T\,\mathbf{d}] + \delta\mathbf{X}^T[\mathbf{d}_x - \mathbf{b}_1^T\,\mathbf{d}] = 0 \tag{7.68}$$

Since $\delta\mathbf{P}$ and $\delta\mathbf{X}$ are arbitrary, nonzero, and independent, Eq. (7.68) can only be satisfied if the expressions in the square brackets are each zero, that is,

$$\mathbf{U} = \mathbf{b}_0^T\mathbf{d} \tag{7.69}$$

and

$$\mathbf{d}_x = \mathbf{b}_1^T\mathbf{d} \tag{7.70}$$

Substituting

$$\mathbf{d} = \mathbf{f}\mathbf{s} = \mathbf{f}(\mathbf{b}_0\,\mathbf{P} + \mathbf{b}_1\mathbf{X}) \tag{7.71}$$

into Eq. (7.70) and noting that $\mathbf{d}_x = 0$ we have

$$\mathbf{X} = -(\mathbf{b}_1^T\mathbf{f}\mathbf{b}_1)^{-1}\mathbf{b}_1^T\mathbf{f}\mathbf{b}_0\,\mathbf{P} = -\mathbf{F}_{11}^{-1}\mathbf{F}_{10}\,\mathbf{P} \tag{7.72}$$

Substituting Eqs. (7.71) and (7.72) into Eq. (7.69) yields

$$\begin{aligned}\mathbf{U} &= (\mathbf{b}_0^T\,\mathbf{f}\,\mathbf{b}_0 - \mathbf{b}_0^T\,\mathbf{f}\,\mathbf{b}_1\mathbf{F}_{11}^{-1}\mathbf{F}_{10})\mathbf{P} = (\mathbf{F}_{00} - \mathbf{F}_{01}\mathbf{F}_{11}^{-1}\mathbf{F}_{10})\mathbf{P}\\ &= \mathbf{F}\mathbf{P}\end{aligned} \tag{7.73}$$

where

$$\mathbf{F} = \mathbf{F}_{00} - \mathbf{F}_{01}\mathbf{F}_{11}^{-1}\mathbf{F}_{10} \tag{7.74}$$

This result is identical to that obtained using the basic equations [see Eq. (7.12)].

7.6 RELATIONSHIPS BETWEEN STATICS AND KINEMATICS

In Ch. 3 we noted that for conjugate vectors (e.g., **P** and **U** or **s** and **d**) there are corresponding contragredient transformations. That is,

$$\mathbf{s} = \mathbf{b}\mathbf{P} \tag{7.75}$$

implies the existence of the corresponding transformation

$$\mathbf{U} = \mathbf{b}^T\mathbf{d} \tag{7.76}$$

Equilibrium for the statically determinate truss of Example 2.5, which is illustrated in Fig. 7.12a, yields (see Fig. 7.12b)

$$\begin{bmatrix} s^{ab} \\ s^{bc} \\ s^{ac} \end{bmatrix} = \begin{bmatrix} -1 & 0 & 0 \\ \frac{5}{3} & 0 & \frac{5}{3} \\ 0 & -1 & 0 \end{bmatrix} \begin{bmatrix} P_{xa} \\ P_{ya} \\ P_{xb} \end{bmatrix} \tag{7.77}$$

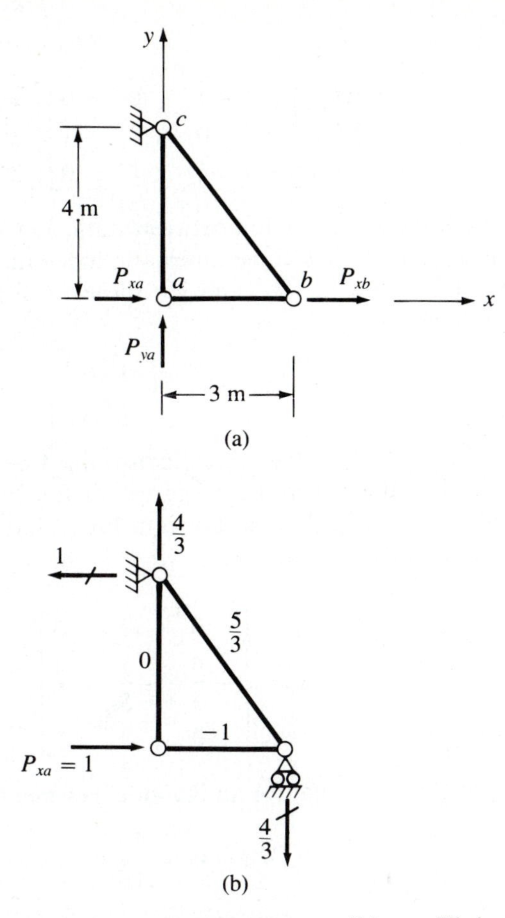

Figure 7.12 (a) Statically determinate truss; (b) equilibrium solution for unit applied force P_{xa}

Therefore, from Eq. (7.76),

$$\begin{bmatrix} U_a \\ V_a \\ U_b \end{bmatrix} = \begin{bmatrix} -1 & \frac{5}{3} & 0 \\ 0 & 0 & -1 \\ 0 & \frac{5}{3} & 0 \end{bmatrix} \begin{bmatrix} d^{ab} \\ d^{bc} \\ d^{ac} \end{bmatrix} \tag{7.78}$$

Similarly, using the contragredient rule, the transformation

$$\mathbf{d} = \mathbf{aU} \tag{7.79}$$

implies

$$\mathbf{P} = \mathbf{a}^T\mathbf{s} \tag{7.80}$$

For the truss in Fig. 7.12a,

$$\begin{bmatrix} d^{ab} \\ d^{bc} \\ d^{ac} \end{bmatrix} = \begin{bmatrix} -1 & 0 & 1 \\ 0 & 0 & \frac{3}{5} \\ 0 & -1 & 0 \end{bmatrix} \begin{bmatrix} U_a \\ V_a \\ U_b \end{bmatrix} \tag{7.81}$$

Therefore,

$$\begin{bmatrix} P_{xa} \\ P_{ya} \\ P_{xb} \end{bmatrix} = \begin{bmatrix} -1 & 0 & 0 \\ 0 & 0 & -1 \\ 1 & \frac{3}{5} & 0 \end{bmatrix} \begin{bmatrix} s^{ab} \\ s^{bc} \\ s^{ac} \end{bmatrix} \tag{7.82}$$

In Ch. 2 we discussed the analog between static and kinematic indeterminacy. The truss in Fig. 7.12a has three kinematic indeterminacies corresponding to the degrees of freedom V_b, U_c, and V_c; therefore Eq. (7.81) could have been written as follows:

$$\mathbf{d} = [\mathbf{a}_0 \quad \mathbf{a}_1] \begin{bmatrix} \mathbf{U}_f \\ \mathbf{U}_s \end{bmatrix} \tag{7.83}$$

where $\mathbf{U}_f = \{U_a \quad V_a \quad U_b\}$ (i.e., the unconstrained degrees of freedom), $\mathbf{U}_s = \{V_b \quad U_c \quad V_c\}$ (i.e., the constrained degrees of freedom), $\mathbf{a}_0$ is the kinematic matrix associated with $\mathbf{U}_f$ [i.e., $\mathbf{a}$ shown in Eq. (7.81)], and $\mathbf{a}_1$ is the kinematic matrix associated with $\mathbf{U}_s$; that is,

$$\mathbf{a}_1 = \begin{bmatrix} 0 & 0 & 0 \\ -\dfrac{4}{5} & -\dfrac{3}{5} & \dfrac{4}{5} \\ 0 & 0 & 1 \end{bmatrix} \tag{7.84}$$

We recall from Eq. (7.55) that for an individual element

$$d = fs = \frac{L}{AE} s \tag{7.85}$$

Therefore

$$s = kd = \frac{AE}{L} d \tag{7.86}$$

Thus for the entire truss of Fig. 7.12a,

$$\begin{bmatrix} s^{ab} \\ s^{bc} \\ s^{ac} \end{bmatrix} = \begin{bmatrix} k^{ab} & 0 & 0 \\ 0 & k^{bc} & 0 \\ 0 & 0 & k^{ac} \end{bmatrix} \begin{bmatrix} d^{ab} \\ d^{bc} \\ d^{ac} \end{bmatrix} \tag{7.87}$$

$$\mathbf{s} \quad = \quad \mathbf{k} \quad \mathbf{d}$$

Substituting Eq. (7.79) into Eq. (7.85) yields

$$\mathbf{s} = \mathbf{kaU} \tag{7.88}$$

From the principle of virtual work (i.e., $\delta W_e = \delta W_i$) and Eqs. (7.79) and (7.88),

$$\boldsymbol{\delta}\mathbf{U}^T\mathbf{P} = \boldsymbol{\delta}\mathbf{d}^T\mathbf{s} = \boldsymbol{\delta}\mathbf{U}^T\mathbf{a}^T\mathbf{kaU}$$

Thus

$$\mathbf{P} = \mathbf{a}^T\mathbf{kaU} \tag{7.89}$$

Operating with $\mathbf{a}$ in the partitioned form shown in Eq. (7.83) we have

$$\mathbf{a}^T\mathbf{k}\mathbf{a} = \begin{bmatrix} \mathbf{a}_0^T \\ \mathbf{a}_1^T \end{bmatrix} \mathbf{k}[\mathbf{a}_0 \quad \mathbf{a}_1] = \begin{bmatrix} \mathbf{a}_0^T\mathbf{k}\mathbf{a}_0 & \mathbf{a}_0^T\mathbf{k}\mathbf{a}_1 \\ \mathbf{a}_1^T\mathbf{k}\mathbf{a}_0 & \mathbf{a}_1^T\mathbf{k}\mathbf{a}_1 \end{bmatrix}$$
$$= \begin{bmatrix} \mathbf{K}_{00} & \mathbf{K}_{01} \\ \mathbf{K}_{10} & \mathbf{K}_{11} \end{bmatrix} \tag{7.90}$$

Using Eq. (7.90) and recalling Eq. (2.30) we note that

$$\begin{aligned} \mathbf{a}_0^T\mathbf{k}\mathbf{a}_0 &= \mathbf{K}_{00} = \mathbf{K}_{ff} \\ \mathbf{a}_0^T\mathbf{k}\mathbf{a}_1 &= \mathbf{K}_{01} = \mathbf{K}_{fs} \\ \mathbf{a}_1^T\mathbf{k}\mathbf{a}_0 &= \mathbf{K}_{10} = \mathbf{K}_{sf} \\ \mathbf{a}_1^T\mathbf{k}\mathbf{a}_1 &= \mathbf{K}_{11} = \mathbf{K}_{ss} \end{aligned} \tag{7.91}$$

Manipulating $\mathbf{a}_0$ and $\mathbf{a}_1$ from Eqs. (7.81) and (7.84) for the truss of Fig. 7.12a we obtain the same global stiffness matrix $\mathbf{K}$ calculated by the direct stiffness method in Example 2.5.

For statically determine structures $\mathbf{b} = \mathbf{b}_0$. Considering only the $\mathbf{a}_0$ partition of the kinematic matrix and substituting Eq. (7.80) into Eq. (7.81), we have the following for statically determinate structures:

$$\mathbf{s} = \mathbf{b}_0\mathbf{P} = \mathbf{b}_0\mathbf{a}_0^T\mathbf{s} \tag{7.92}$$

Thus

$$\mathbf{b}_0\mathbf{a}_0^T = \mathbf{I} \tag{7.93}$$

and

$$\mathbf{a}_0^T = \mathbf{b}_0^{-1} \tag{7.94}$$

From Eqs. (7.77) and (7.82) we note that Eq. (7.94) is valid for the statically determinate truss of Fig. 7.12a.

Recall from Eqs. (2.72a) and (2.72b) that for planar trusses the number of static redundancies (NOS) and the number of kinematic redundancies (NOK) are, respectively,

$$\text{NOS} = \text{NE} + \text{NR} - 2\text{NN}$$

and

$$\text{NOK} = 2\text{NN} - \text{NR}$$

where NE is the number of elements, NR is the number of support conditions, and NN is the number of nodes. In general, $\mathbf{a}_0$ is NE by NR and $\mathbf{a}_1$ is NE by NOK; therefore $\mathbf{a}$ is NE by 2NN. In contrast, $\mathbf{b}_1$ is NE by NOS, but while the number of rows in $\mathbf{b}_0$ is NE, the number of columns can be variable, depending upon the number of displacements to be calculated. At most $\mathbf{b}_0$ has (2NN − NR) columns, and the largest size of $\mathbf{b}$ is NE by NE. Therefore, Eq. (7.92) is not true in general; that is,

$$\mathbf{s} \neq \mathbf{b}\mathbf{a}^T\mathbf{s}$$

and

$$\mathbf{b}\mathbf{a}^T \neq \mathbf{I}$$

Since the basic matrices are not conformable for multiplication, the kinematic matrix cannot be derived from the statics matrix nor vice versa. An alternative interrelationship of the kinematic and statics matrices appears in Sec. 7.9.

7.7 THE ELEMENT FLEXIBILITY-STIFFNESS TRANSFORMATION

The element flexibility matrix characterizes a constrained element, whereas the element stiffness matrix contains all degrees of freedom, including rigid-body modes, and has no unique inverse; that is,

$$\begin{bmatrix} \mathbf{p}_f \\ \mathbf{p}_s \end{bmatrix} = \begin{bmatrix} \mathbf{k}_{ff} & \mathbf{k}_{fs} \\ \mathbf{k}_{sf} & \mathbf{k}_{ss} \end{bmatrix} \begin{bmatrix} \mathbf{u}_f \\ \mathbf{u}_s \end{bmatrix} \tag{7.95}$$

where the subscript f refers to free degrees of freedom and the subscript s indicates supported degrees of freedom. If $\mathbf{u}_s$ is null,

$$\begin{bmatrix} \mathbf{p}_f \\ \mathbf{p}_s \end{bmatrix} = \begin{bmatrix} \mathbf{k}_{ff} \\ \mathbf{k}_{sf} \end{bmatrix} \mathbf{u}_f \tag{7.96}$$

Therefore,

$$\mathbf{p}_f = \mathbf{k}_{ff}\,\mathbf{u}_f \tag{7.97}$$

or

$$\mathbf{u}_f = \mathbf{k}_{ff}^{-1}\,\mathbf{p}_f = \mathbf{f}\mathbf{p}_f \tag{7.98}$$

where

$$\mathbf{k}_{ff}^{-1} = \mathbf{f} \tag{7.99}$$

or

$$\mathbf{k}_{ff} = \mathbf{f}^{-1} \tag{7.100}$$

The transformation matrix $\mathbf{T}$ relates the free and supported forces as follows:

$$\mathbf{p}_s = \mathbf{T}\mathbf{p}_f \tag{7.101}$$

Substituting Eqs. (7.97) and (7.100) into Eq. (7.101) yields

$$\mathbf{p}_s = \mathbf{T}\mathbf{f}^{-1}\mathbf{u}_f \tag{7.102}$$

Comparing Eqs. (7.95) and (7.102) reveals that

$$\mathbf{k}_{sf} = \mathbf{T}\mathbf{f}^{-1} \tag{7.103}$$

Since the element stiffness matrix is symmetrical,

$$\mathbf{k}_{fs} = \mathbf{k}_{sf}^{T} = \mathbf{f}^{-1}\mathbf{T}^{T} \tag{7.104}$$

Each column of the stiffness matrix represents an equilibrated set of forces; therefore, these forces should be related in a fashion similar to the transformation of Eq. (7.101). That is,

$$\mathbf{k}_{ss} = \mathbf{T}\mathbf{k}_{fs} = \mathbf{T}\mathbf{f}^{-1}\mathbf{T}^{T} \tag{7.105}$$

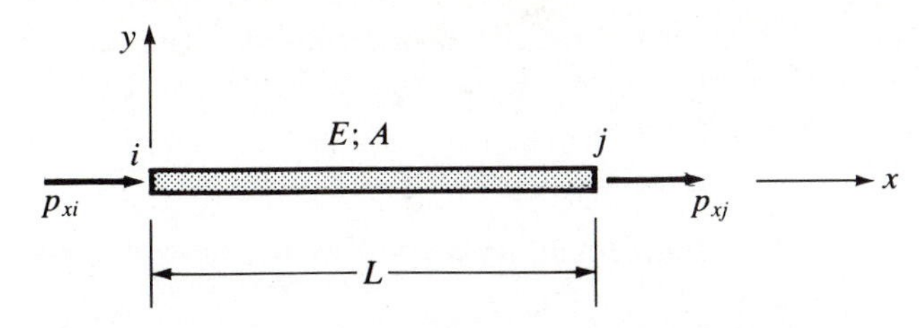

Figure 7.13 Axial force element with end forces

The stiffness matrix, constructed from Eqs. (7.100), (7.103), (7.104), and (7.105), is

$$\mathbf{k} = \begin{bmatrix} \mathbf{f}^{-1} & \mathbf{f}^{-1}\mathbf{T}^T \\ \mathbf{T}\mathbf{f}^{-1} & \mathbf{T}\mathbf{f}^{-1}\mathbf{T}^T \end{bmatrix} \tag{7.106}$$

For the axial force element of Fig. 7.13, $\mathbf{p}_f = p_{xi}$ and $\mathbf{p}_s = p_{xj}$; therefore $\mathbf{T} = -1$. Noting that $\mathbf{f}^{-1} = AE/L$ and substituting into Eq. (7.106) gives the same 2 by 2 matrix of Eq. (2.26). We can also obtain the stiffness matrix for the beam element shown in Fig. 7.14 by applying the force transformation of Eq. (7.101) with

$$\mathbf{p}_f = \begin{bmatrix} m_i \\ m_j \end{bmatrix}; \quad \mathbf{p}_s = \begin{bmatrix} p_{yi} \\ p_{yj} \end{bmatrix}; \quad \text{and} \quad \mathbf{T} = \frac{1}{L}\begin{bmatrix} 1 & 1 \\ -1 & -1 \end{bmatrix}$$

Substituting into Eq. (7.106) yields the 4 by 4 stiffness matrix for the prismatic beam element shown in Eq. (4.19).

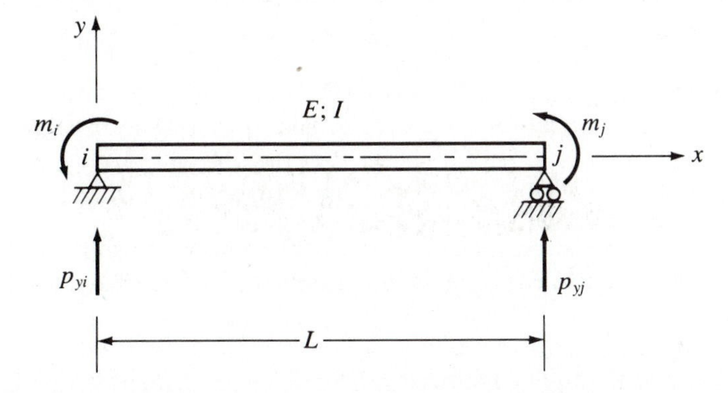

Figure 7.14 Simply supported prismatic beam element with end forces

7.8 INITIAL STRAINS, THERMAL STRAINS, AND DISTRIBUTED LOADS

Temperature changes, fabrication errors, and prestrains create initial element deformations. These effects induce nodal displacements, but element forces are introduced only for statically indeterminate structures. The initial deformations, $\mathbf{d}^o$, combine with the elastic deformations to yield the total element deformations; that is,

$$\mathbf{d} = \mathbf{fs} + \mathbf{d}^o \tag{7.107}$$

Therefore, the complementary virtual work associated with the internal forces is

$$\begin{aligned} \delta W_i^* &= \delta\mathbf{s}^T\,\mathbf{d} = (\mathbf{b}_0\,\delta\mathbf{P} + \mathbf{b}_1\,\delta\mathbf{X})^T\,\mathbf{d} \\ &= \delta\mathbf{P}^T\mathbf{b}_0^T(\mathbf{fs} + \mathbf{d}^o) + \delta\mathbf{X}^T\mathbf{b}_1^T(\mathbf{fs} + \mathbf{d}^o) \end{aligned} \tag{7.108}$$

The principle of complementary virtual work (i.e., $\delta W_e^* = \delta W_i^*$) produces

$$\delta\mathbf{P}^T\mathbf{U} + \delta\mathbf{X}^T\,\mathbf{d}_x = \delta\mathbf{P}^T\mathbf{b}_0^T(\mathbf{fs} + \mathbf{d}^o) + \delta\mathbf{X}^T\mathbf{b}_1^T(\mathbf{fs} + \mathbf{d}^o)] \tag{7.109}$$

Thus

$$\delta\mathbf{P}^T[\mathbf{U} - \mathbf{b}_0^T(\mathbf{fs} + \mathbf{d}^o)] + \delta\mathbf{X}^T[\mathbf{d}_x - \mathbf{b}_1^T(\mathbf{fs} + \mathbf{d}^0)] = 0 \tag{7.110}$$

Since $\delta\mathbf{P}$ and $\delta\mathbf{X}$ are arbitrary, nonzero, and independent, Eq. (7.110) can only be satisfied if each of the expressions in the square brackets is zero; that is,

$$\mathbf{U} = \mathbf{b}_0^T(\mathbf{fs} + \mathbf{d}^o) \tag{7.111}$$

$$\mathbf{d}_x = \mathbf{b}_1^T(\mathbf{fs} + \mathbf{d}^o) \tag{7.112}$$

Substituting Eq. (7.1a) into Eq. (7.112) and noting that compatibility requires $\mathbf{d}_x = \mathbf{0}$, we have

$$\mathbf{X} = -\mathbf{F}_{11}^{-1}\mathbf{F}_{10}\mathbf{P} - \mathbf{F}_{11}^{-1}\mathbf{b}_1^T\,\mathbf{d}^o \tag{7.113}$$

where $\mathbf{F}_{11}$ and $\mathbf{F}_{10}$ are defined in Eq. (7.7). Substituting Eqs. (7.1a) and (7.113) into Eq. (7.111), the nodal displacements are

$$\mathbf{U} = \mathbf{FP} + (-\mathbf{F}_{01}\mathbf{F}_{11}^{-1}\mathbf{b}_1^T + \mathbf{b}_0)\,\mathbf{d}^o \tag{7.114}$$

where $\mathbf{F}_{01} = \mathbf{F}_{10}^T$ and $\mathbf{F}$ is defined in Eq. (7.12).

For an axial force element subjected to a uniform increase in temperature of ΔT,

$$\mathbf{d}^o = \alpha\Delta TL \tag{7.115}$$

where α is the coefficient of linear thermal expansion and L is the original element length. For an axial force element fabricated ΔL too long,

$$\mathbf{d}^o = \Delta L \tag{7.116}$$

For a simply supported prismatic beam element with a linear temperature gradient from the upper to the lower surface (see Fig. 7.15),

$$\mathbf{d}^o = \begin{bmatrix} \theta_i^o \\ \theta_j^o \end{bmatrix} = \frac{\alpha(T_u - T_l)}{2h}\begin{bmatrix} 1 \\ -1 \end{bmatrix} \tag{7.117}$$

where T_u and T_l are the upper and lower surface temperatures and h is the depth of the beam.

The approach outlined above can also be used to model loads between the nodes. That is, initial element deformations induced by intermediate

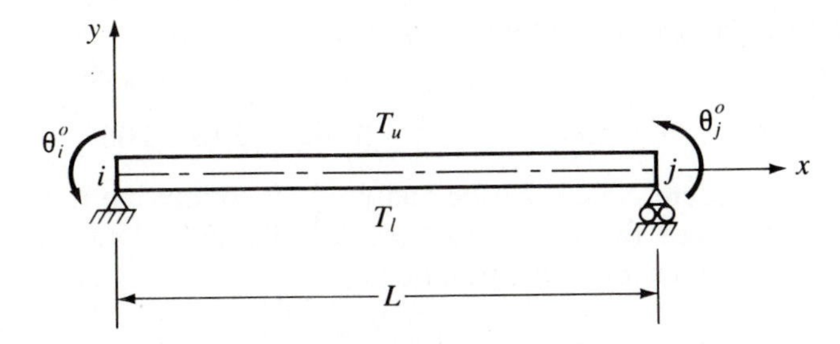

Figure 7.15 Simply supported prismatic beam element with linear temperature gradient through the depth

loading are characterized as $\mathbf{d}^o$. For the simply supported uniformly loaded prismatic beam element in Fig. 7.16,

$$\mathbf{d}^o = \begin{bmatrix} \theta_i^o \\ \theta_j^o \end{bmatrix} = \frac{qL^3}{24EI}\begin{bmatrix} 1 \\ -1 \end{bmatrix} \tag{7.118}$$

The truss of Example 2.7 (see Fig. 7.3a) was investigated in Example 2.13 for a temperature increase of 100°F for element *cd*. We will calculate the horizontal displacement of node *d* and the element forces using the flexibility approach as outlined above. With $\alpha = 6.5 \times 10^{-6}$ in./(in. · °F) and the basic matrices derived in Sec. 7.1, along with

$$\begin{aligned} d^{o(cd)} &= \alpha \Delta T L = [6.5 \times 10^{-6} \text{ in./(in.} \cdot {}^\circ\text{F)}](100^\circ\text{F})(144 \text{ in.}) \\ &= 0.0936 \text{ in.} \end{aligned}$$

we have, from Eq. (7.113) with $\mathbf{P} = \mathbf{0}$,

$$\mathbf{X} = \begin{bmatrix} s^{bc} \\ P_{xb} \end{bmatrix} = -\mathbf{F}_{11}^{-1}\mathbf{b}_1^T\, \mathbf{d}^o = \begin{bmatrix} 3.05 \\ 2.16 \end{bmatrix} \text{kips}$$

Using Eq. (7.114), the nodal displacement is

$$\mathbf{U} = U_d = (-\mathbf{F}_{01}\mathbf{F}_{11}^{-1}\mathbf{b}_1^T + \mathbf{b}_0^T)\, \mathbf{d}^o = 0.0414 \text{ in.}$$

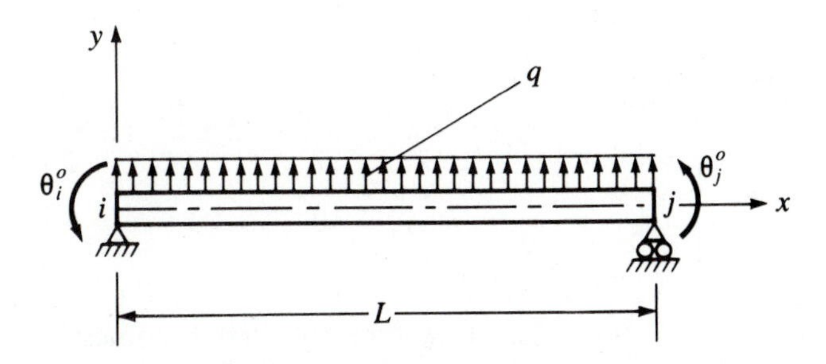

Figure 7.16 Simply supported prismatic beam element with uniform loading between the ends

From Eq. (7.1a) the element forces are

$$\mathbf{s} = \{s^{ab} \quad s^{cd} \quad s^{ac} \quad s^{bd} \quad s^{ad} \quad s^{bc}\}$$
$$= \{0.00 \quad -2.16 \quad -2.16 \quad -2.16 \quad 3.06 \quad 3.06\} \text{ kips}$$

Settlement effects can be analyzed by computing initial element displacements induced by the support movements and using the approach outlined above for other initial deformations.

7.9 AUTOMATIC SELECTION OF REDUNDANTS

The basic equilibrium matrices $\mathbf{b}_0$ and $\mathbf{b}_1$ can be constituted logically by operating upon the kinematic matrix $\mathbf{a}$ with Gauss elimination. Appendix A contains a discussion of the Gauss algorithm. We automatically select redundants starting with Eq. (7.80):

$$\mathbf{a}^T\mathbf{s} - \mathbf{IP} = \mathbf{0} \tag{7.119}$$

or in partitioned form

$$[\mathbf{a}^T \quad -\mathbf{I}]\begin{bmatrix}\mathbf{s}\\ \mathbf{P}\end{bmatrix} = \mathbf{0} \tag{7.120}$$

Operating on Eq. (7.120) with Gauss elimination, we have

$$[\mathbf{I} \quad -\mathbf{C}_1 \quad -\mathbf{C}_0]\begin{bmatrix}\mathbf{s}_s\\ \mathbf{X}\\ \mathbf{P}_f\end{bmatrix} = \mathbf{0} \tag{7.121}$$

Thus

$$\mathbf{s}_s = \mathbf{C}_0\mathbf{P}_f + \mathbf{C}_1\mathbf{X} \tag{7.122}$$

where $\mathbf{s}_s$ contains the element forces plus the statically determinate reaction forces. Therefore, $\mathbf{C}_0$ contains the same entries as $\mathbf{b}_0$ augmented by the statically determinate reactions, and $\mathbf{C}_1$ resembles $\mathbf{b}_1$ similarly augmented.

For the truss of Example 2.7 (see Fig. 7.3a) Eq. (7.83) is

$$\begin{bmatrix} d^{ab}\\ d^{cd}\\ d^{ac}\\ d^{bd}\\ d^{ad}\\ d^{bc}\end{bmatrix} = \begin{bmatrix} 0 & 0 & 0 & 0 & -1 & 0 & 1 & 0\\ -1 & 0 & 1 & 0 & 0 & 0 & 0 & 0\\ 0 & 1 & 0 & 0 & 0 & -1 & 0 & 0\\ 0 & 0 & 0 & 1 & 0 & 0 & 0 & -1\\ 0 & 0 & \frac{\sqrt{2}}{2} & \frac{\sqrt{2}}{2} & -\frac{\sqrt{2}}{2} & -\frac{\sqrt{2}}{2} & 0 & 0\\ -\frac{\sqrt{2}}{2} & \frac{\sqrt{2}}{2} & 0 & 0 & 0 & 0 & \frac{\sqrt{2}}{2} & -\frac{\sqrt{2}}{2} \end{bmatrix}\begin{bmatrix} U_c\\ V_c\\ U_d\\ V_d\\ U_a\\ V_a\\ U_b\\ V_b\end{bmatrix} \tag{7.123}$$

Note that $\mathbf{U} = \{\mathbf{U}_f \quad \mathbf{U}_s\}$; therefore Eq. (7.80) becomes

$$\begin{bmatrix}\mathbf{P}_f\\ \mathbf{P}_s\end{bmatrix} = \begin{bmatrix}\mathbf{a}_0^T\\ \mathbf{a}_1^T\end{bmatrix}\mathbf{s} \tag{7.124}$$

It is not necessary to include all unconstrained degrees of freedom; since only $\mathbf{P}_{xd}$ is nonzero we have the following form of Eq. (7.124):

	s^{ab}	s^{cd}	s^{ac}	s^{bd}	s^{ad}	s^{bc}	P_{xa}	P_{ya}	P_{xb}	P_{yb}	P_{xd}
F_{xc}	0	-1	0	0	0	$-\frac{\sqrt{2}}{2}$	0	0	0	0	0
F_{yc}	0	0	1	0	0	$\frac{\sqrt{2}}{2}$	0	0	0	0	0
F_{xd}	0	1	0	0	$\frac{\sqrt{2}}{2}$	0	0	0	0	0	-1
F_{yd}	0	0	0	1	$\frac{\sqrt{2}}{2}$	0	0	0	0	0	0
F_{xa}	-1	0	0	0	$-\frac{\sqrt{2}}{2}$	0	-1	0	0	0	0
F_{ya}	0	0	-1	0	$-\frac{\sqrt{2}}{2}$	0	0	-1	0	0	0
F_{xb}	1	0	0	0	0	$\frac{\sqrt{2}}{2}$	0	0	-1	0	0
F_{yb}	0	0	0	-1	0	$-\frac{\sqrt{2}}{2}$	0	0	0	-1	0

(7.125)

Equation (7.125) is in the form of Eqs. (7.120) and (7.124) for the truss of Fig. 7.3a, with columns and rows labeled and matrix brackets omitted for clarity. The first step in Gauss elimination is to select the largest entry in each row and place it on the diagonal by rearranging the columns (this process is called pivoting). Note that the last column (P_{xd}) represents the right-hand side of the equation; therefore, it cannot be involved in this rearrangement scheme. Doing this for Eq. (7.125) we have

	s^{cd}	s^{ac}	s^{ad}	s^{bd}	P_{xa}	P_{ya}	s^{ab}	P_{yb}	s^{bc}	P_{xb}	P_{xd}
[1]	-1	0	0	0	0	0	0	0	$-\frac{\sqrt{2}}{2}$	0	0
[2]	0	1	0	0	0	0	0	0	$\frac{\sqrt{2}}{2}$	0	0
[3]	1	0	$\frac{\sqrt{2}}{2}$	0	0	0	0	0	0	0	-1
[4]	0	0	$\frac{\sqrt{2}}{2}$	1	0	0	0	0	0	0	0
[5]	0	0	$-\frac{\sqrt{2}}{2}$	0	-1	0	-1	0	0	0	0
[6]	0	-1	$-\frac{\sqrt{2}}{2}$	0	0	-1	0	0	0	0	0
[7]	0	0	0	0	0	0	1	0	$\frac{\sqrt{2}}{2}$	-1	0
[8]	0	0	0	-1	0	0	0	-1	$-\frac{\sqrt{2}}{2}$	0	0

(7.126)

The rows in Eq. (7.126) are in the same order as in Eq. (7.125), but each row is numbered for reference. Dividing each row by the diagonal term and combining rows to obtain an 8 by 8 identity matrix in the first eight columns gives

$$
\begin{array}{lccccccccccc}
 & s^{cd} & s^{ac} & s^{ad} & s^{bd} & P_{xa} & P_{ya} & s^{ab} & P_{yb} & s^{bc} & P_{xb} & P_{xd} \\
[1'] & 1 & 0 & 0 & 0 & 0 & 0 & 0 & 0 & \frac{\sqrt{2}}{2} & 0 & 0 \\
[2'] & 0 & 1 & 0 & 0 & 0 & 0 & 0 & 0 & \frac{\sqrt{2}}{2} & 0 & 0 \\
[3'] & 0 & 0 & 1 & 0 & 0 & 0 & 0 & 0 & -1 & 0 & -\sqrt{2} \\
[4'] & 0 & 0 & 0 & 1 & 0 & 0 & 0 & 0 & \frac{\sqrt{2}}{2} & 0 & 1 \\
[5'] & 0 & 0 & 0 & 0 & 1 & 0 & 0 & 0 & 0 & 1 & 1 \\
[6'] & 0 & 0 & 0 & 0 & 0 & 1 & 0 & 0 & 0 & 0 & 1 \\
[7'] & 0 & 0 & 0 & 0 & 0 & 0 & 1 & 0 & \frac{\sqrt{2}}{2} & -1 & 0 \\
[8'] & 0 & 0 & 0 & 0 & 0 & 0 & 0 & 1 & 0 & 0 & -1
\end{array}
\qquad (7.127)
$$

where the new equations (i.e, primed numbers in square brackets) were obtained as follows:

$$
\begin{aligned}
[1'] &= -[1] \\
[2'] &= [2] \\
[3'] &= \sqrt{2}\{-[1'] + [3]\} \\
[4'] &= -\frac{\sqrt{2}}{2}[3'] + [4] \\
[5'] &= -\frac{\sqrt{2}}{2}[3'] - [5] - [7] \\
[6'] &= -[2'] - \frac{\sqrt{2}}{2}[3'] - [6] \\
[7'] &= [7] \\
[8'] &= -[4'] - [8]
\end{aligned}
$$

Therefore Eq. (7.127) is in the form of Eq. (7.121), and we observe that

$$
\begin{bmatrix} s^{cd} \\ s^{ac} \\ s^{ad} \\ s^{bd} \\ P_{xa} \\ P_{ya} \\ s^{ab} \\ P_{yb} \end{bmatrix}
=
\begin{bmatrix} 0 \\ 0 \\ \sqrt{2} \\ -1 \\ -1 \\ -1 \\ 0 \\ 1 \end{bmatrix} P_{xd}
+
\begin{bmatrix} -\frac{\sqrt{2}}{2} & 0 \\ -\frac{\sqrt{2}}{2} & 0 \\ 1 & 0 \\ -\frac{\sqrt{2}}{2} & 0 \\ 0 & -1 \\ 0 & 0 \\ -\frac{\sqrt{2}}{2} & 1 \\ 0 & 0 \end{bmatrix}
\begin{bmatrix} s^{bc} \\ P_{xb} \end{bmatrix}
\qquad (7.128)
$$

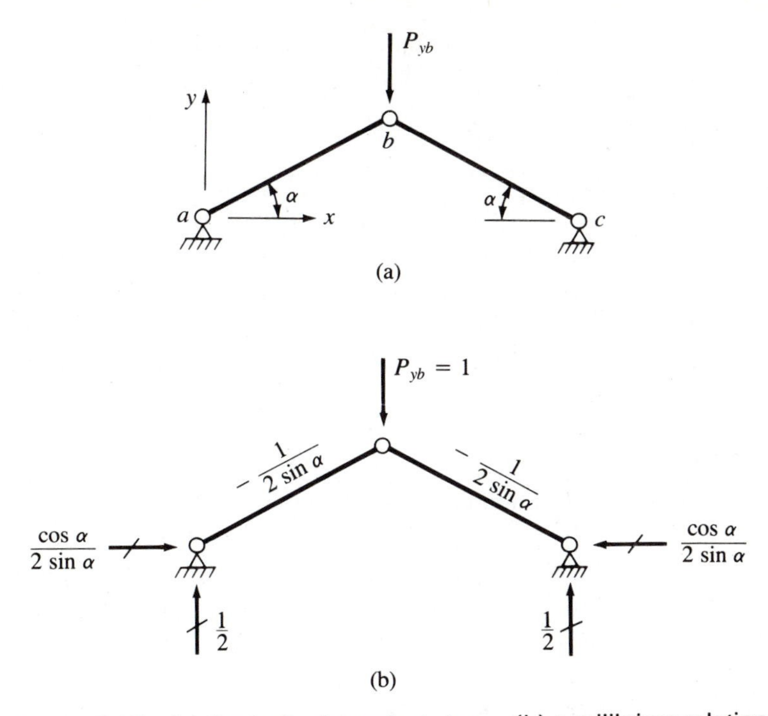

Figure 7.17 (a) Statically determinate truss; (b) equilibrium solution

Thus, the above $\mathbf{C}_0$ and $\mathbf{C}_1$ matrices contain the information of the $\mathbf{b}_0$ and $\mathbf{b}_1$ matrices, respectively, plus the statically determinate reactions.

Obtaining the $\mathbf{C}_0$ and $\mathbf{C}_1$ from the kinematic matrix $\mathbf{a}$ by operating upon Eq. (7.120) with Gauss elimination can also be used to examine the stability of a structure. If the process fails to produce the identity matrix from the 2NN columns of the augmented matrix $[\mathbf{a}^T \quad -\mathbf{I}]$, there is no unique equilibrium solution for the structure. That is, the structure is not stable. The truss of Example 5.3 was examined for stability by testing the consistency of the force-displacement equations formulated by the stiffness method. We found that the system was unstable for $\alpha = 0$. The truss in Fig. 7.17a has the reactions and element forces shown in Fig. 7.17b. As an example of a stability investigation, we will obtain the equilibrium equations by operating on the kinematic equations [see Eq. (7.120)] using Gauss elimination as suggested by Eq. (7.121). For a general configuration of this truss,

$$\begin{bmatrix} d^{ab} \\ d^{bc} \end{bmatrix} = \begin{bmatrix} -\cos\alpha & -\sin\alpha & \cos\alpha & \sin\alpha & 0 & 0 \\ 0 & 0 & -\cos\alpha & \sin\alpha & \cos\alpha & -\sin\alpha \end{bmatrix} \begin{bmatrix} U_a \\ V_a \\ U_b \\ V_b \\ U_c \\ V_c \end{bmatrix} \tag{7.129}$$

Writing the equilibrium equations corresponding to the kinematic equations of Eq. (7.129) and labeling the columns and rows as in Eq. (7.125) we have

$$\begin{array}{c|ccccccc} & s^{ab} & s^{bc} & P_{xa} & P_{ya} & P_{xc} & P_{yc} & P_{yb} \\ F_{xa} & -\cos\alpha & 0 & -1 & 0 & 0 & 0 & 0 \\ F_{ya} & -\sin\alpha & 0 & 0 & -1 & 0 & 0 & 0 \\ F_{xb} & \cos\alpha & -\cos\alpha & 0 & 0 & 0 & 0 & 0 \\ F_{yb} & \sin\alpha & \sin\alpha & 0 & 0 & 0 & 0 & 1 \\ F_{xc} & 0 & \cos\alpha & 0 & 0 & -1 & 0 & 0 \\ F_{yc} & 0 & -\sin\alpha & 0 & 0 & 0 & -1 & 0 \end{array} \tag{7.130}$$

Rearranging the columns so that the largest term is on the diagonal yields

$$\begin{array}{c|ccccccc} & P_{xa} & P_{ya} & s^{ab} & s^{bc} & P_{xc} & P_{yc} & P_{yb} \\ [1] & -1 & 0 & -\cos\alpha & 0 & 0 & 0 & 0 \\ [2] & 0 & -1 & -\sin\alpha & 0 & 0 & 0 & 0 \\ [3] & 0 & 0 & \cos\alpha & -\cos\alpha & 0 & 0 & 0 \\ [4] & 0 & 0 & \sin\alpha & \sin\alpha & 0 & 0 & 1 \\ [5] & 0 & 0 & 0 & \cos\alpha & -1 & 0 & 0 \\ [6] & 0 & 0 & 0 & -\sin\alpha & 0 & -1 & 0 \end{array} \tag{7.131}$$

Dividing each row by the diagonal term and combining rows to obtain a 6 by 6 identity matrix in the first six rows yields

$$\begin{array}{c|ccccccc} & P_{xa} & P_{ya} & s^{ab} & s^{bc} & P_{xc} & P_{yc} & P_{yb} \\ [1'] & 1 & 0 & 0 & 0 & 0 & 0 & -\dfrac{\cos\alpha}{2\sin\alpha} \\ [2'] & 0 & 1 & 0 & 0 & 0 & 0 & -\dfrac{1}{2} \\ [3'] & 0 & 0 & 1 & 0 & 0 & 0 & \dfrac{1}{2\sin\alpha} \\ [4'] & 0 & 0 & 0 & 1 & 0 & 0 & \dfrac{1}{2\sin\alpha} \\ [5'] & 0 & 0 & 0 & 0 & 1 & 0 & \dfrac{\cos\alpha}{2\sin\alpha} \\ [6'] & 0 & 0 & 0 & 0 & 0 & 1 & -\dfrac{1}{2} \end{array} \tag{7.132}$$

where the new equations were obtained as follows:

$$
\begin{aligned}
[1'] &= -[1] - \frac{1}{2}[3] - \frac{\cos\alpha}{2\sin\alpha}[4] \\
[2'] &= -[2] - \frac{\sin\alpha}{2\cos\alpha}[3] - \frac{1}{2}[4] \\
[3'] &= \frac{[3]}{2\cos\alpha} + \frac{[4]}{2\sin\alpha} \\
[4'] &= -[3'] + \frac{[4]}{\sin\alpha} \\
[5'] &= [4']\cos\alpha - [5] \\
[6'] &= -[4']\sin\alpha - [6]
\end{aligned}
\tag{7.133}
$$

The last column of Eq. (7.132) is $-\mathbf{C}_0$ ordered so that $\mathbf{s}_s = \{P_{xa} \;\; P_{ya} \;\; s^{ab} \;\; s^{bc} \;\; P_{xc} \;\; P_{yc}\}$. Comparing Eq. (7.132) with Fig. 7.17b reveals that we have obtained the correct solution.

For the structure of Fig. 7.17a with $\alpha = 0$, the augmented kinematic matrix of Eq. (7.120) with the columns rearranged is

$$
\begin{array}{c|ccccccc}
 & s^{ab} & s^{bc} & P_{xa} & P_{ya} & P_{xc} & P_{yc} & P_{yb} \\
F_{xa} & -1 & 0 & 0 & -1 & 0 & 0 & 0 \\
F_{ya} & 0 & -1 & 0 & 0 & 0 & 0 & 0 \\
F_{xb} & 1 & 0 & -1 & 0 & 0 & 0 & 0 \\
F_{yb} & 0 & 0 & 0 & 0 & 0 & 0 & 1 \\
F_{xc} & 0 & 0 & 1 & 0 & 0 & 0 & 0 \\
F_{yc} & 0 & 0 & 0 & 0 & 0 & -1 & 0
\end{array}
\tag{7.134}
$$

Since the first six entries of row four are zeros, there is no structural resistance for the applied force P_{yb}, and the structure is unstable. That is, there is no inverse of the augmented kinematic matrix; thus the element forces and reactions are unbounded for any imposed loads. This, in essence, is the zero-load test.

7.10 DISCUSSION

Casting classical compatibility methods in matrix form yields the flexibility method. The basic matrices consist of the flexibility matrix for all elements $\mathbf{f}$, plus the equilibrium matrices $\mathbf{b}_0$ and $\mathbf{b}_1$. The stiffness method can also be envisaged as composed of comparable matrices: $\mathbf{k}$ for all elements, with the kinematic matrices $\mathbf{a}_0$ and $\mathbf{a}_1$. Viewed from this perspective there is a duality of the force and stiffness methods. That is, the global equilibrium equations, $\mathbf{P} = \mathbf{KU}$, are intrinsically formulated in a fashion analogous to the compatibility equations, $\mathbf{U} = \mathbf{FP}$. While the global equilibrium equations can be formulated efficiently using the direct stiffness approach, there is no similar analog for the flexibility method. Therefore, computer efficiencies associated with the flexibility method are limited to automatic selection of the redundants. This key difference accounts for the overwhelming popularity of the stiffness method over the flexibility method.

7.11 Problems

Calculate nodal displacements and element forces for the original problems using the flexibility method.

7.1 Problem 2.12

7.2 Problem 2.29

7.3 Problem 6.4

7.4 Problem 4.5

7.5 Problem 4.9

7.6 Problem 4.10

7.7 Problem 5.2

7.8 Problem 5.3

7.9 Problem 2.16

7.10 Problem 2.26

7.11 Problem 2.19

7.12 Problem 2.28

7.13 Problem 2.18

7.14 Problem 2.32

7.15 Problem 4.6

7.16 Problem 4.7

7.17 Problem 6.7

7.18 Problem 4.26

7.19 We are attempting to determine experimentally the stiffness matrix for the truss in Fig. P7.19a. First, the structure is supported as shown in Fig. 7.19b with $U_a =$ 0.50 in. imposed, which requires $P_{xa} = 20$ kips and $P_{ya} = 40$ kips. Next, the structure is supported as in Fig. P7.19c, with $V_a = 0.25$ in. imposed, which requires $P_{xa} = 20$ kips and $P_{ya} = 60$ kips. Construct the matrix $\mathbf{K}_{ff}$ from these data and comment on the quality of the data.

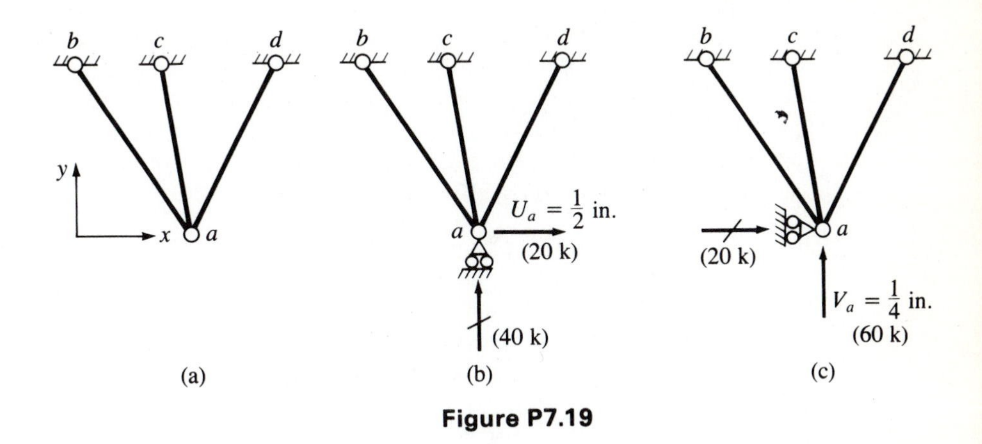

Figure P7.19

7.20 We do not know the properties of the beam in Fig. P7.20; therefore we load it with two distinct load cases as shown. Using the data from these figures calculate the stiffness and flexibility matrices. Is there any evidence that the data are consistent? Is the stiffness matrix obtained $\mathbf{K}$, $\mathbf{K}_{ff}$, or something else (identify if the latter is your answer).

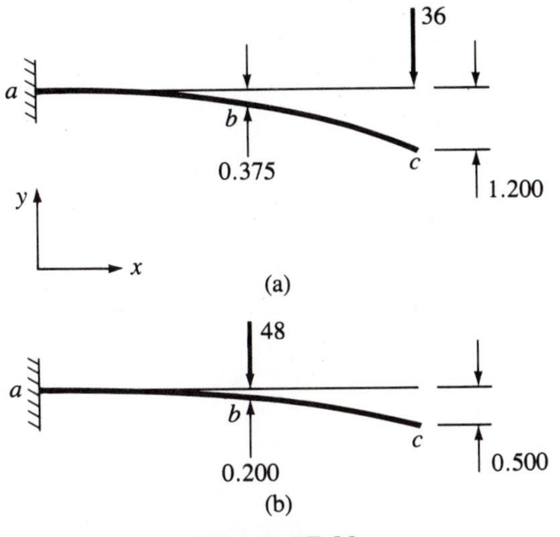

Figure P7.20

7.21 Discuss the concepts of virtual work and complementary virtual work (use the terms admissible, real, and actual displacements, plus clarify conditions on force equilibrium). Under these two broad categories, classify the following: equilibrium method; compatibility method; method of consistent displacements; three-moment equation; slope-deflection method; moment distribution; the stiffness method; and the flexibility method.

7.22 Using the techniques developed in Sec. 7.7 transform the beam flexibility matrices shown in Eqs. (7.21) and (7.22) to the beam stiffness matrix of Eq. (4.19).

A SOLUTION OF THE EQUILIBRIUM EQUATIONS

The task that consumes the most computer time in executing the stiffness method of analysis is that of solving the equilibrium equations for the unknown generalized displacements. In the case of a dynamic analysis the percentage of solution time is even larger than that for a static solution. The structural engineer is typically concerned with: idealization of a structure into its individual structural elements; efficient assembly of the element stiffness matrices into the global stiffness matrix of the system; and interpreting the calculated nodal displacements and element forces. The feasibility of a structural analysis depends upon an efficient algorithm for solving the algebraic equations of equilibrium. In the late 1960s an airframe manufacturer pushed the state of the art of equation solvers with an investigation of the wing-body intersection of the 747 airplane; this involved approximately 2500 equations. Today, structures with 10,000 equilibrium equations are routinely solved.

Computer solution algorithms are usually formulated by numerical analysts, but the structural engineer must be aware of the cost, accuracy, and sources of errors associated with the solution process. Equipped with a knowledge of a solution algorithm we can: minimize analysis time by appropriately numbering the nodes or elements; interpret errors; and understand the limitations of the results.

Sets of simultaneous linear algebraic equations can be solved using either direct elimination methods or iterative solutions. During the emergence of the stiffness method of analysis the latter solution algorithms were widely used; today these have been largely supplanted by Gauss elimination. The structural equilibrium equations have special qualities that make them amenable to efficient solution algorithms. That is, the coefficient matrix: is symmetrical; includes many zeros; contains nonzero entries clustered about the main diagonal; and has positive terms on the main diagonal. These respective characteristics can be succinctly described by stating that the global stiffness matrix is symmetric, sparse, banded, and positive definite. Solution errors are introduced by the computer, the algorithm, and the structural idealization. An added benefit of a well-executed solution package is that it can frequently detect an unstable structure.

A.1 DIRECT SOLUTIONS BASED ON GAUSS ELIMINATION

We are interested in obtaining the generalized displacements associated with the static nodal equilibrium equations of a structure. In this section we will

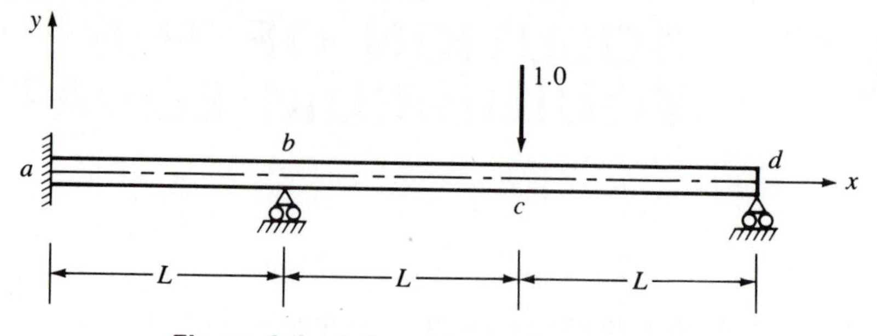

Figure A.1 A three-element beam structure

discuss methods of direct solutions based upon Gauss elimination. For the sake of simplicity we will write the force-displacement equations for a beam element [see (Eq. (4.19)] as follows:

$$\begin{bmatrix} \bar{p}_{\bar{y}i} \\ \bar{m}_i/L \\ \bar{p}_{yj} \\ \bar{m}_j/L \end{bmatrix} = \frac{EI}{L^3} \begin{bmatrix} 12 & 6 & -12 & 6 \\ 6 & 4 & -6 & 2 \\ -12 & -6 & 12 & -6 \\ 6 & 2 & -6 & 4 \end{bmatrix} \begin{bmatrix} \bar{v}_i \\ \bar{\theta}_i L \\ \bar{v}_j \\ \bar{\theta}_j L \end{bmatrix} \tag{A.1}$$

Thus we can write the force-displacement equations for the supported beam of Fig. A.1 with $EI/L^3 = 1$ and $L = 1$ as follows:

$$\begin{bmatrix} M_b \\ P_{yc} \\ M_c \\ M_d \end{bmatrix} = \begin{bmatrix} 8 & -6 & 2 & 0 \\ -6 & 24 & 0 & 6 \\ 2 & 0 & 8 & 2 \\ 0 & 6 & 2 & 4 \end{bmatrix} \begin{bmatrix} \Theta_b \\ V_c \\ \Theta_c \\ \Theta_d \end{bmatrix} = \begin{bmatrix} 0 \\ -1 \\ 0 \\ 0 \end{bmatrix} \tag{A.2}$$

A.1.1 SOLUTION BY ELIMINATION We will solve the equilibrium equations in Eq. (A.2) using Gauss elimination as follows:

STEP 1 Subtract a multiple of the first equation from the second and third equations so that the first column of **K** contains all zeros; that is, we multiply the first equation by $-3/4$ and subtract it from the second, and the third equation is replaced by 1/4 times the first subtracted from the third. This gives

$$\begin{bmatrix} 8 & -6 & 2 & 0 \\ 0 & \frac{39}{2} & \frac{3}{2} & 6 \\ 0 & \frac{3}{2} & \frac{15}{2} & 2 \\ 0 & 6 & 2 & 4 \end{bmatrix} \begin{bmatrix} \Theta_b \\ V_c \\ \Theta_c \\ \Theta_d \end{bmatrix} = \begin{bmatrix} 0 \\ -1 \\ 0 \\ 0 \end{bmatrix} \tag{A.3}$$

STEP 2 Using Eq. (A.3), we multiply the second equation by 1/13 and subtract it from the third. Furthermore, replacing the fourth equation by 4/13 times the second subtracted from the fourth equation gives

$$\begin{bmatrix} 8 & -6 & 2 & 0 \\ 0 & \frac{39}{2} & \frac{3}{2} & 6 \\ 0 & 0 & \frac{96}{13} & \frac{20}{13} \\ 0 & 0 & \frac{20}{13} & \frac{28}{13} \end{bmatrix} \begin{bmatrix} \Theta_b \\ V_c \\ \Theta_c \\ \Theta_d \end{bmatrix} = \begin{bmatrix} 0 \\ -1 \\ \frac{1}{13} \\ \frac{4}{13} \end{bmatrix} \tag{A.4}$$

STEP 3 Replace the fourth equation of Eq. (A.4) by the difference of the fourth equation and 5/24 times the third. The final result is

$$\begin{bmatrix} 8 & -6 & 2 & 0 \\ 0 & \frac{39}{2} & \frac{3}{2} & 6 \\ 0 & 0 & \frac{96}{13} & \frac{20}{13} \\ 0 & 0 & 0 & \frac{11}{6} \end{bmatrix} \begin{bmatrix} \Theta_b \\ V_c \\ \Theta_c \\ \Theta_d \end{bmatrix} = \begin{bmatrix} 0 \\ -1 \\ \frac{1}{13} \\ \frac{7}{24} \end{bmatrix} \tag{A.5}$$

Using Eq. (A.5) we can solve for the displacements by backsubstitution to give

$$\begin{aligned} \Theta_d &= \frac{6}{11}\left(\frac{7}{24}\right) = \frac{7}{44} \\ \Theta_c &= \frac{13}{96}\left[\frac{1}{13} - \frac{20}{13}\left(\frac{7}{44}\right)\right] = -\frac{1}{44} \\ V_c &= \frac{2}{39}\left[-1 - \frac{3}{2}\left(-\frac{1}{44}\right) - 6\left(\frac{7}{44}\right)\right] = -\frac{13}{132} \\ \Theta_b &= \frac{1}{8}\left[6\left(-\frac{13}{132}\right) - 2\left(-\frac{1}{44}\right)\right] = -\frac{3}{44} \end{aligned} \tag{A.6}$$

The objective of Gauss elimination is to convert the stiffness matrix to an upper triangular form as in Eq. (A.5). We accomplished this in $n-1$ distinct steps, where n is the size of the stiffness matrix. At the ith step we replaced equations $i+1, i+2, \ldots, n$ by the respective equation minus a multiple of the ith equation so that the elements $(i+1, i)$, $(i+2, i)$, $\ldots$, (n, i) became zero. Finally, we used backsubstitution to calculate the unknowns.

The physical significance of Gauss elimination is apparent if we compare the process with static condensation (cf. Sec. 6.1). By eliminating a degree of

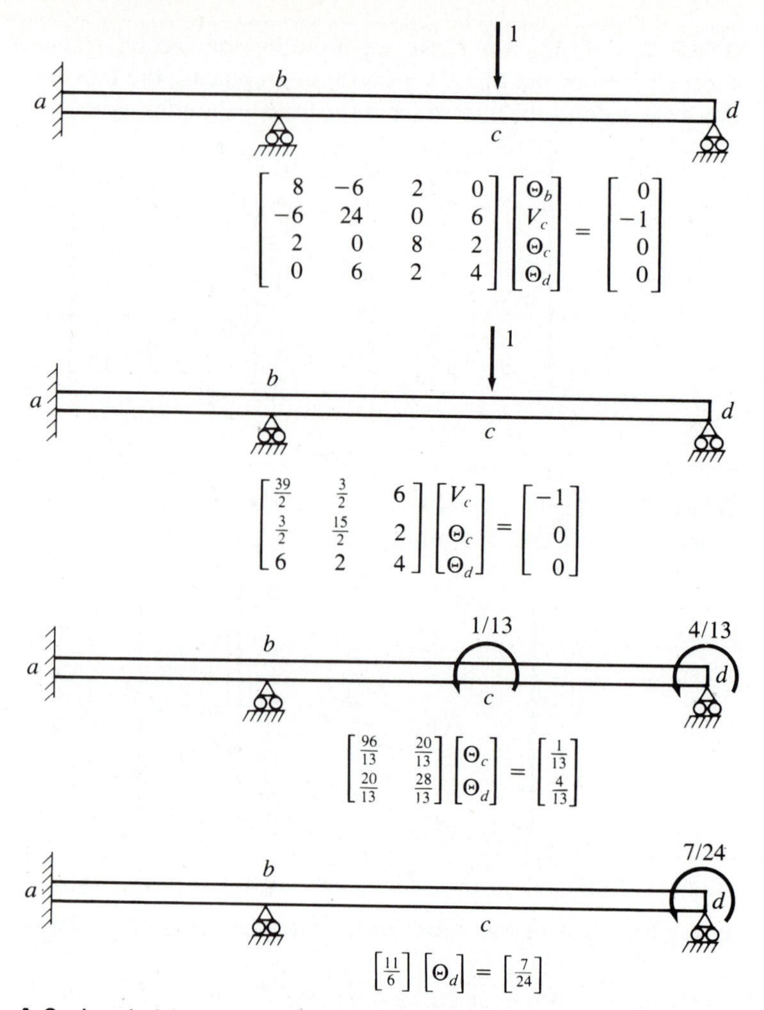

Figure A.2 Loaded beams implied by the steps of a Gauss elimination of the beam of Fig. A.1

freedom from a set of equations we retain its effect while suppressing the variable (i.e., the degree of freedom is implicit in the equations). In Eqs. (A.3) through (A.5) the matrix partition in the lower right corner of the stiffness matrix enclosed within the dotted lines is the condensed stiffness matrix; furthermore, the corresponding matrix of loads is also condensed. Figure A.2 shows the active degrees of freedom for the structure and the associated loads for each step of the Gauss elimination procedure.

In general, k_{ij} is the force at the ith degree of freedom required to maintain a unit displacement of the jth degree of freedom. Thus, the global stiffness matrix must have positive terms on the main diagonal. Furthermore, since the diagonal elements of the stiffness matrix at any point in the Gauss elimination process are the condensed stiffness coefficients, they must always be positive. If

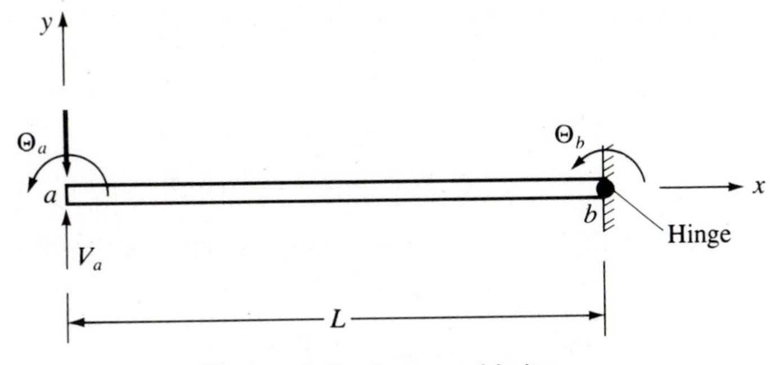

Figure A.3 An unstable beam

a negative or zero element occurs on the diagonal during solution, the structure is unstable. The unstable beam of Fig. A.3 has the following stiffness matrix (with support conditions imposed and degrees of freedom ordered as V_a, Θ_a, and Θ_b):

$$\begin{bmatrix} 12 & 6 & -12 \\ 6 & 4 & -6 \\ -12 & -6 & 12 \end{bmatrix}$$

After operating on the second and third rows using Gauss elimination we have

$$\begin{bmatrix} 12 & 6 & -12 \\ 0 & 1 & 0 \\ 0 & 0 & 0 \end{bmatrix}$$

which reveals that we have no stiffness term to resist rotation at point b. That is, the beam of Fig. A.3 is unstable.

A.1.2 SOLUTION BY DECOMPOSITION Gauss elimination as performed in the previous section is limited to hand calculations. In this section we will show a systematic procedure for carrying out the two basic steps: decomposition of the stiffness matrix and solution for the displacements. The operations of combining the equations to obtain zero elements in the stiffness matrix below the diagonal can be represented by the matrix product:

$$\mathbf{L}_{n-1}^{-1} \cdots \mathbf{L}_2^{-1}\mathbf{L}_1^{-1}\mathbf{K} = \mathbf{R} \tag{A.7}$$

where **R** is the matrix with nonzero elements above the diagonal and

$$\mathbf{L}_i^{-1} = \begin{bmatrix} 1 & & & & \\ & \ddots & & & \\ & & 1 & & \\ & & -l_{i+1,i} & \ddots & \\ & & -l_{i+2,i} & & 1 \\ & & \vdots & & \\ & & -l_{ni} & & \end{bmatrix} \qquad l_{i+j,i} = \frac{k_{i+j,i}^i}{k_{ii}^i} \tag{A.8}$$

Elements not shown in the matrix of Eq. (A.8) are zero. Elements $l_{i+j,\,i}$ are the Gauss multiplying factors, and the superscripts i on the stiffness terms indicate that an element of the matrix $\mathbf{L}_{i-1}^{-1} \cdots \mathbf{L}_2^{-1}\mathbf{L}_1^{-1}\mathbf{K}$ is used. Multiplying Eq. (A.7) by the various lower triangular matrices $\mathbf{L}_i$, we have

$$\mathbf{K} = \mathbf{L}_1\mathbf{L}_2 \cdots \mathbf{L}_{n-1}\mathbf{R} \tag{A.9}$$

We can obtain the matrix $\mathbf{L}_i$ by reversing the signs of the off-diagonal terms of $\mathbf{L}_i^{-1}$; thus

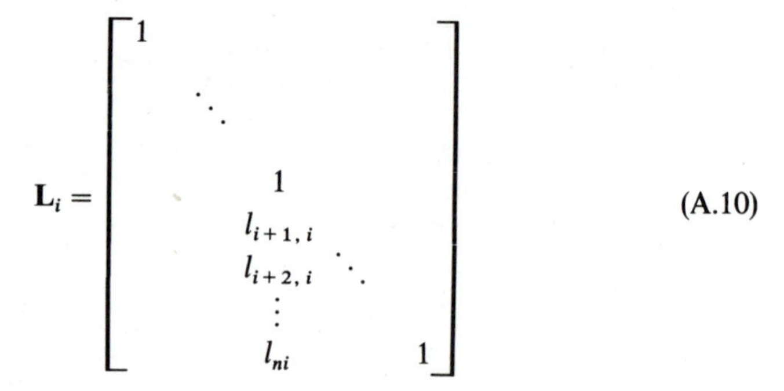

$$\mathbf{L}_i = \begin{bmatrix} 1 & & & & \\ & \ddots & & & \\ & & 1 & & \\ & & l_{i+1,\,i} & & \\ & & l_{i+2,\,i} & \ddots & \\ & & \vdots & & \\ & & l_{ni} & & 1 \end{bmatrix} \tag{A.10}$$

The lower diagonal matrix, $\mathbf{L} = \mathbf{L}_1\mathbf{L}_2 \cdots \mathbf{L}_{n-1}$, is formulated as follows:

$$\mathbf{L} = \begin{bmatrix} 1 & & & & & \\ l_{21} & 1 & & & & \\ l_{31} & l_{32} & 1 & & & \\ l_{41} & l_{42} & l_{43} & \ddots & & \\ \vdots & \vdots & \vdots & & & \\ & & & & 1 & \\ l_{n1} & l_{n2} & \cdots & & l_{n,\,n-1} & 1 \end{bmatrix} \tag{A.11}$$

Thus

$$\mathbf{K} = \mathbf{L}\mathbf{R} \tag{A.12}$$

Since $\mathbf{L}$ has unit values on the diagonal, $\mathbf{R}$ contains entries on the main diagonal, which are the so-called pivot elements of the Gauss elimination process. Making the substitution:

$$R = \mathbf{D}\hat{\mathbf{R}} \tag{A.13}$$

where $\mathbf{D}$ is a diagonal matrix containing the diagonal elements of $\mathbf{R}$ (i.e., $D_{ii} = R_{ii}$). By substituting Eq. (A.13) into Eq. (A.12) and noting that $\mathbf{K}$ is symmetric and the decomposition is unique, we have

$$\hat{\mathbf{R}} = \mathbf{L}^T \tag{A.14}$$

and

$$\mathbf{K} = \mathbf{L}\mathbf{D}\mathbf{L}^T \tag{A.15}$$

Using this $\mathbf{LDL}^T$ decomposition of the stiffness matrix we can solve the equilibrium equations as follows:

$$\mathbf{K}\mathbf{U} = \mathbf{L}\mathbf{D}\mathbf{L}^T\mathbf{U} = \mathbf{P} \tag{A.16}$$

Let

$$\mathbf{LY} = \mathbf{P} \tag{A.17}$$

Comparing Eqs. (A.16) and (A.17) we note that

$$\mathbf{Y} = \mathbf{DL}^T\mathbf{U} \tag{A.18}$$

Since the coefficient matrix in Eq. (A.17) is a lower diagonal form, $\mathbf{Y}$ can be obtained by forward substitution, that is,

$$\mathbf{Y} = \mathbf{L}_{n-1}^{-1} \cdots \mathbf{L}_2^{-1}\mathbf{L}_1^{-1}\mathbf{P} \tag{A.19}$$

and from Eq. (A.18),

$$\mathbf{L}^T\mathbf{U} = \mathbf{D}^{-1}\mathbf{Y} \tag{A.20}$$

Equation (A.20) is solved by backsubstitution because $\mathbf{L}^T$ is an upper triangular matrix.

We will obtain the solution of the equilibrium equations for the beam of Fig. A.1 using decomposition. The equilibrium equations are

$$\begin{bmatrix} M_b \\ P_{yc} \\ M_c \\ M_d \end{bmatrix} = \begin{bmatrix} 8 & -6 & 2 & 0 \\ -6 & 24 & 0 & 6 \\ 2 & 0 & 8 & 2 \\ 0 & 6 & 2 & 4 \end{bmatrix} \begin{bmatrix} \Theta_b \\ V_c \\ \Theta_c \\ \Theta_d \end{bmatrix} = \begin{bmatrix} 0 \\ -1 \\ 0 \\ 0 \end{bmatrix}$$

Using Eq. (A.8) we have

$$\mathbf{L}_1^{-1} = \begin{bmatrix} 1 & & & \\ \frac{3}{4} & 1 & & \\ -\frac{1}{4} & 0 & 1 & \\ 0 & 0 & 0 & 1 \end{bmatrix}$$

Thus

$$\mathbf{L}_1^{-1}\mathbf{K} = \begin{bmatrix} 8 & -6 & 2 & 0 \\ 0 & \frac{39}{2} & \frac{3}{2} & 6 \\ 0 & \frac{3}{2} & \frac{15}{2} & 2 \\ 0 & 6 & 2 & 4 \end{bmatrix}$$

Using Eq. (A.8) and the above product,

$$\mathbf{L}_2^{-1} = \begin{bmatrix} 1 & & & \\ 0 & 1 & & \\ 0 & -\frac{1}{13} & 1 & \\ 0 & -\frac{4}{13} & 0 & 1 \end{bmatrix}$$

Similarly, from the product $\mathbf{L}_2^{-1}\mathbf{L}_1^{-1}\mathbf{K}$ we have

$$\mathbf{L}_3^{-1} = \begin{bmatrix} 1 & & & \\ 0 & 1 & & \\ 0 & 0 & 1 & \\ 0 & 0 & -\frac{5}{24} & 1 \end{bmatrix}$$

and

$$\boldsymbol{R} = \begin{bmatrix} 8 & -6 & 2 & 0 \\ 0 & \frac{39}{2} & \frac{3}{2} & 6 \\ 0 & 0 & \frac{96}{13} & \frac{20}{13} \\ 0 & 0 & 0 & \frac{11}{6} \end{bmatrix} \tag{A.21}$$

The matrix **D** contains the pivot elements on its diagonal; thus from **R**,

$$\mathbf{D} = \begin{bmatrix} 8 & & & \\ & \frac{39}{2} & & \\ & & \frac{96}{13} & \\ & & & \frac{11}{6} \end{bmatrix} \tag{A.22}$$

From Eq. (A.11),

$$\mathbf{L} = \begin{bmatrix} 1 & & & \\ -\frac{3}{4} & 1 & & \\ \frac{1}{4} & \frac{1}{13} & 1 & \\ 0 & \frac{4}{13} & \frac{5}{24} & 1 \end{bmatrix} \tag{A.23}$$

we obtain the following vector **Y** using Eq. (A.19):

$$\mathbf{Y} = \begin{bmatrix} 0 \\ -1 \\ \frac{1}{13} \\ \frac{7}{24} \end{bmatrix}$$

and

$$\mathbf{D}^{-1}\mathbf{Y} = \begin{bmatrix} 0 \\ -\frac{2}{39} \\ \frac{1}{96} \\ \frac{7}{44} \end{bmatrix}$$

Finally, we can solve Eq. (A.20) for **U**; this gives the same displacements obtained previously.

A.1.3 COMPUTER IMPLEMENTATION

The stiffness matrix is symmetric (i.e., $k_{ij} = k_{ji}$) with positive terms on the diagonal, and generally all nonzero terms are clustered about the diagonal. This latter property is dependent upon how the nodes are numbered. For example, the node numbers in Fig. A.4a spread the nonzero entries (shown as $\times$'s) in a much wider strip about the diagonal than does the numbering scheme of Fig. A.4b. Note that $k_{ij} = 0$ for $j > i + m_k$, where m_k is the half-bandwidth of the matrix. In Fig. A.4a, $m_k = 13$ and $m_k = 7$ in Fig. A.4b. The zero entries outside the half-bandwidth reduce the number of operations required to solve the equations; therefore, the analyst should be careful in numbering the nodes.

The $\mathbf{LDL}^T$ decomposition is most efficiently executed by operating upon the stiffness matrix column by column. Because **K** is banded we need not store most of the zero entries; to optimize the process we define the skyline of **K** as the profile of the various column heights (see Figs. A.4a and A.4b). The column heights are $i - m_i$, where m_i is the row with the first nonzero entry in the ith column. Thus **K** for the node numbering of Fig. A.4a has $m_2 = 1$, $m_3 = m_4 = 3$, $m_5 = m_6 = 5$, etc. Note that the maximum column height is the half-bandwidth (i.e., in Fig. A.4a, $m_{14} = 1$ and $i - m_i = 13$). With $d_{11} = k_{11}$, the intermediate entries of the decomposed matrix are

$$\begin{aligned} g_{mj,\,j} &= k_{mj,\,j} \\ g_{ij} &= k_{ij} - \sum_{s=m_m}^{i-1} \hat{r}_{si}\, g_{sj}, \qquad i = m_j + 1, \ldots, j - 1 \end{aligned} \tag{A.24}$$

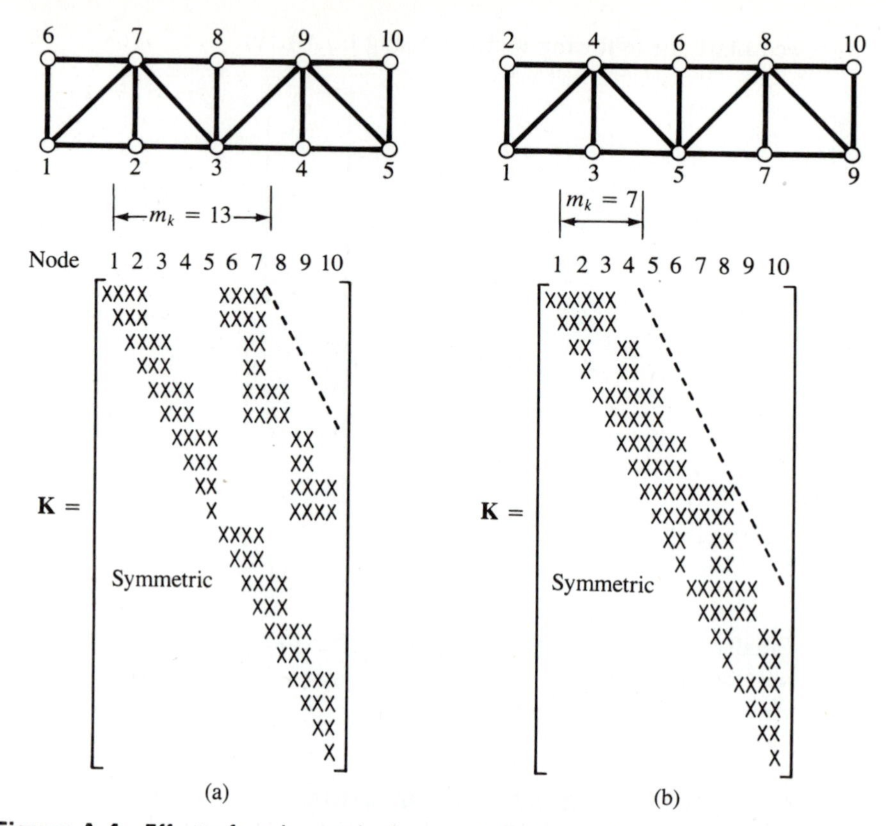

Figure A.4 Effect of node numbering on half-bandwidth (omitted entries in **K** above diagonal are zero)

and finally,

$$\hat{r}_{ij} = \frac{g_{ij}}{d_{ii}}, \qquad i = m_j, \ldots, j-1 \tag{A.25}$$

$$d_{jj} = k_{jj} - \sum_{s=m_j}^{j-1} \hat{r}_{sj} g_{sj} \tag{A.26}$$

Note that $\hat{r}_{ij}$ are elements of the matrix $\mathbf{L}^T$. This solution algorithm is the active column solution or the skyline reduction method.

To illustrate the process, we calculate $\mathbf{D}$ and $\mathbf{L}^T$ for the beam of Fig. A.1. The upper beam portion of the stiffness matrix is

$$\begin{bmatrix} 8 & -6 & 2 & \\ & 24 & 0 & 6 \\ & & 8 & 2 \\ & & & 4 \end{bmatrix}$$

The skyline is defined by $m_1 = 1$, $m_2 = 1$, $m_3 = 1$, and $m_4 = 2$. Using $d_{11} = k_{11} = 8$ from Eqs. (A.24) through (A.26) with $j = 2$ yields $g_{12} = k_{12} = -6$, and

$$\hat{r}_{12} = \frac{g_{12}}{d_{11}} = -\frac{6}{8}$$

$$d_{22} = k_{22} - \hat{r}_{12} g_{22} = 24 - \frac{3}{4}(-6) = \frac{39}{2}$$

Therefore the matrix becomes, with the dashed line separating reduced and unreduced columns,

$$\left[\begin{array}{cc:cc} 8 & -\frac{3}{4} & 2 & \\ & \frac{39}{2} & 0 & 6 \\ & & 8 & 2 \\ & & & 4 \end{array}\right]$$

For $j = 3$, $m_3 = 1$, $g_{13} = k_{13} = 2$; therefore

$$g_{23} = k_{23} - \hat{r}_{12} g_{13} = 0 - \left(-\frac{3}{4}\right)(2) = \frac{3}{2}$$

$$\hat{r}_{13} = \frac{g_{13}}{d_{11}} = \frac{2}{8}$$

$$\hat{r}_{23} = \frac{g_{23}}{d_{22}} = \frac{3/2}{39/2} = \frac{3}{39}$$

$$\begin{aligned} d_{33} &= k_{33} - \hat{r}_{13} g_{13} - \hat{r}_{23} g_{23} \\ &= 8 - \left(\frac{2}{8}\right)(2) - \left(\frac{1}{13}\right)\left(\frac{3}{2}\right) = \frac{96}{13} \end{aligned}$$

and the matrix becomes

$$\left[\begin{array}{ccc:c} 8 & -\frac{3}{4} & \frac{1}{4} & \\ & \frac{39}{2} & \frac{1}{13} & 6 \\ & & \frac{96}{13} & 2 \\ & & & 4 \end{array}\right]$$

For $j = 4$, $m_j = 2$, $g_{24} = k_{24} = 6$, and

$$g_{34} = k_{34} - \hat{r}_{23} g_{24} = \frac{20}{13}$$

Thus

$$\hat{r}_{24} = \frac{g_{24}}{d_{22}} = \frac{4}{13}$$

$$\hat{r}_{34} = \frac{g_{34}}{d_{33}} = \frac{20}{96}$$

$$d_{44} = k_{44} - \sum_{s=2}^{3} \hat{r}_{sj} g_{sj} = 4 - \frac{24}{13} - \frac{25}{78} = \frac{11}{6}$$

and the final elements are

$$\begin{bmatrix} 8 & -\frac{3}{4} & \frac{1}{4} & \\ & \frac{39}{2} & \frac{1}{13} & \frac{4}{13} \\ & & \frac{96}{13} & \frac{5}{24} \\ & & & \frac{11}{16} \end{bmatrix}$$

Hence we have the elements of $\mathbf{D}$ stored on the diagonal and the elements $\hat{r}_{ij}$ have replaced k_{ij} where $j > i$.

A.1.4 CHOLESKI DECOMPOSITION A variation on Gauss elimination is the Choleski decomposition wherein the stiffness matrix is factored as follows:

$$\mathbf{K} = \tilde{\mathbf{L}}\tilde{\mathbf{L}}^T \tag{A.27}$$

where

$$\tilde{\mathbf{L}} = \mathbf{L}\mathbf{D}^{1/2} \tag{A.28}$$

Thus the Choleski decomposition factors can be obtained from the basic matrices of Gauss elimination, but $\tilde{\mathbf{L}}$ is usually calculated independently since the two approaches are rarely used together. For the beam of Fig. A.1, with the stiffness matrix of Eq. (A.1), we can deduce the Choleski factors from $\mathbf{L}$ and $\mathbf{D}$ shown in Eqs. (A.22) and (A.23), respectively. Therefore,

$$\tilde{\mathbf{L}} = \begin{bmatrix} 2.828 & & & \\ -2.121 & 4.416 & & \\ 0.707 & 0.340 & 2.718 & \\ 0 & 1.359 & 0.566 & 1.354 \end{bmatrix}$$

Choleski decomposition is limited to symmetric matrices, but Gauss elimination can be extended to more general categories of matrices.

A.1.5 FRONTAL SOLUTION By successively eliminating degrees of freedom during the assembly process we can avoid storing the entire structural stiffness matrix. This algorithm, known as the frontal or wavefront solution, eliminates a particular degree of freedom from the structural stiffness matrix after all contributing element stiffness terms have been added. Either Gauss elimination or static condensation (see Sec. 6.1) yields the condensed structural stiffness matrix. Thus, assembly of the structural stiffness matrix is interspersed with calculations to selectively eliminate (i.e., condense) degrees of freedom.

Consider the frontal solution for the beam of Fig. A.1. We designate degrees of freedom as $\Theta_b \to 1$, $V_c \to 2$, $\Theta_c \to 3$, and $\Theta_d \to 4$, and process elements in the order *ab*, *bc*, and *cd*. After merging elements *ab* and *bc* and imposing the support conditions at node *a* the nodal equilibrium equations are

$$\begin{bmatrix} 8 & -6 & 2 \\ -6 & 12 & -6 \\ 2 & -6 & 4 \end{bmatrix} \begin{bmatrix} \Theta_b \\ V'_c \\ \Theta'_c \end{bmatrix} = \begin{bmatrix} 0 \\ -1 \\ 0 \end{bmatrix} \tag{A.29}$$

where primes on a degree of freedom denote that all stiffness contributions have not been included. Eliminating Θ_b (by either static condensation or Gauss elimination) yields

$$\begin{bmatrix} \frac{15}{2} & -\frac{9}{2} \\ -\frac{9}{2} & \frac{7}{2} \end{bmatrix} \begin{bmatrix} V'_c \\ \Theta'_c \end{bmatrix} = \begin{bmatrix} -1 \\ 0 \end{bmatrix} \tag{A.30}$$

We can envisage the process physically in Fig. A.5a, where the active degrees of freedom are enclosed between the two dashed lines. Thus V_c and Θ_c remain active after Θ_b is eliminated. Finally, we combine the stiffness matrix for element *cd* with the previous result; thus

$$\mathbf{k}^{cd} = \begin{bmatrix} 12 & 6 & 6 \\ 6 & 4 & 2 \\ 6 & 2 & 4 \end{bmatrix}$$

Note that the stiffness terms for V_d are not included since that degree of freedom is supported. The final equilibrium equations are

$$\begin{bmatrix} \frac{39}{2} & \frac{3}{2} & 6 \\ \frac{3}{2} & \frac{15}{2} & 2 \\ 6 & 2 & 4 \end{bmatrix} \begin{bmatrix} V_c \\ \Theta_c \\ \Theta_d \end{bmatrix} = \begin{bmatrix} -1 \\ 0 \\ 0 \end{bmatrix} \tag{A.31}$$

Solving Eq. (A.31) yields the displacements for V_c, Θ_c, and Θ_d shown in Eq. (A.6). Thus from Eq. (A.29) we have

$$8\Theta_b - 6V'_c + 2\Theta'_c = 0 \tag{A.32}$$

Recall that the primes indicate that all stiffness terms have not been included for those degrees of freedom throughout the equilibrium equations, but

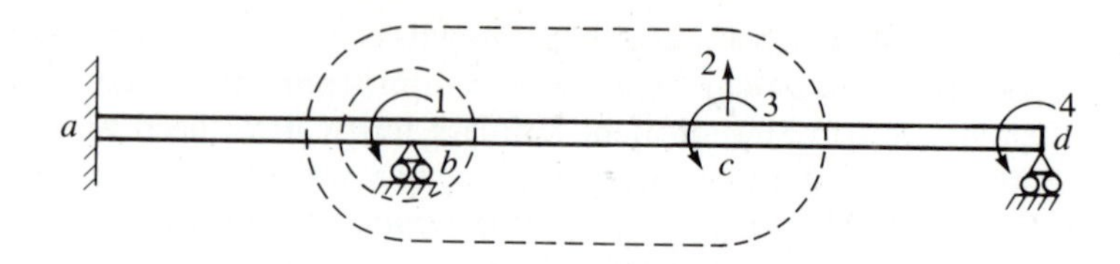

Figure A.5 Frontal solution concept with active degrees of freedom after reducing *ab* and *bc*

Eq. (A.32) is complete nonetheless. Therefore, solving Eq. (A.32) yields Θ_b, as shown in Eq. (A.6).

Thus the frontal solution is in principal Gauss elimination, wherein we successively eliminate degrees of freedom after assembling only those element stiffness matrices that contribute to a given equilibrium equation. Computational efficiency of the frontal solution depends upon the element numbering, which, in turn, determines the size of the wavefront (i.e., the half-bandwidth). By comparing the solution in Sec. A.1.4 with that in this section we observe that the same number of basic operations is involved in the active column and frontal solution methods provided the element and nodal point numbering correspond. Since nodal points need not be renumbered to minimize the bandwidth, the frontal solution is advantageous in situations where elements are added. We can envisage that if the wavefront is sufficiently large so that the active freedoms cannot be accommodated in high-speed storage, the inefficiencies of storing and retrieving information from low-speed storage will significantly diminish the power of the approach.

Most structural analysis programs use a direct method, stemming from Gauss elimination, to solve the equilibrium equations, but this has not always been the case. Previously, iterative methods were seriously studied and used as solution algorithms; this is not surprising when we recall that the moment distribution method involves solving the slope-deflection equations by Gauss-Seidel iteration. Solution times for iterative methods cannot be accurately predicted because the number of iterations required for convergence depends upon the condition number of the stiffness matrix.

A.2 SOLUTION ACCURACY

Numerical solutions are rarely exact; thus, results from matrix structural analysis are an approximation of actual behavior. The novice is tempted to quote displacements and stresses to as many decimal places as the computer displays. We know that dimensions, material properties, and applied forces input to the computer carry an implicit precision that is usually less than the number of digits in the computer output. Therefore, the designer must interpret the accuracy of computer results without being duped into believing every displayed digit. The manner in which we numerically describe a structure for analysis is an additional source of potential error. For example, beam elements must be defined and distributed loads represented as concentrated nodal loads; this task, known as modeling or idealization, can introduce errors if improperly done. In addition, the computer hardware and software establish limits on the solution accuracy. We will describe several sources of errors; the reader is urged to consult one of the many excellent texts on numerical analysis for a more detailed discussion.

A.2.1 NUMERICAL ACCURACY The computer output from the analysis of a truss displays the axial force in an element as +.181342E + 05 lb. We know that this floating point number was printed to conform to a Fortran format statement (e.g., E12.6). If engineers design the member for an axial tension of 18,134.2 lb, they believe that the number has a *precision* of 0.1 lb with an *accuracy* of 1/180,000. Since the applied forces are probably known with a precision of 10 lb (or possibly 100 lb), the designer should interpret the axial force to be 18,130 lb (or 18,100 lb). That is, working with the last two (or three) digits of the printed value of the axial force is not defensible. Results can be easily interpreted by relying on significant digits.

The numbers 142, 2.61×10^5, 0.00732, and 4.58 each have three *significant digits*. It is not clear how many significant digits there are in 92,300, but by writing the quantity as 9.23×10^4 we imply that it also has three significant digits. In division or multiplication, the result contains the same number of significant digits as the quantity with the least number of significant digits. Thus the product of 3.14159 and 0.736 is 2.31. The sum or difference of quantities has no more significant digits than the least precise number. That is, 253.1582 added to 34 is 287; it should not be written with four numbers after the decimal place.

If we input 3.141592654 (i.e., π) to a computer that uses six digits to represent a floating point number, it will *truncate* the quantity to six digits and use 3.14159 for the value. That is, digits seven and beyond are discarded. If we are asked to *round off* the original quantity to five digits, the value is 3.1416, whereas π becomes 3.14 when rounded off to three digits and 3.142 when rounded off to four digits. The rules for rounding off a number to n digits are: (a) if the $n + 1$ digit is less than 5 drop it; (b) if the $n + 1$ digit is greater than 5 increase the nth digit by one; and (c) if the $n + 1$ digit is exactly 5 the nth digit remains the same if it is even and the nth digit is increased by one if it is odd. Incidentally, many calculators round off by simply increasing the nth digit by one if the $n + 1$ digit is exactly 5.

Truncation and round-off can have significant effects on the results of matrix structural analysis solutions. Consider the assemblage shown in Fig. A.6, where the middle spring is much stiffer than either of the other two elements. The equilibrium equations for the constrained assemblage are

$$\begin{bmatrix} k_1 + k_2 & -k_2 \\ -k_2 & k_2 + k_3 \end{bmatrix} \begin{bmatrix} U_b \\ U_c \end{bmatrix} = \begin{bmatrix} P_b \\ P_c \end{bmatrix} \tag{A.33}$$

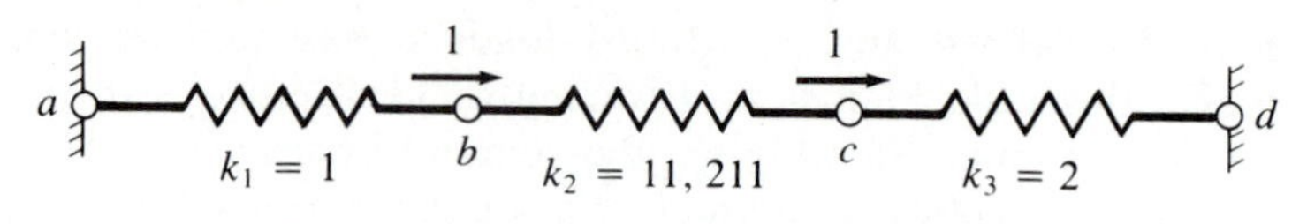

Figure A.6 Three-spring assemblage

If $k_1 = 1$, $k_2 = 11{,}211$, $k_3 = 2$, and $P_b = P_c = 1$, the equations become

$$\begin{bmatrix} 11212 & -11211 \\ -11211 & 11213 \end{bmatrix} \begin{bmatrix} U_b \\ U_c \end{bmatrix} = \begin{bmatrix} 1 \\ 1 \end{bmatrix} \tag{A.34}$$

Assume we are using a hypothetical computer that represents floating point numbers by truncating to four digits; therefore, the stored form of the equilibrium equations is

$$\begin{bmatrix} 11210 & -11210 \\ -11210 & 11210 \end{bmatrix} \begin{bmatrix} U_b \\ U_c \end{bmatrix} = \begin{bmatrix} 1 \\ 1 \end{bmatrix} \tag{A.35}$$

After the first step of Gauss elimination we have

$$\begin{bmatrix} 11210 & -11210 \\ 0 & 0 \end{bmatrix} \begin{bmatrix} U_b \\ U_c \end{bmatrix} = \begin{bmatrix} 1 \\ 2 \end{bmatrix} \tag{A.36}$$

Therefore, there is no unique solution because $d_{22} = 0$.

Assume that we will solve Eq. (A.34) by Gauss elimination with a computer that uses five significant digits to represent a floating point number, and that this hypothetical computer truncates after each addition, subtraction, multiplication, and division. Furthermore, the operational hierarchy of operations is multiplication/division followed by addition/subtraction. The first step of decomposition yields

$$\begin{aligned} \mathbf{L}_1^{-1}\mathbf{K} &= \begin{bmatrix} 1.0000 & 0.0000 \\ 0.99991 & 1.0000 \end{bmatrix} \begin{bmatrix} 11212. & -11211. \\ -11211. & 11213. \end{bmatrix} \\ &= \begin{bmatrix} 11212. & -11211. \\ 0.0000 & 4.0000 \end{bmatrix} \end{aligned} \tag{A.37}$$

We assume that the computer algorithm sets off-diagonal entries in the decomposed matrix to zero because performing the arithmetic with truncation will not yield an upper triangular matrix. That is, (0.99991)(11212.) = 11210.99092, which becomes 11210.9 when truncated to six significant digits. Thus

$$\mathbf{L} = \begin{bmatrix} 1.0000 & 0.0000 \\ -0.99991 & 1.0000 \end{bmatrix} \tag{A.38}$$

and

$$\mathbf{D} = \begin{bmatrix} 11212. & 0.0000 \\ 0.0000 & 4.0000 \end{bmatrix} \tag{A.39}$$

Table A.1 Solutions for three-spring assemblage of Fig. A.6

		Number of significant digits used					
	Exact	*Four*	*Five*	*Five + R/O*	*Six*	*Six + R/O*	*Seven*
U_b	0.666686	No solution	0.50002	0.66666	0.645162	0.665869	0.6666667
U_c	0.666657	No solution	0.49999	0.66663	0.645132	0.666637	0.6666370

Therefore

$$\mathbf{Y} = \mathbf{L}^{-1}\mathbf{P} = \begin{bmatrix} 1.0000 \\ 1.9999 \end{bmatrix} \tag{A.40}$$

and

$$\mathbf{D}^{-1}\mathbf{Y} = \begin{bmatrix} 0.000089190 \\ 0.49999 \end{bmatrix} \tag{A.41}$$

Solving $\mathbf{L}^T\mathbf{U} = \mathbf{D}^{-1}\mathbf{Y}$ yields

$$\begin{bmatrix} U_b \\ U_c \end{bmatrix} = \begin{bmatrix} 0.50002 \\ 0.49999 \end{bmatrix} \tag{A.42}$$

If our hypothetical computer performs arithmetic operations by rounding off the fifth significant digit and then truncating, we observe the following: **L** remains unchanged; D_{22} becomes 3.0000, while other entries in **D** are unaltered;

$$\mathbf{D}^{-1}\mathbf{Y} = \begin{bmatrix} 0.000089190 \\ 0.66663 \end{bmatrix} \tag{A.43}$$

and $\mathbf{L}^T\mathbf{U} = \mathbf{D}^{-1}\mathbf{Y}$ yields

$$\begin{bmatrix} U_b \\ U_c \end{bmatrix} = \begin{bmatrix} 0.66666 \\ 0.66663 \end{bmatrix} \tag{A.44}$$

The displacements for the assemblage of Fig. A.6 using various calculation strategies are summarized in Table A.1.

B MATRICES

B.1 DEFINITIONS

A matrix is a rectangular array of elements arranged in rows and columns as follows:

$$\mathbf{A} = [a_{ij}] = \begin{bmatrix} a_{11} & a_{12} & \cdots & a_{1n} \\ a_{21} & a_{22} & \cdots & a_{2n} \\ \vdots & \vdots & \vdots\vdots & \\ a_{m1} & a_{m2} & \cdots & a_{mn} \end{bmatrix} \tag{B.1}$$

where a_{ij} is the element in the ith row and jth column. The above matrix is *order* (i.e., size) $m \times n$ (i.e., m by n). The row index is always given first. If $m = n$, $\mathbf{A}$ is a square matrix of order n; the elements $a_{11}, a_{22}, \ldots, a_{nn}$ lie on and form the *main* (or principal) *diagonal.* If $n = 1$, $\mathbf{A}$ is a *column matrix* of order $m \times 1$, and if $m = 1$, $\mathbf{A}$ is a *row matrix* of order $1 \times n$.

B.2 EQUALITY

Two matrices $\mathbf{A} = [a_{ij}]$ and $\mathbf{B} = [b_{ij}]$ are *equal* if they have the same order and $a_{ij} = b_{ij}$; that is,

$$\begin{bmatrix} x + 2y \\ 3x + 4y \end{bmatrix} = \begin{bmatrix} 2 \\ 8 \end{bmatrix}$$

implies that $x + 2y = 2$ and $3x + 4y = 8$.

B.3 ADDITION AND SUBTRACTION

The *sum* of two matrices $\mathbf{A} = [a_{ij}]$ and $\mathbf{B} = [b_{ij}]$ is

$$\mathbf{A} + \mathbf{B} = [a_{ij} + b_{ij}] = [c_{ij}] = \mathbf{C} \tag{B.2}$$

where it is implied that $\mathbf{A}$ and $\mathbf{B}$ are the same order. Thus,

$$\begin{bmatrix} 1 & 8 \\ 3 & 4 \\ 7 & 2 \end{bmatrix} + \begin{bmatrix} 6 & 9 \\ 8 & 5 \\ 12 & 3 \end{bmatrix} = \begin{bmatrix} 7 & 17 \\ 11 & 9 \\ 19 & 5 \end{bmatrix}$$

In general, matrix addition has the following properties:

$$\mathbf{A} + \mathbf{B} = \mathbf{B} + \mathbf{A} \qquad \text{(commutative law)} \tag{B.3a}$$

$$\mathbf{A} + (\mathbf{B} + \mathbf{C}) = (\mathbf{A} + \mathbf{B}) + \mathbf{C} \qquad \text{(associative law)} \tag{B.3b}$$

The *null* (or zero) *matrix* **0**, consisting of only zero elements, implies that

$$\mathbf{A} + \mathbf{0} = \mathbf{A} \qquad \text{and} \qquad \mathbf{A} + (-\mathbf{A}) = \mathbf{0} \tag{B.4}$$

where $-\mathbf{A} = [-a_{ij}]$. Note that $-\mathbf{A} = [-a_{ij}] = (-1)[a_{ij}]$, where we multiply each element of the matrix by the scalar -1. In general multiplication of a matrix by a scalar implies that

$$k\mathbf{A} = k[a_{ij}] = [ka_{ij}] \tag{B.5}$$

Thus matrix *subtraction* is similar to matrix addition since

$$\mathbf{A} - \mathbf{B} = \mathbf{A} + (-1)\mathbf{B} = \mathbf{A} + -\mathbf{B} \tag{B.6}$$

For example,

$$\begin{bmatrix} 1 & 8 \\ 3 & 4 \\ 7 & 2 \end{bmatrix} - \begin{bmatrix} 6 & 9 \\ 8 & 5 \\ 12 & 3 \end{bmatrix} = \begin{bmatrix} -5 & -1 \\ -5 & -1 \\ -5 & -1 \end{bmatrix}$$

B.4 MULTIPLICATION

If **A** is a matrix of order $m \times n$ and **B** a matrix of order $n \times p$, then the matrix product of **A** and **B** is a matrix of order $m \times p$, that is,

$$\underset{(m \times n)}{\mathbf{A}} \; \underset{(n \times p)}{\mathbf{B}} = \underset{(m \times p)}{\mathbf{C}} = [c_{ij}] = \left[\sum_{r=1}^{n} a_{ir} b_{rj} \right] \tag{B.7}$$

where $i = 1, 2, \ldots, m$; $j = 1, 2, \ldots, p$. Note that for matrices **A** and **B** to conform for multiplication in **AB**, the number of columns in **A** must equal the number of rows of **B**. When the product is written **AB**, **A** premultiplies **B**, and **B** postmultiplies **A**. For example, if

$$\mathbf{A} = \begin{bmatrix} 1 & 3 & 5 \\ 2 & 4 & 6 \end{bmatrix} \qquad \text{and} \qquad \mathbf{B} = \begin{bmatrix} -7 & 10 \\ 8 & -11 \\ -9 & 0 \end{bmatrix}$$

then

$$\begin{aligned} \mathbf{AB} &= \begin{bmatrix} 1 & 3 & 5 \\ 2 & 4 & 6 \end{bmatrix} \begin{bmatrix} -7 & 10 \\ 8 & -11 \\ -9 & 0 \end{bmatrix} \\ &= \begin{bmatrix} 1(-7) + 3(8) + 5(-9) & 1(10) + 3(-11) + 5(0) \\ 2(-7) + 4(8) + 6(-9) & 2(10) + 4(-11) + 6(0) \end{bmatrix} \\ &= \begin{bmatrix} -28 & -23 \\ -36 & -24 \end{bmatrix} \end{aligned}$$

Note that **B** and **A** are conformable for the product **BA** since **B** has two columns and **A** has two rows. The product is of order 3 × 3; that is,

$$\mathbf{BA} = \begin{bmatrix} -7 & 10 \\ 8 & -11 \\ -9 & 0 \end{bmatrix} \begin{bmatrix} 1 & 3 & 5 \\ 2 & 4 & 6 \end{bmatrix}$$

$$= \begin{bmatrix} (-7)1 + 10(2) & (-7)3 + 10(4) & (-7)5 + 10(6) \\ 8(1) + (-11)2 & 8(3) + (-11)4 & 8(5) + (-11)6 \\ (-9)1 + 0(2) & (-9)3 + 0(4) & (-9)5 + 0(6) \end{bmatrix}$$

$$= \begin{bmatrix} 13 & 19 & 25 \\ -14 & -20 & -26 \\ -9 & -27 & -45 \end{bmatrix}$$

In this case $\mathbf{AB} \neq \mathbf{BA}$; that is, matrix multiplication is not commutative. While matrices in general do not commute, there are some matrices that do commute. We urge the reader to demonstrate the unusual situation in which $\mathbf{AB} = \mathbf{BA}$ if

$$\mathbf{A} = \begin{bmatrix} 3 & 0 \\ 1 & 4 \end{bmatrix} \quad \text{and} \quad \mathbf{B} = \begin{bmatrix} 1 & 0 \\ 1 & 2 \end{bmatrix}$$

The following three fundamental properties of scalar multiplication that are not applicable in matrix algebra are:

$$\mathbf{AB} = \mathbf{BA} \text{ is not true in general} \tag{B.8a}$$

$$\mathbf{AB} = \mathbf{0} \text{ does not imply that either } \mathbf{A} \text{ or } \mathbf{B} \text{ are null matrices} \tag{B.8b}$$

$$\mathbf{AB} = \mathbf{AC} \text{ does not imply that } \mathbf{B} = \mathbf{C} \text{ even if } \mathbf{A} \neq \mathbf{0} \tag{B.8c}$$

In general, matrix multiplication has the following properties:

$$\mathbf{A}(\mathbf{BC}) = (\mathbf{AB})\mathbf{C} \quad \text{(associative law)} \tag{B.9a}$$

$$\mathbf{A}(\mathbf{B} + \mathbf{C}) = \mathbf{AB} + \mathbf{AC} \quad \text{(left distributive law)} \tag{B.9b}$$

$$(\mathbf{B} + \mathbf{C})\mathbf{A} = \mathbf{BA} + \mathbf{CA} \quad \text{(right distributive law)} \tag{B.9c}$$

The unit matrix of order n is defined as follows:

$$\mathbf{I}_n = \mathbf{I} = [\delta_{ij}] \tag{B.10}$$

where $\delta_{ij} = 1$ for $i = j$ and $\delta_{ij} = 0$ for $i \neq j$. Thus

$$\mathbf{AI} = \mathbf{A} \quad \text{and} \quad \mathbf{IB} = \mathbf{B} \tag{B.11}$$

for every matrix **A** of order $m \times n$ and every matrix **B** of order $n \times r$. The matrix **I** commutes with any square matrix of the same order.

B.5 TRANSPOSE

We obtain the transpose of a matrix **A**, denoted by $\mathbf{A}^T$, by changing all the rows of **A** into the columns of $\mathbf{A}^T$. That is, the first row of **A** becomes the first column of $\mathbf{A}^T$, the second row of **A** becomes the second column of $\mathbf{A}^T$, and the last row of **A** becomes the last column of $\mathbf{A}^T$. Thus, if

$$\mathbf{A} = \begin{bmatrix} 1 & 3 & 5 \\ 2 & 4 & 6 \end{bmatrix} \quad \text{then} \quad \mathbf{A}^T = \begin{bmatrix} 1 & 2 \\ 3 & 4 \\ 5 & 6 \end{bmatrix}$$

Therefore, if $\mathbf{A} = [a_{ij}]$ is an $n \times p$ matrix, then the transpose of **A**, denoted by $\mathbf{A}^T = [a_{ij}^T]$ is a $p \times n$ matrix with $a_{ij}^T = a_{ji}$. The transpose possesses the following properties:

$$(\mathbf{A}^T)^T = \mathbf{A} \tag{B.12a}$$

$$(k\mathbf{A})^T = k\mathbf{A}^T \qquad \text{where } k \text{ is a scalar} \tag{B.12b}$$

$$(\mathbf{A} + \mathbf{B})^T = \mathbf{A}^T + \mathbf{B}^T \tag{B.12c}$$

$$(\mathbf{AB})^T = \mathbf{B}^T\mathbf{A}^T \tag{B.12d}$$

$$(\mathbf{ABC})^T = \mathbf{C}^T\mathbf{B}^T\mathbf{A}^T \tag{B.12e}$$

B.6 SPECIAL MATRICES

A matrix **A** is *symmetric* if $\mathbf{A} = \mathbf{A}^T$; it is *skew symmetric* if $\mathbf{A} = -\mathbf{A}^T$. Note that all diagonal elements of a skew symmetric matrix must be zero.

A matrix **A** is *lower triangular* if $a_{ij} = 0$ for $j > i$ (i.e., all elements above the main diagonal are zero); it is *upper triangular* if $a_{ij} = 0$ for $i > j$ (i.e., all elements below the main diagonal are zero).

B.7 SUBMATRICES AND PARTITIONING

A *submatrix* of matrix **A** is a matrix obtained by removing any number of rows or columns from **A**. For example, if

$$\mathbf{A} = \begin{bmatrix} 16 & 15 & 14 & 13 \\ 12 & 11 & 10 & 9 \\ 8 & 7 & 6 & 5 \\ 4 & 3 & 2 & 1 \end{bmatrix}$$

we create the submatrix $\mathbf{C} = [16]$ by removing a_{11}. Similarly,

$$\mathbf{D} = [15 \quad 14 \quad 13]; \qquad \mathbf{E} = \begin{bmatrix} 12 \\ 8 \\ 4 \end{bmatrix}; \qquad \text{and} \qquad \mathbf{F} = \begin{bmatrix} 11 & 10 & 9 \\ 7 & 6 & 5 \\ 3 & 2 & 1 \end{bmatrix}$$

are submatrices created by removing the appropriate rows and columns from **A**.

A matrix is *partitioned* if it is divided into submatrices. We can partition a matrix in many different ways. For example, referring to **A** and its submatrices **C**, **D**, **E**, and **F**, we can write

$$\mathbf{A} = \left[\begin{array}{c|c} \mathbf{C} & \mathbf{D} \\ \hline \mathbf{E} & \mathbf{F} \end{array}\right]$$

where **C**, **D**, **E**, and **F** are of order 1×1, 1×3, 3×1, and 3×3, respectively.

Matrix partitioning can be very useful in some situations (e.g., for mixed boundary conditions); operations with partitioned matrices are straightforward. For example, consider matrix **A** above and

$$\mathbf{B} = \left[\begin{array}{c|c} \mathbf{G} & \mathbf{H} \\ \hline \mathbf{J} & \mathbf{K} \end{array}\right]$$

where **G**, **H**, **J**, and **K** are submatrices. We obtain the product **AB** by regarding the submatrices as elements and observing the rules of matrix multiplication, that is,

$$\mathbf{AB} = \left[\begin{array}{c|c} \mathbf{CG} + \mathbf{DJ} & \mathbf{CH} + \mathbf{DK} \\ \hline \mathbf{EG} + \mathbf{FJ} & \mathbf{EH} + \mathbf{FK} \end{array}\right]$$

providing the various partitions are conformable for multiplication.

B.8 DETERMINANTS

Every square matrix has associated with it a scalar called its *determinant*. The determinant of a square matrix $\mathbf{A} = [a_{ij}]$ of order n is defined by

$$|\mathbf{A}| = \begin{vmatrix} a_{11} & a_{12} & \cdots & a_{1n} \\ a_{21} & a_{22} & \cdots & a_{2n} \\ \vdots & \vdots & \vdots\vdots & \\ a_{m1} & a_{m2} & \cdots & a_{mn} \end{vmatrix}$$

$$= \sum \varepsilon_{ij \cdots k} a_{i1} a_{j2} \cdots a_{kn} \tag{B.13}$$

where we take the summation over all values $i, j, \ldots, k = 1, 2, \ldots, n$, and $\varepsilon_{ij \cdots k}$ is the alternating symbol defined as follows:

$$\varepsilon_{ij \cdots k} = \begin{cases} +1, & \text{if } (i, j, \ldots, k) \text{ is an even permutation of } (1, 2, \ldots, n) \\ -1, & \text{if } (i, j, \ldots, k) \text{ is an odd permutation of } (1, 2, \ldots, n) \\ 0, & \text{otherwise} \end{cases} \tag{B.14}$$

While the definition shown in Eq. (B.13) is rigorously correct, in practice it is difficult to use this approach; therefore we will describe another procedure that yields the determinant in a more straightforward manner.

For a square matrix $\mathbf{A} = [a_{ij}]$, the *cofactor* of the element a_{ij} (denoted as A_{ij}) is a scalar obtained by multiplying together the term $(-1)^{i+j}$ and the minor obtained from $\mathbf{A}$ after removing the ith row and jth column. The minor of a square matrix $\mathbf{A}$ is the determinant of any square submatrix of $\mathbf{A}$; that is, if

$$\mathbf{A} = \begin{bmatrix} 1 & 4 & 7 \\ 2 & 5 & 8 \\ 3 & 6 & 9 \end{bmatrix}$$

then

$$\begin{vmatrix} 4 & 7 \\ 6 & 9 \end{vmatrix} \quad \text{and} \quad \begin{vmatrix} 2 & 5 \\ 3 & 6 \end{vmatrix}$$

are both minors since

$$\begin{bmatrix} 4 & 7 \\ 6 & 9 \end{bmatrix} \quad \text{and} \quad \begin{bmatrix} 2 & 5 \\ 3 & 6 \end{bmatrix}$$

are both square submatrices of $\mathbf{A}$. Thus, to obtain the cofactor of the element 4 [i.e., element (1, 2)] in the matrix

$$\mathbf{A} = \begin{bmatrix} 1 & 4 & 7 \\ 2 & 5 & 8 \\ 3 & 6 & 9 \end{bmatrix}$$

we cross out the first row and second column and use the remaining submatrix. Thus,

$$A_{12} = (-1)^{1+2}\begin{vmatrix} 2 & 8 \\ 3 & 9 \end{vmatrix} = (-1)\{2(9) - 3(8)\} = 6$$

To find the determinant of a matrix $\mathbf{A}$ of arbitrary order: (a) select any single row (or column) of the matrix; (b) for each element in the row (or column) selected, find its cofactor; and (c) multiply each element in the row (or column) selected by its cofactor and sum the results. This sum is the determinant of the matrix. We suggest that the reader use expansion by cofactors to demonstrate that the determinant of the following matrix is 2:

$$\mathbf{E} = \begin{bmatrix} 2 & 0 & 1 \\ 1 & 1 & 0 \\ 1 & 1 & 1 \end{bmatrix}$$

The following are useful properties of determinants:

1. If one row of a matrix consists entirely of zeros, then the determinant is zero.
2. If two rows of a matrix are interchanged, the determinant changes sign.

3. If two rows of a determinant are identical, the determinant is zero.
4. If the matrix **B** is obtained from the matrix **A** by multiplying every element in one row of **A** by the scalar k, then $|\mathbf{B}| = k|\mathbf{A}|$.
5. For an $n \times n$ matrix **A** and any scalar k, $|k\mathbf{A}| = k^n|\mathbf{A}|$.
6. If a matrix **B** is obtained from a matrix **A** by adding to one row of **A** a scalar times another row of **A**, then $|\mathbf{A}| = |\mathbf{B}|$.

B.9 SIMULTANEOUS LINEAR ALGEBRAIC EQUATIONS

Consider the following set of simultaneous linear algebraic equations:

$$\begin{aligned} x_1 + 3x_2 + 2x_3 &= 1 \\ 2x_1 + 8x_2 + 6x_3 &= 4 \\ 4x_1 + 4x_2 + 2x_3 &= 2 \end{aligned} \tag{B.15}$$

Using the rules of matrix multiplication we can write these equations as follows:

$$\underset{\mathbf{A}}{\begin{bmatrix} 1 & 3 & 2 \\ 2 & 8 & 6 \\ 4 & 4 & 2 \end{bmatrix}} \underset{\mathbf{X}}{\begin{bmatrix} x_1 \\ x_2 \\ x_3 \end{bmatrix}} = \underset{\mathbf{B}}{\begin{bmatrix} 1 \\ 4 \\ 2 \end{bmatrix}} \tag{B.16}$$

The inverse of an $n \times n$ matrix **A** is an $n \times n$ matrix, denoted by $\mathbf{A}^{-1}$, with the property

$$\mathbf{A}\mathbf{A}^{-1} = \mathbf{A}^{-1}\mathbf{A} = \mathbf{I} \tag{B.17}$$

If a square matrix **A** has an inverse, it is *nonsingular* or invertible. If **A** has no inverse, it is *singular*. Note that an inverse is only defined for a square matrix. Therefore, premultiplying both sides of Eq. (B.16) by $\mathbf{A}^{-1}$ yields

$$\mathbf{A}^{-1}\mathbf{A}\mathbf{X} = \mathbf{A}^{-1}\mathbf{B}$$

and substituting from Eq. (B.17) we have

$$\mathbf{X} = \mathbf{A}^{-1}\mathbf{B} \tag{B.18}$$

The *cofactor matrix* associated with an $n \times n$ matrix **A** is an $n \times n$ matrix $\mathbf{A}^c$ obtained from **A** by replacing each element of **A** by its cofactor. For example, $\mathbf{A}^c$ for the coefficient matrix of Eq. (B.16) is

$$\mathbf{A}^c = \begin{bmatrix} -8 & 20 & -24 \\ 2 & -6 & 8 \\ 2 & -2 & 2 \end{bmatrix}$$

The *adjoint* of an $n \times n$ matrix A is the transpose of the cofactor matrix of **A**. Designating the adjoint of **A** by $\mathbf{A}^a$, we have $\mathbf{A}^a = (\mathbf{A}^c)^T$. Thus for the coefficient matrix of Eq. (B.16),

$$\mathbf{A}^a = \begin{bmatrix} -8 & 2 & 2 \\ 20 & -6 & -2 \\ -24 & 8 & 2 \end{bmatrix}$$

The inverse of an $n \times n$ matrix $\mathbf{A}$ is obtained as follows:

$$\mathbf{A}^{-1} = \frac{1}{|\mathbf{A}|}\mathbf{A}^a \qquad \text{if} \qquad |\mathbf{A}| \neq 0 \tag{B.19}$$

The determinant of the coefficient matrix of Eq. (B.16) is 4; therefore the inverse is

$$\mathbf{A}^{-1} = \begin{bmatrix} -2.0 & 0.5 & 0.5 \\ 5.0 & -3.5 & -0.5 \\ -6.0 & 2.0 & 0.5 \end{bmatrix}$$

We urge the reader to verify that $\mathbf{A}^{-1}\mathbf{A} = \mathbf{I}$ for this example. Consequently, the solution of Eq. (B.16) is

$$\begin{bmatrix} x_1 \\ x_2 \\ x_3 \end{bmatrix} = \begin{bmatrix} -2.0 & 0.5 & 0.5 \\ 5.0 & -3.5 & -0.5 \\ -6.0 & 2.0 & 0.5 \end{bmatrix}\begin{bmatrix} 1 \\ 4 \\ 2 \end{bmatrix} = \begin{bmatrix} 1 \\ -10 \\ 3 \end{bmatrix}$$

The inverse possesses the following properties:

$$(\mathbf{A}^{-1})^{-1} = \mathbf{A} \tag{B.20a}$$

$$(\mathbf{AB})^{-1} = \mathbf{B}^{-1}\mathbf{A}^{-1} \tag{B.20b}$$

$$(\mathbf{ABC})^{-1} = \mathbf{C}^{-1}\mathbf{B}^{-1}\mathbf{A}^{-1} \tag{B.20c}$$

$$(\mathbf{A}^T)^{-1} = (\mathbf{A}^{-1})^T \tag{B.20d}$$

The *rank* of a matrix $\mathbf{A}$, designated $r(\mathbf{A})$, is the order of the largest nonzero minor of $\mathbf{A}$. For example, if

$$\mathbf{A} = \begin{bmatrix} 12 & 11 & 10 & 9 \\ 8 & 7 & 6 & 5 \\ 4 & 3 & 2 & 1 \end{bmatrix}$$

all possible minors of order 3 are

$$\begin{vmatrix} 12 & 11 & 10 \\ 8 & 7 & 6 \\ 4 & 3 & 2 \end{vmatrix}; \qquad \begin{vmatrix} 11 & 10 & 9 \\ 7 & 6 & 5 \\ 3 & 2 & 1 \end{vmatrix}; \qquad \begin{vmatrix} 12 & 10 & 9 \\ 8 & 6 & 5 \\ 4 & 2 & 1 \end{vmatrix}; \qquad \begin{vmatrix} 12 & 11 & 9 \\ 8 & 7 & 5 \\ 4 & 3 & 1 \end{vmatrix}$$

Evaluating these minors reveals that they are all zero; therfore $\mathbf{A}$ is not of rank 3. However, $\mathbf{A}$ is of rank 2 because it has at least one nonzero minor of order 2 (actually all minors of order 2 are nonzero). For example, examining the upper left 2×2 submatrix gives

$$\begin{vmatrix} 12 & 11 \\ 8 & 7 \end{vmatrix} = (12)(7) - (8)(11) = 84 - 88 = -4$$

B.10 DIFFERENTIATION

The operation $d\mathbf{A}/dx$ indicates that we must differentiate the matrix $\mathbf{A}$ element by element with respect to x. For example, if

$$\mathbf{A} = \begin{bmatrix} 2x^3 & 4x^2 & 5x \\ \sin x & \cos^2 x & 2 \sin x \\ 8x & 3x^2 & x^5 \end{bmatrix}$$

therefore,

$$\frac{dA}{dx} = \begin{bmatrix} 6x^2 & 8x & 5 \\ \cos x & -2 \cos x \sin x & 2 \cos x \\ 8 & 6x & 5x^4 \end{bmatrix}$$

Partial differentiation is also performed by operating upon each element.

B.11 INTEGRATION

We integrate a matrix by operating upon each element. Thus if

$$\mathbf{A} = \begin{bmatrix} 2x^3 & 4x^2 & 5x \\ \sin x & \cos^2 x & 2 \sin x \\ 8x & 3x^2 & x^5 \end{bmatrix}$$

therefore,

$$\int_0^L \mathbf{A}\, dx = \begin{bmatrix} \dfrac{x^2}{4} & \dfrac{4x^3}{3} & \dfrac{5x^2}{2} \\ -\cos x & \dfrac{1}{4}(2x + \sin 2x) & -2 \cos x \\ 4x^2 & x^3 & \dfrac{x^6}{6} \end{bmatrix}_0^L$$

$$= \begin{bmatrix} \dfrac{L^2}{4} & \dfrac{4L^3}{3} & \dfrac{5L^2}{2} \\ 1 - \cos L & \dfrac{1}{4}(2L + \sin Lx) & 2(1 - \cos L) \\ 4L^2 & L^3 & \dfrac{x^6}{6} \end{bmatrix}$$

B.12 BILINEAR AND QUADRATIC FORMS

We define a *bilinear form* in the $m + n$ variables $x_1, x_2, \ldots, x_m$ and $y_1, y_2, \ldots, y_n$ to be a scalar function of the type

$$\mathbf{X}^T\mathbf{A}\mathbf{X} = \sum_{i=1}^{m} \sum_{j=1}^{n} a_{ij} x_i x_j \tag{B.21}$$

where $\mathbf{A}$ is an $m \times n$ matrix, and $\mathbf{X}$ and $\mathbf{Y}$ are column matrices of order $m \times 1$ and $n \times 1$, respectively. The rank of $\mathbf{A}$ is the rank of the form. When $\mathbf{A}$ is symmetric, the form itself is called a *symmetric bilinear form.* When $\mathbf{X}$ and $\mathbf{Y}$ of a bilinear form $\mathbf{X}^T\mathbf{A}\mathbf{Y}$, in which $\mathbf{A}$ is square and order n, are subjected to transformations $\mathbf{X} = (\mathbf{T}^{-1})^T\mathbf{U}$ and $\mathbf{Y} = \mathbf{T}\mathbf{V}$, the bilinear form is said to be transformed *contragrediently.*

A *quadratic form* in the n variables $x_1, x_2, \ldots, x_n$ is a scalar function of the form

$$\mathbf{X}^T\mathbf{A}\mathbf{X} = \sum_{i=1}^{m} \sum_{j=1}^{n} a_{ij} x_j y_j \tag{B.22}$$

C GENERAL REFERENCES

Bergan, P. G., G. Horrigmoe, and T. G. Syvertsen, *Matrisestatikk*, 2nd ed., Tapir, Trondheim, Norway, 1982.

Davies, G. A. O., *Virtual Work in Structural Analysis*, Wiley Interscience, New York, 1982.

Fleming, J. F., *Structural Engineering Analysis on Personal Computers*, McGraw-Hill, New York, 1986.

Harrison, H. B., *Structural Analysis and Design*, Parts 1 and 2, Pergamon, New York, 1980.

Holzer, S. M., *Computer Analysis of Structures*, Elsevier, New York, 1985.

Huston, R. L., and C. E. Passerello, *Finite Element Methods*, Marcel Dekker, Inc., New York, 1984.

Kanchi, M. B., *Matrix Methods of Structural Analysis*, Halstad (Wiley), New York, 1981.

Kardestuncer, H., *Elementary Matrix Analysis of Structures*, McGraw-Hill, New York, 1974.

Livesley, R. K., *Matrix Methods of Structural Analysis*, 2nd ed., Pergamon, New York, 1975.

McGuire, W., and R. H. Gallagher, *Matrix Structural Analysis*, Wiley, New York, 1979.

Martin, H. C., *Introduction to Matrix Methods of Structural Analysis*, McGraw-Hill, New York, 1966.

Meek, J. L., *Matrix Structural Analysis*, McGraw-Hill, New York, 1971.

Meyers, V. J., *Matrix Analysis of Structures*, Harper and Row, New York, 1983.

Rubinstein, M. F., *Matrix Computer Analysis of Structures*, Prentice-Hall, Englewood Cliffs, N.J., 1966.

Smith, T. R. G., *Linear Analysis of Frameworks*, Ellis Horwood Ltd., Chichester, U.K., 1983.

Vanderbilt, M. D., *Matrix Structural Analysis*, Quantum, New York, 1974.

Wang, C. K., *Matrix Methods of Structural Analysis*, 2nd ed., Intext, Scranton, Pa., 1970.

Wang, C. K., *Structural Analysis on Microcomputers*, Macmillan, New York, 1986.

Weaver, W., Jr., *Computer Programs for Structural Analysis*, Van Nostrand, Princeton, N.J., 1967.

Weaver, W., Jr., and J. M. Gere, *Matrix Analysis of Framed Structures*, 2nd ed., Van Nostrand, New York, 1980.

D SYSTEMS OF MEASUREMENT

Although the structural engineer in the United States must work with products, documents, institutes, and industries that use U.S. Customary System (USCS) units almost exclusively, the civil engineering profession is working toward the official adoption and use of the System International (SI) units. It therefore behooves the student to develop a facility for working with both systems. Tables D.1 through D.3 provide a ready reference for the SI units used in this book. More complete information on the entire SI system is available in many references.

Table D.1 Selected SI units

Quantity	*Unit*	*SI symbol*
Base units:		
Length	meter	m
Mass	kilogram	kg
Time	second	s
Derived units:		
Area	meter^2	m^2
Density, mass	kilogram/meter^3	kg/m^3
Force	newton	N(=kg · m/s^2)
Moment of force	newton-meter	N · m
Moment of inertia, area	meter^4	m^4
Pressure, stress	pascal	Pa(=N/m^2)
Temperature	Celsius	°C
Work, energy	joule	J(=N · m)
Supplementary units:		
Plane angle	radian	rad

Table D.2 SI unit prefixes

Prefix	*Symbol*	*Multiplication factor*	*Prefix*	*Symbol*	*Multiplication factor*
exa	E	10^{18}	deci[a]	d	10^{-1}
peta	P	10^{15}	centi[a]	c	10^{-2}
tera	T	10^{12}	milli	m	10^{-3}
giga	G	10^{9}	micro	μ	10^{-6}
mega	M	10^{6}	nano	n	10^{-9}
kilo	k	10^{3}	pico	p	10^{-12}
hecto[a]	h	10^{2}	femto	f	10^{-15}
deka[a]	da	10^{1}	atto	a	10^{-18}

[a] Use should generally be avoided.

Table D.3 Some conversion factors

Length	
1 in. = 25.40 mm	1 mm = 0.03937 in.
= 0.0254 m	1 m = 39.37 in.
1 ft. = 304.8 mm	1 mm = 0.003281 ft
= 0.3048 m	1 m = 3.281 ft
Area	
1 in.^{2} = 6.452×10^2 mm^2	1 mm^2 = 1.550×10^{-3} in.^{2}
= 6.452×10^{-4} m^2	1 m^2 = 1.550×10^3 in.^{2}
1 ft^2 = 9.290×10^4 mm^2	1 mm^2 = 1.076×10^{-5} ft^2
= 9.290×10^{-2} m^2	1 m^2 = 1.076×10^1 ft^2
Moment of inertia	
1 in.^{4} = 4.162×10^5 mm^4	1 mm^4 = 2.402×10^{-6} in.^{4}
= 4.162×10^{-7} m^4	1 m^4 = 2.402×10^6 in.^{4}
Force and force per unit length	
1 lbf = 4.448 N	1 N = 0.2248 lbf
1 kip = 4.448×10^3 N	1 kN = 2.248×10^2 lbf
1 lbf/ft = 14.59 N/m	1 N/m = 6.853×10^{-2} lbf/ft
1 kip/ft = 14.59×10^3 N/m	1 kN/m = 6.853×10^1 lbf/ft
Bending moment	
1 lbf · in. = 0.1130 N · m	1 N · m = 8.851 lbf · in.
1 lbf · ft = 1.3558 N · m	1 N · m = 0.7376 lbf · ft
1 kip · ft = 1.3558×10^3 N · m	1 kN · m = 0.7376×10^3 lbf · ft
Stress	
1 lbf/in.^{2} (psi) = 6.895×10^3 Pa	1 Pa = 0.1450×10^{-3} psi
1 kip/in.^{2} (ksi) = 6.895×10^6 Pa	1 kPa = 0.1450 psi
1 lbf/ft^2 (psi) = 47.88 Pa	1 Pa = 2.089×10^{-2} psf
Temperature	
$t_{°F} = (9/5)t_{°C} + 32$	$t_{°C} = (5/9)(t_{°F} - 32)$

E ANSWERS TO SELECTED PROBLEMS

Chapter 2

2.2 **(a)** $V_b = -10.0$ mm; **(b)** $P_{yb} = 216$ kN

2.3 **(a)** $V_a = 0.40$ in.; **(b)** $P_{yb} = 165$ kN

2.4 **(a)** and **(b)**

$$\mathbf{K} = 3 \times 10^3 \begin{bmatrix} 16 & -16 & 0 & 0 \\ -16 & 24 & -8 & 0 \\ 0 & -8 & 29 & -21 \\ 0 & 0 & -21 & 21 \end{bmatrix} \text{kN/m}$$

(c)

$$\begin{bmatrix} U_b \\ U_c \\ U_d \end{bmatrix} = \begin{bmatrix} 12 \\ 28 \\ 31 \end{bmatrix} \text{mm}$$

$$P_{xa} = -576 \text{ kN}$$

(d)

$$\begin{bmatrix} U_c \\ U_b \end{bmatrix} = \begin{bmatrix} 32 \\ 37 \end{bmatrix} \text{mm}$$

$$\begin{bmatrix} P_{xa} \\ P_{xd} \end{bmatrix} = \begin{bmatrix} -177.6 \\ -201.6 \end{bmatrix} \text{kN}$$

(e) $U_c = 8.8$ mm

$$\begin{bmatrix} P_{xa} \\ P_{xb} \\ P_{xd} \end{bmatrix} = \begin{bmatrix} -1152 \\ 1517 \\ -554 \end{bmatrix} \text{kN}$$

2.7

$$\mathbf{K} = 2 \times 10^4 \begin{bmatrix} 0.36 & 0.48 & 0.00 & 0.00 & -0.36 & -0.48 \\ 0.48 & 1.64 & 0.00 & -1.00 & -0.48 & -0.64 \\ 0.00 & 0.00 & 1.00 & 0.00 & -1.00 & 0.00 \\ 0.00 & -1.00 & 0.00 & 1.00 & 0.00 & 0.00 \\ -0.36 & -0.48 & -1.00 & 0.00 & 1.36 & 0.48 \\ -0.48 & -0.64 & 0.00 & 0.00 & 0.48 & 0.64 \end{bmatrix} \text{kN/m}$$

2.8 **(a)** Equilibrium in x and y directions is satisfied.

(b) The forces P_{xa}, P_{ya}, P_{xb}, P_{yb}, P_{xc}, P_{yc} due to a $V_b = 1$, while all other nodal displacements are zero.

(c) The diagonal terms are the forces required to impose a unit displacement at the corresponding degree of freedom.

(d) The truss is not restrained against rigid-body motion.

2.9 If $V_a = 1$, while $U_a = 0$, $P_{xa} = (\sqrt{3}AE/2L)$ and $P_{ya} = 2AE/L$:

$$\begin{bmatrix} U_a \\ V_a \end{bmatrix} = \frac{PL}{AE}\begin{bmatrix} 0.0824 \\ 0.9643 \end{bmatrix}$$

$$s^{ca} = 0.55P \text{ (tension)}$$

2.12

$$\mathbf{K}_{ff} = 200\begin{bmatrix} 1.000 & 0.000 & -1.000 & 0.000 \\ 0.000 & 1.000 & 0.000 & 0.000 \\ -1.000 & 0.000 & 1.354 & 0.354 \\ 0.000 & 0.000 & 0.354 & 1.354 \end{bmatrix}$$

2.13 (a)

$$\mathbf{K} = 2 \times 10^4\begin{bmatrix} 1.72 & 0.00 & -0.36 & -1.00 & 0.00 \\ 0.00 & 1.28 & 0.48 & 0.00 & 0.00 \\ -0.36 & 0.48 & 1.72 & -0.36 & -0.48 \\ -1.00 & 0.00 & -0.36 & 1.36 & 0.48 \\ 0.00 & 0.00 & -0.48 & 0.48 & 0.36 \end{bmatrix} \text{kN/m}$$

(b)

$$\mathbf{K} = 2 \times 10^4\begin{bmatrix} 1.72 & 0.00 & -0.36 & -1.00 & 0.00 \\ 0.00 & 1.28 & 0.48 & 0.00 & 0.00 \\ -0.36 & 0.48 & 1.72 & -0.36 & -0.48 \\ -1.00 & 0.00 & -0.36 & 2.20 & 0.85 \\ 0.00 & 0.00 & -0.48 & 0.85 & 0.80 \end{bmatrix} \text{kN/m}$$

2.15 $s^{ac} = 22.5$ kips (c); $s^{be} = 75$ kips (c)

2.16 (a)

$$\mathbf{K}_{ff} = 8 \times 10^4\begin{bmatrix} 1.5 & -0.5 & -1.0 & 0.0 \\ -0.5 & 1.5 & 0.0 & 0.0 \\ -1.0 & 0.0 & 1.5 & 0.5 \\ 0.0 & 0.0 & 0.5 & 1.5 \end{bmatrix}$$

(b) $\mathbf{K}_{ff}$ same as in part (a).

(c) $s^{cd} = 386$ kN (c); $s^{bd} = 218$ kN (c)

2.17 (a)

$$\mathbf{K} = \frac{AE}{L}\begin{bmatrix} 2.0 & 0.0 & 0.5 & -0.5 \\ 0.0 & 2.0 & -0.5 & 0.5 \\ 0.5 & -0.5 & 1.5 & 0.0 \\ -0.5 & 0.5 & 0.0 & 1.5 \end{bmatrix}$$

(b) $s^{cd} = 11.79$ kips (t)

2.19

$$\begin{bmatrix} V_b \\ V_c \end{bmatrix} = \begin{bmatrix} 13.72 \\ 8.10 \end{bmatrix} \text{mm}$$

$$s^{bc} = 90.0 \text{ kN (c)}$$

2.21

$$\begin{bmatrix} U_d \\ V_d \end{bmatrix} = \begin{bmatrix} 0.00 \\ 0.32 \end{bmatrix} \text{in.}$$

$$s^{ad} = 6.0 \text{ kips (c)}$$

2.23

$$\begin{bmatrix} U_a \\ V_a \end{bmatrix} = \frac{PL}{AE}\begin{bmatrix} 0.7155 \\ -0.5822 \end{bmatrix}$$

2.25 $\{U_a \quad V_a \quad U_b \quad V_b \quad U_c \quad V_c \quad U_d \quad V_d \quad U_f \quad V_f \quad U_g \quad V_g \quad U_h\} =$
$10^{-4}\{114 \quad -1009 \quad 56 \quad -347 \quad -197 \quad -346 \quad -293 \quad -49 \quad -98 \quad 157 \quad -51 \quad 168 \quad -113\}$ (in.)

$s^{ab} = 1.73$ kips (c); $s^{ac} = 2.00$ kips (t); $s^{cd} = 2.08$ kips (t); $s^{be} = 1.66$ kips (c); $s^{de} = 1.89$ kips (c); $s^{df} = 2.13$ kips (t); $s^{eg} = 1.52$ kips (c); $s^{fh} = 2.11$ kips (t); $s^{gh} = 1.83$ kips (c)

2.26

$$\begin{bmatrix} V_a \\ U_b \\ V_b \\ U_d \\ V_d \\ U_e \\ V_e \end{bmatrix} = 10^{-3} \begin{bmatrix} 1.281 \\ 6.734 \\ 0.497 \\ 1.538 \\ -2.616 \\ 2.893 \\ 0.415 \end{bmatrix} \text{ in.}$$

$s^{ab} = 0.260$ kips (c); $s^{bc} = 0.165$ kips (c); $s^{ad} = 0.372$ kips (t); $s^{bd} = 0.127$ kips (c); $s^{de} = 1.005$ kips (t); $s^{be} = 0.928$ kips (c); $s^{ce} = 0.442$ kips (t); $s^{ae} = 0.315$ kips (t)

2.28 **(a)**

$$\begin{bmatrix} U_a \\ V_a \end{bmatrix} = \begin{bmatrix} 0.109 \\ -2.126 \end{bmatrix} \text{ mm}$$

(d)

$$\begin{bmatrix} U_a \\ V_a \end{bmatrix} = \begin{bmatrix} -0.517 \\ -7.942 \end{bmatrix} \text{ mm}$$

(e) All displacements are divided by two; element forces are the same as in part (a).

2.30 **(a)** $U_a = 0.0690$ in.; $s^{ac} = 33.33$ kips (c)
(b) $U_a = -0.020$ in.; $s^{ac} = 19.66$ kips (c)
(c) $U_a = 0.000$ in.; $s^{ac} = 0.0$ kips

2.31 **(a)** $U_e = 0.084$ in.; $s^{ec} = 29.8$ kips (c)
(b) $U_e = -0.45$ in.; $s^{ec} = 15.9$ kips (t)
(c) $U_e = 0.00$ in.; $s^{ec} = 35.4$ kips (t)

2.32 **(a)** $\{V_b \quad V_c \quad U_d \quad V_d \quad U_e \quad V_e \quad U_f \quad V_f\} =$
$\{-1.04 \quad -4.12 \quad 2.41 \quad -6.45 \quad 5.12 \quad -4.41 \quad 0.00 \quad -4.41\}$ mm
$s^{ab} = 13.9$ kN (c); $s^{bc} = 41.0$ kN (c); $s^{ad} = 36.2$ kN (t); $s^{bd} = 45.0$ kN (c); $s^{de} = 27.1$ kN (t); $s^{be} = 73.0$ kN (c); $s^{ce} = 68.4$ kN (t); $s^{ef} = 0.0$ kN

(b) $\{V_b \quad V_c \quad U_d \quad V_d \quad U_e \quad V_e \quad U_f \quad V_f\} =$
$\{0.38 \quad 0.00 \quad 0.90 \quad -1.65 \quad 0.00 \quad -0.88 \quad 0.00 \quad -0.88\}$ mm
$s^{ab} = 5.1$ kN (t); $s^{bc} = 5.1$ kN (c); $s^{ad} = 13.6$ kN (t); $s^{bd} = 17.0$ kN (c); $s^{de} = 10.2$ kN (t); $s^{be} = 0.0$ kN; $s^{ce} = 8.5$ kN (t); $s^{ef} = 0.0$ kN

Chapter 3

3.2 $P_{ya} = 2.75$ kips; $P_{xb} = 1.00$ kips; $P_{yb} = 0.25$ kips

3.4 $P_{ya} = 7.5$ kips; $P_{yb} = -22.5$ kips; $P_{ye} = 105.0$ kips; $P_{yh} = 40.0$ kips

3.6 $s^{bd} = 27.9$ kN (c)

3.8 $s^{bd} = 75$ kips (c)

3.9 $A(x) = A_i(1 - x/L) + A_j x/L$

(a)

$$\mathbf{k} = \frac{2E}{L}\begin{bmatrix} A_1 & 0 & -A_1 \\ 0 & A_2 & -A_2 \\ -A_1 & -A_2 & A_1 + A_2 \end{bmatrix}$$

where $\mathbf{u} = \{u_i \quad u_j \quad u_m\}$; $u_m = u(x = L/2)$; $A_m(A_i + A_j)/2$; $A_1 = (A_i + A_m)/2$; and $A_2 = (A_m + A_j)/2$. Using static condensation

$$\mathbf{k} = \frac{2A_1 A_2 E}{L(A_1 + A_2)}\begin{bmatrix} 1 & -1 \\ -1 & 1 \end{bmatrix}$$

where $\mathbf{u} = \{u_i \quad u_j\}$.

(b) Using $u = a_1 x + a_2$,

$$\mathbf{k} = \frac{E(A_i + A_j)}{2L}\begin{bmatrix} 1 & -1 \\ -1 & 1 \end{bmatrix}$$

(c) With a quadratic function, choose either: $u = a_2 x^2 + a_1 x + a_0$ (three nodes) and condense the third node; with two nodes use $u = a_2 x^2 + a_0$ or $u = a_2 x^2 + a_1 x$.

(d) $A(x) = A_i(1 + mx/L)$. Choose a value for m and use ln.

3.14 $N_i = 1 - 11x/2L + 9x^2/L^2 - 9x^3/2L^3$; $N_j = x/L - 9x^2/2L^2 + 9x^3/2L^3$; $N_k = 9x/L - 45x^2/2L^2 + 27x^3/2L^3$; $N_l = -9x/2L + 18x^2/L^2 - 27x^3/2L^3$. It is a compatible displacement function.

$$k_{11} = 3.70AE/L.$$

If A = constant, the stiffness matrix will not be exact since σ is actually constant whereas the admissible displacements give a quadratic σ.

3.15 (a) $\{p_{xi} \quad p_{xj}\} = (qL/2)\{1 \quad 1\}$

(b) $\{p_{xi} \quad p_{xj}\} = (qL/2)\{1 \quad 1\}$

(c) $\{p_{xi} \quad p_{xj}\} = (qL/6)\{1 \quad 2\}$

3.17 See Prob. 2.9.

3.19 See Prob. 2.19.

3.20 See Prob. 2.21.

3.21 See Prob. 2.23.

3.23 See Prob. 2.17.

3.25 See Prob. 2.28.

3.26 See Prob. 2.30.

3.27 See Prob. 2.31.

Chapter 4

4.1 $v_a = +2P_a L^3/3\pi^4 EI$

4.3 K_{ij} is the force at degree of freedom (dof) i required to sustain a unit displacement at dof j.

4.5 $\mathbf{U} = \{\Theta_a \quad V_b \quad \Theta_b \quad V_c \quad \Theta_c \quad \Theta_d\}$

$$\mathbf{K}_{ff} = \frac{EI}{L}\begin{bmatrix} 4 & -\frac{6}{L} & 2 & 0 & 0 & 0 \\ -\frac{6}{L} & \frac{48}{L^2} & \frac{12}{L} & -\frac{36}{L^2} & \frac{18}{L} & 0 \\ 2 & \frac{12}{L} & 16 & -\frac{18}{L} & 6 & 0 \\ 0 & -\frac{36}{L^2} & -\frac{18}{L} & \frac{60}{L^2} & -\frac{6}{L} & \frac{12}{L} \\ 0 & \frac{18}{L} & 6 & -\frac{6}{L} & 20 & 4 \\ 0 & 0 & 0 & \frac{12}{L} & 4 & 8 \end{bmatrix}$$

4.7 **(a)**

$$\mathbf{K}_{ff} = \begin{bmatrix} 12 & -6 & 2 \\ -6 & 12 & -6 \\ 2 & -6 & 4 \end{bmatrix}$$

(b)

$$\begin{bmatrix} \bar{y}_a \\ \bar{m}_a \\ \bar{m}_b \end{bmatrix} = -\frac{P}{2}\begin{bmatrix} 3 \\ 1 \\ 2 \end{bmatrix}$$

(c)

$$\mathbf{K}_{ff} = \begin{bmatrix} 12 & -6 & 2 \\ -6 & 15 & -6 \\ 2 & -6 & 4 \end{bmatrix}$$

4.9

$$\begin{bmatrix} \Theta_a \\ V_b \\ \Theta_b \\ \Theta_c \end{bmatrix} = -\frac{PL^2}{9EI}\begin{bmatrix} 5 \\ 4L \\ 2 \\ -4 \end{bmatrix}; \qquad \mathbf{s}^{ab} = \frac{2P}{3}\begin{bmatrix} 1 \\ 0 \\ L \end{bmatrix}$$

4.11 $\Theta_b = -\Theta_c = 0.0013$ rad

4.13 $\{\Theta_b \quad \Theta_c\} = (1/EI)\{45.45 \quad -22.72\}$

4.17

$$\mathbf{F} = 10^{-5}\begin{bmatrix} 75 & 100 \\ 100 & 250 \end{bmatrix}; \qquad \mathbf{K} = \begin{bmatrix} 2857 & -1143 \\ -1143 & 857 \end{bmatrix}$$

4.19 $\{\Theta_a \quad V_b \quad \Theta_b \quad V_c \quad \Theta_c \quad \Theta_d \quad V_e \quad \Theta_e \quad V_f \quad \Theta_f \quad \Theta_g\} =$
$10^{-5}\{26 \quad 3781 \quad 18 \quad 4726 \quad -9 \quad -53 \quad -12254$
$-96 \quad -29886 \quad -119 \quad -127\}$ (in. and rad)

$\{m_i \quad m_j \quad y_i\}^{ij} = \{0 \quad -30.1 \quad -0.186\}^{ab}; \quad \{30.1 \quad -60.1 \quad -0.186\}^{bc};$
$\{60.1 \quad -90.2 \quad -0.186\}^{cd}; \quad \{90.2 \quad -58.0 \quad 0.199\}^{de};$
$\{58.0 \quad -25.7 \quad 0.199\}^{ef}; \quad \{25.7 \quad 0.0 \quad 0.0\}^{fg}$

4.20 **(a)** $\{\Theta_b \quad \Theta_c\} = \{0.000458 \quad -0.00183\}$ rad
$\{y_i \quad m_i \quad m_j\}^{ij} = \{2.98 \quad 143 \quad 286\}^{ab}; \{-8.95 \quad -286 \quad -1000\}^{bc};$
$\{17.9 \quad 1000 \quad 1580\}^{cd}$ (kips and kip · in.)

(b) $\Theta_b = \Theta_c = 0$
$\{y_i \quad m_i \quad m_j\} = 548\{0 \quad -1 \quad 1\}$ (kips and kip · in.) for all elements

4.25 $\Theta_b = LM_b/24EI$; $\bar{m}_a^{ab} = M_b/12$; $\bar{m}_b^{ab} = M_b/6$; $\bar{m}_b^{bc} = M_b/3$; $\bar{m}_c^{bc} = M_b/6$; $\bar{m}_b^{db} = M_b/2$; $\bar{m}_d^{db} = M_b/4$

4.27 $\Theta_b = -1.441 \times 10^{-4}$ rad
$\{\bar{y}_a \quad \bar{m}_a \quad \bar{m}_b\}^{ab} = \{3.12 \text{ kN} \quad 3.32 \text{ kN}\cdot\text{m} \quad -10.24 \text{ kN}\cdot\text{m}\}$
$\{\bar{y}_b \quad \bar{m}_b \quad \bar{m}_c\}^{bc} = \{21.84 \text{ kN} \quad 10.24 \text{ kN}\cdot\text{m} \quad -18.88 \text{ kN}\cdot\text{m}\}$

4.29 (a) $\{\bar{y}_a \quad \bar{m}_a \quad \bar{m}_b\}^{ab} = \{-17.6 \text{ k} \quad -113 \text{ k}\cdot\text{ft} \quad -135 \text{ k}\cdot\text{ft}\}$
$\{\bar{y}_b \quad \bar{m}_b \quad \bar{m}_c\}^{bc} = \{13.5 \text{ k} \quad 135 \text{ k}\cdot\text{ft} \quad 135 \text{ k}\cdot\text{ft}\}$

Chapter 5

5.2 $\{U_a \quad V_a \quad W_a\} = \{0.076 \quad 0.093 \quad 0.103\}$ in.
$s^{ab} = 93$ kips (t)

5.4 $s^{ad} = 28.28$ kN (t); $s^{bd} = s^{cf} = 28.28$ kN (c); $s^{df} = 20$ kN (t); all other element forces zero

5.6 $s^{a1b2} = 1.94$ kN (c); $s^{a1h2} = 0.65$ kN (c); $s^{a1a2} = 2.83$ kN (c);
$s^{a2a3} = 3.64$ kN (c); $s^{a2b2} = s^{a2h2} = 0.94$ kN (t); $s^{a3b2} = 3.64$ kN (t);
$s^{a3h2} = s^{a3g3} = 0$; $s^{a3c3} = 3.75$ kN (c)

5.8 $\{W_2 \quad W_5 \quad W_6 \quad W_9 \quad W_{13}\} =$
$\{-0.134 \quad -0.105 \quad -0.150 \quad -0.168 \quad -0.168\}$ in.

5.13 $\{W_2 \quad W_6\} = \{-0.549 \quad -0.825\}$ in.

5.15 $\{\Theta_{xa} \quad \Theta_{ya} \quad \Theta_{za}\}10^{-3} = \{5.000 \quad -0.865 \quad -6.667\}$ rad
$\{p_{\bar{x}a} \quad p_{\bar{y}a} \quad p_{\bar{z}a} \quad m_{\bar{x}a} \quad m_{\bar{y}a} \quad m_{\bar{z}a}\} =$
$\{0 \quad 5.4 \quad 5.0 \quad 30 \quad -67 \quad 283\}$ (kips and kip · in.)

5.17 $\{V_b \quad \Theta_{xb} \quad \Theta_{zb}\} = \{-0.508 \text{ in.} \quad 0.00365 \text{ rad} \quad -0.00480 \text{ rad}\}$

5.19 (a) $\{V_b \quad \Theta_{xb} \quad \Theta_{zb} \quad V_c \quad \Theta_{xc}\} =$
$10^{-3}\{-2.257 \quad 1.1285 \quad -0.871 \quad -4.128 \quad 1.128\}$ (m and rad)

$$\begin{bmatrix} \bar{y}_i \\ \bar{m}_{\bar{x}i} \\ \bar{m}_{\bar{z}i} \\ \bar{m}_{\bar{z}j} \end{bmatrix}^{ij} = \begin{bmatrix} 5.0 \\ 1.71 \\ 15.0 \\ 0 \end{bmatrix}^{ab}; \begin{bmatrix} 5.0 \\ 0 \\ 1.71 \\ 13.29 \end{bmatrix}^{bc}; \begin{bmatrix} 5.0 \\ 0 \\ 1.71 \\ 13.29 \end{bmatrix}^{dc} \begin{bmatrix} 5.0 \\ -1.71 \\ 15.0 \\ 0 \end{bmatrix}^{ed} \text{(kN and kN}\cdot\text{m)}$$

(local $\bar{y}$ axis coincides with y axis)

(b) $\{V_b \quad \Theta_{xb} \quad \Theta_{zb} \quad V_c \quad \Theta_{xc}\} =$
$10^{-3}\{-2.708 \quad 1.354 \quad -0.697 \quad -4.092 \quad 1.354\}$ (m and rad)

$$\begin{bmatrix} \bar{y}_i \\ \bar{m}_{\bar{x}i} \\ \bar{m}_{\bar{z}i} \\ \bar{m}_{\bar{z}j} \end{bmatrix}^{ij} = \begin{bmatrix} 6.0 \\ 1.37 \\ 18.0 \\ 0 \end{bmatrix}^{ab}; \begin{bmatrix} 6.0 \\ 0 \\ 1.37 \\ 7.63 \end{bmatrix}^{bc}; \begin{bmatrix} 6.0 \\ 0 \\ 1.37 \\ 7.63 \end{bmatrix}^{dc}; \begin{bmatrix} 6.0 \\ -1.37 \\ 18.0 \\ 0 \end{bmatrix}^{ed} \text{(kN and kN}\cdot\text{m)}$$

(local $\bar{y}$ axis coincides with y axis)

Chapter 6

6.1 $$\mathbf{K}_{cc} = \frac{AE}{4L}\begin{bmatrix} 3 & -2 \\ -2 & 6 \end{bmatrix}; \quad \begin{bmatrix} U_b \\ V_b \end{bmatrix} = \frac{2PL}{AE}\begin{bmatrix} 2 \\ 1 \end{bmatrix}$$

6.2 $\mathbf{U}_\beta = -K_{\beta\beta}^{-1}\mathbf{K}_{\beta\alpha}U_\alpha$, where $\mathbf{U}_\alpha = \{U_b \quad V_b\}$ and $\mathbf{U}_\beta = \{U_c \quad V_c\}$

$$\begin{bmatrix} U_c \\ V_c \end{bmatrix} = -\frac{AE}{L}\begin{bmatrix} 0.7929 & -0.2071 \\ -0.2071 & 0.7929 \end{bmatrix}\frac{L}{AE}\begin{bmatrix} -1.0 & 0.0 \\ 0.0 & 0.0 \end{bmatrix}\frac{PL}{AE}\begin{bmatrix} 4.828 \\ 2.000 \end{bmatrix}$$

$$= \frac{PL}{AE}\begin{bmatrix} 3.828 \\ -1.100 \end{bmatrix}$$

6.3 Place the hinge on element ab. Therefore, $\mathbf{K}_{ff} = 4 \times 10^4$ kN · m; $\mathbf{P}_f^o = 16$ kN · m; $\Theta_{b+} = -4 \times 10^{-4}$ rad; ($\Theta_{b-} = 1.7578 \times 10^{-4}$ rad if hinge is on bc). Element forces: $\{\bar{y}_a \quad \bar{m}_a\}^{ab} = \{6.82 \text{ kN} \quad 8.44 \text{ kN} \cdot \text{m}\}$ and $(\bar{y}_b \quad \bar{m}_b \quad \bar{m}_c\} = \{18 \text{ kN} \quad 0 \text{ kN} \cdot \text{m} \quad -24 \text{ kN} \cdot \text{m}\}$

6.5 Place the hinge on element ab. $\{V_b \quad \Theta_{b+}\} = \{-0.053 \text{ in.} \quad 0.000 \text{ rad}\}$; $p_{yc} = 13.33$ kips; $m_c = -400$ kip · in.

6.7 $U_b = U_c = PL^3/12EI$; $m_d = PL/2$

6.9 See Prob. 4.27.

6.10 See Prob. 4.29(a).

6.11 $$\begin{bmatrix} U_c \\ V_c \\ U_d \\ V_d \end{bmatrix} = 10^{-6}\begin{bmatrix} 1.832 & -2.992 & -0.168 \\ -1.160 & 6.962 & 0.840 \\ -0.168 & 3.008 & 1.832 \\ -0.840 & 7.038 & 1.160 \end{bmatrix}\begin{bmatrix} P_{xa} \\ P_{ya} \\ P_{xb} \end{bmatrix}$$

6.12 See Prob. 2.12.

6.13 See Prob. 2.13.

6.15 See Prob. 2.26.

6.16 See Prob. 2.32.

6.17 Geometrically unstable, but truss can support loading because equations are consistent

6.20 $\{U_a \quad V_a \quad V_c\} = \{0 \quad 0 \quad 4.53 \text{ mm}\}$; $s^{ab} = s^{bc} = 0$; $s^{ac} = 72.5$ kN (t)

6.21 $N_1 = 2(x/L)^3 - 3(x/L)^2 + 1$; $B_1 = 6y\{2(x/L) - 1\}/L^2$; $k_{11} = 6.45Ebd^3/L^3$

Chapter 7

7.1 See Prob. 2.12.

7.4 See Prob. 4.5.

7.5 See Prob. 4.9.

7.7 See Prob. 5.2.

7.9 See Prob. 2.16.

7.11 See Prob. 2.19.

7.12 See Prob. 2.28.

7.14 See Prob. 2.32.

7.15 $V_b = L^3/EI$; $\Theta_b = L^2/EI$; $\Theta_c = -2L^2/EI$
$\{m_a^{ab} \quad m_{b-}^{ab} \quad m_{b+}^{bc} \quad m_c^{bc}\} = \{-20 \quad -10 \quad 30 \quad 0\}$ kN·m

7.16 $\Theta_b = -P/8$; $V_c = -11P/24$; $\Theta_c = -5P/8$
$\{m_a^{ab} \quad m_{b-}^{ab} \quad m_{b+}^{bc} \quad m_c^{bc}\} = (P/2)\{-1 \quad -2 \quad 2 \quad 0\}$

7.17 See Prob. 6.7.

7.20
$$\mathbf{F} = 10^{-3}\begin{bmatrix} 4.167 & 10.417 \\ 10.417 & 33.333 \end{bmatrix}; \qquad \mathbf{K}_{ff} = \begin{bmatrix} 10.97 & -342.9 \\ -342.9 & 137.2 \end{bmatrix}$$

INDEX